Power Electronics Handbook

Volume I

Power Electronics Handbook
Volume I

Edited by **Annie Kent**

NY RESEARCH PRESS

New York

Published by NY Research Press,
23 West, 55th Street, Suite 816,
New York, NY 10019, USA
www.nyresearchpress.com

Power Electronics Handbook: Volume I
Edited by Annie Kent

International Standard Book Number: 978-1-63238-367-9 (Hardback)

Printed in China.

Contents

Preface

Power electronics is a field which deals with concepts related to application of solid state electronics. It basically deals with control and conversion of electric power. Electrical and electronic engineering are also referred to as a part of the subject which conceptually deals with designing, control computational area and integrated nonlinear, and time varying energy processing electronic systems. The very first power electronic device invented was mercury arc valves. But in present day systems, the process of conversions are performed by the use of semiconductors switching devices like diodes and transistors, etc.

In simple words, power electronics systems are concerned with transmission and processing of signals and data. In power electronics, a huge amount of electrical energy is consumed as well as produced. The origins of power electronics can be traced to the development of the mercury arc rectifier, which was invented by Peter Cooper Hewitt in the year 1902. This mercury arc rectifier served the purpose of converting alternating current (AC) into direct current (DC).

There are a few power conversion systems that are classified as follows; AC to DC (rectifier), DC to AC (inverter), DC to DC (DC-to-DC converter), AC to AC (AC-to-AC converter), and they are classified on the basis of types of input and output. There are numerous uses of power electronic devices. They may be used as switches or even as amplifiers.

I especially wish to acknowledge the contributing authors, without whom a work of this magnitude would clearly not have been realizable. I appreciate their participation and their adherence as a group to the time parameters set for this publication. I hope that this book proves to be a resourceful guide for both basic and advanced concepts in power electronics.

Editor

1

Using Averaged Modeling for Capacitors Voltages Observer in NPC Inverter

Bassem Omri, Kaiçar Ammous, and Anis Ammous

Power Electronics Group, Université de Sfax, ENIS-Sfax, BP. 1173, 3038 Sfax, Tunisia

Correspondence should be addressed to Bassem Omri, omri_bessem@yahoo.fr

Academic Editor: C. M. Liaw

This paper developed an adaptive observer to estimate capacitors voltages of a three-level neutral-point-clamped (NPC) inverter. A robust estimated method using one parameter is proposed, which eliminates the voltages sensors. An averaged modeling of the inverter was used to develop the observer. This kind of modeling allows a good trade-off between simulation cost and precision. Circuit model of the inverter (implemented in Simpower Matlab simulator) associated to the observer algorithm was used to validate the proposed algorithm.

1. Introduction

Multilevel inverters can provide an effective alternative of high power applications, providing a high quality voltage, increasing efficiency and robustness, and reducing interference electromagnetics [1–4]. There are three main topologies of multilevel inverters: diode-clamped inverter [2, 5, 6], flying capacitor inverter [7, 8], and cascade multilevel inverter [6, 9]. Among the multilevel inverters, the most popular topology is the diode-clamped inverter which is called as a neutral-point-clamped (NPC) inverter in three levels. It is proposed by Nabae et al. [10]. This type of inverter avoids the complexity associated with the series connection of semiconductor switches or the bulky coupling transformer, produces low distortion harmonics, and has an average cost.

However, in NPC inverter, the DC-link voltage is divided by capacitors, and each capacitor is composed of series connection construction. Therefore, if the voltage unbalancing occurs between each capacitor, the line-to-line output voltage waveform has many harmonic components and the power devices in NPC inverter cannot guarantee the safe operation. So, one NPC inverter requirement is to produce a good control maintaining the optimal balance of capacitor voltages [11–13]. Thus, regulation of these capacitors voltages requires information about such voltages. For that there are two possibilities either measuring or estimating such voltages.

Regarding the first possibility, it presents several disadvantages: besides the known difficulties when measuring a high-voltage level, if the number of levels increases the number of capacitors increases in NPC inverter. Then it will be necessary to use multiple-voltage sensors to measure the capacitor voltages, which reduces the reliability. Furthermore, the use of the voltages sensors presents many difficulties of establishment in front of the use of the current sensors. On the other hand, the second alternative uses software sensors, called observers, which replace the physical sensors (the voltage sensors). It is suitable to estimate the capacitors voltages through only the phase currents measurement. So, the observer is independent of the number of the level of NPC inverter. Unlike the traditional direct voltage sensing depends on the levels number.

An observer is a data-processing algorithm to reconstruct the variable state system from the input, mathematical model, and output measurement of the real system. It is an attractive solution because it benefits from a mathematical model of the system. It is less expensive and more reliable because it is applied by a digital computer. Actually, the calculation of the digital signal processors (DSP) is the execution of such sophisticated and complex algorithms

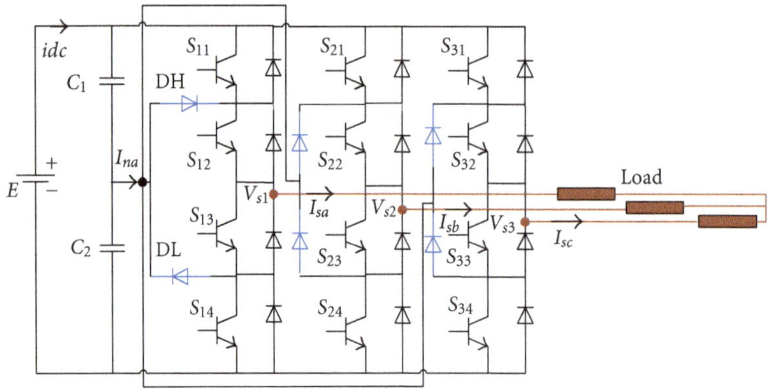

FIGURE 1: A 3-level NPC converter topologies.

with the high-degree precision. Many works related to the observer design for the nonlinear systems were published [14–16].

Recently, the authors [17] designed a robust observer based on adaptive backstepping approach for three-level NPC inverters. It is used for capacitor voltages estimation. This approach gave good results. However, the authors used an instantaneous reference model of NPC inverter. Therefore, the simulation cost is significant. Moreover, the adaptive backstepping algorithm used many parameters (adaptation gains, design constant) to design the observer. The algorithm becomes so complex when the number of levels increases.

This paper proposes an adaptive observer in order to estimate the voltages capacitors of 3-levels inverter NPC (Figure 1). The adapive algorithm is robust, simple to implement, and uses one parameter called forgetting factor. An averaged reference model of three-level NPC inverter is used for simulation. This kind of modeling allows a good tradeoff between simulation cost and precision.

This paper is organized as follows. In Section 2, the PWM control strategies are presented. An averaged reference model of three-level inverter NPC is given in Section 3. Section 4 shows the observer design for the NPC inverter. Finally, simulation results are given in order to illustrate the performance of the proposed observer to estimate NPC capacitors voltages.

2. The PWM Control Strategies of Multilevel Inverters

The proliferation of power electronic devices has led to a demand for more effective pulse width modulation strategies. The current waveform can be improved by increasing the frequency of the carrier wave; this approach reduces copper losses at the expense of increased switching losses. To overcome the limitations of existing switching strategies the technique known as space vector pulse width modulation (SVPWM) is becoming widely used in industry [18, 19].

SVPWM is a highly efficient method of generating the six-pulsed signals for the inverter stage of the motor drive. Conventional switching techniques treat each phase as a separately generated sinusoid that is displaced by 120 degrees. However, a change in the voltage of one-half bridge due to switching invariably influences the other two-phase voltages [20]. SVPWM evaluates the switching schema as a whole, which results in better use of the DC bus and generates significantly less harmonic distortion than the sine triangle method [20].

The SVPWM control of the 3-level NPC inverter consists to control it in the (α,β) coordinates which can generate 27 possible voltages vectors (Figure 2). They are divided in 6 large vectors (PNN,PPN,NPN,NPP,NNP,PNP), 6 medium vectors (PON,OPN,NPO,NOP,ONP,PNO), 12 small vectors (with the redundancy: (PPO,OON), (OPO,NON), (OPP,NOO), (OOP,NNO), (ONO,POP), (POO,ONN)) and 3 zero vectors (OOO,NNN,PPP). When the power devices, S_{k1} and S_{k2}, are connected to the positive DC-link rail the state is noted by "P." In contrast, when the output is connected to the negative DC-rail by turning both S_{k3} and S_{k4} on, the state is called "N." Moreover, the switching state is defined as "O" when S_{k2}, and S_{k3} both power devices, are on.

The objective of SVPWM technique is to approximate a reference space vector V_s somewhere within the hexagon of Figure 2 using a combination of the 27 switching vectors. Indeed, when the reference voltage is in one sector, several couples of vector are possible.

For example in Sector 1 (Figure 3), the V_s vector can be divided on V_{1a}, V_{12} or V_{1b}, V_{12}. We choose the second solution, since V_{1b} is greater in amplitude than V_{1a}. At each commutation period of the inverter, the vector V_s, projected on two adjacent vectors, ensures the calculation of the dwelling time.

Let T_{mod} be a modulation period, T_{1b} dwelling time of the first vector V_{1b}, T_{12} dwelling time of the second vector V_{12}, and $T_{\text{nul}} = T_{\text{mod}} - T_{1b} - T_{12}$ dwelling time of zero vectors. We define the duty cycles ρ_1 and ρ_2 related to times T_{1b} and T_{12}, respectively, of the vectors V_{1b} and V_{1a} by:

$$\rho_1 = \frac{T_1}{T_{\text{mod}}}, \qquad \rho_2 = \frac{T_{12}}{T_{\text{mod}}}. \tag{1}$$

For one modulation period T_{mod}, the output voltage of the inverter is given by:

$$T_{\text{mod}} V_S = (T_{1b} V_{1b} + T_{12} V_{12} + T_3 V_{\text{Null}}). \tag{2}$$

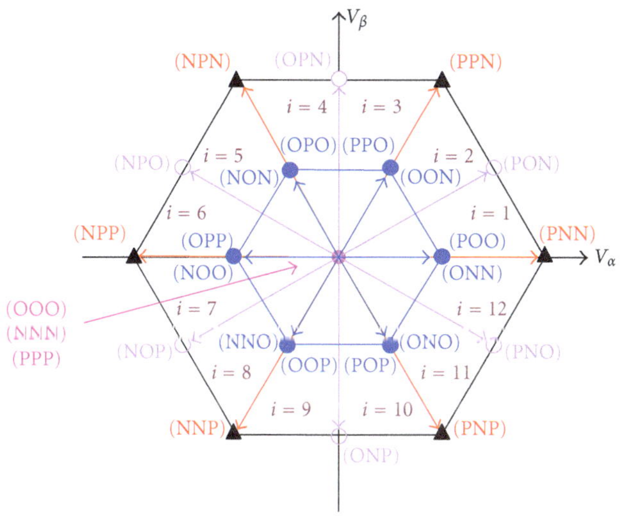

Figure 2: Voltage space vectors in NPC inverter.

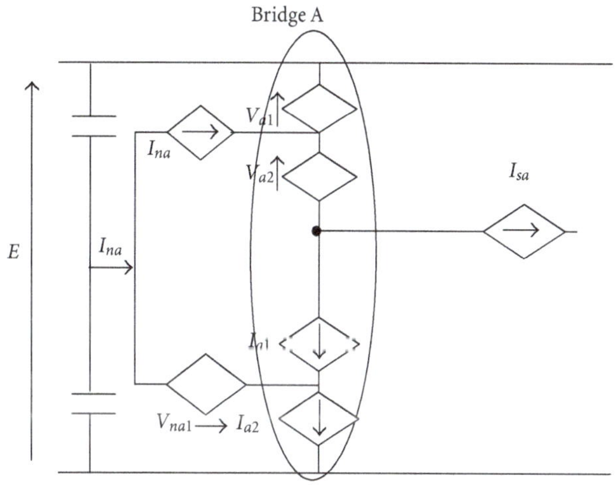

Figure 4: The proposed average model of the 3-level PWM converter.

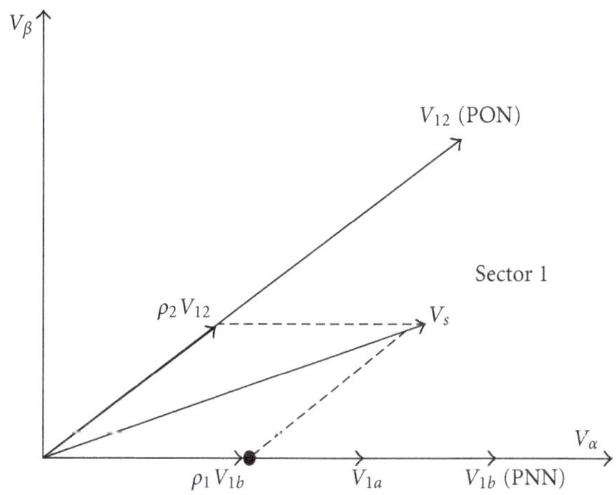

Figure 3: Sector 1.

And the dwelling time of each vector is given by

$$T_{1b} = \frac{2\sqrt{3}T_{\text{mod}}}{E}\left(\frac{1}{2\sqrt{2}}V_{s\alpha} - \frac{\sqrt{3}}{2\sqrt{2}}V_{s\beta}\right),$$
$$T_{12} = \frac{2\sqrt{2}T_{\text{mod}}}{E}V_{s\beta}. \quad (3)$$
$$T_{\text{null}} = T_{\text{mod}} - T_{12} - T_{1b}.$$

For more details see [21].

3. The Nonideal Average Model

Based on the averaged representation of the PWM switch [22], Figure 4 shows the proposed averaged model of the one-leg three-level multilevel converters. I_{sa} is the average value of the load current over the switching period. This model applies a controlled voltage source (V_{a1} and V_{a2}) and a controlled current source (I_{a1} and I_{a2}). Each three-level converter leg has four controlled switches and six uncontrolled switches. The four controlled switches are directly governed by external control signals. The uncontrolled switches are indirectly governed by the state of the controlled switches and the circuit conditions.

The average values of the different voltage and current sources of the proposed averaged model of the bridge (A) are given by [22].

Consider that T_{mod} is the switching period of the active switch and ρ (ρ_1, ρ_2, ρ_3) the duty cycle, which is the ratio of the on-time of the active couple switches ((S_{i1}, S_{i2}), (S_{i2}, S_{i3}), or (S_{i3}, S_{i4})) by the switching period T_{mod}, respectively (Figure 5).

V_d and V_{ds} represent respectively, the voltage across the diode and the MOSFET in on-state. These values can be deduced from the devices static characteristics $I(V)$:

$$\rho_i = \frac{T_{\text{on}\,i}}{T_{pwm}}; \qquad \rho_{Mi} = \frac{T_{\text{off}\,i}}{T_{pwm}}, \quad (4)$$
$$\rho_M = \rho_{M1} + \rho_{M2} + \rho_{M3}.$$

When only output current behavior is considered, a simplified representation of the averaged model (Figure 6) can be used, where

$$\langle I_{na}\rangle = (I_{na1} + I_{a2} - I_{a1}),$$
$$\langle V_a\rangle = (V_{a1} + V_{a2}), \quad (5)$$
$$\langle I_a\rangle = \langle I_{a2}\rangle.$$

The averaged values of the voltage source V_a and the current source I_a of the bridge (A) are given as follows.

If the current I_{sa} is positive,

$$\langle V_a\rangle = 2Vds\rho_1 + (E + 2Vd)\rho_3$$
$$+ \left(\frac{E}{2} + Vd + Vds\right)\rho_2 + (E + 2Vd)\rho_M, \quad (6)$$
$$\langle I_a\rangle = -I_{sa}\rho_3,$$
$$\langle I_{na}\rangle = I_{sa}\rho_2.$$

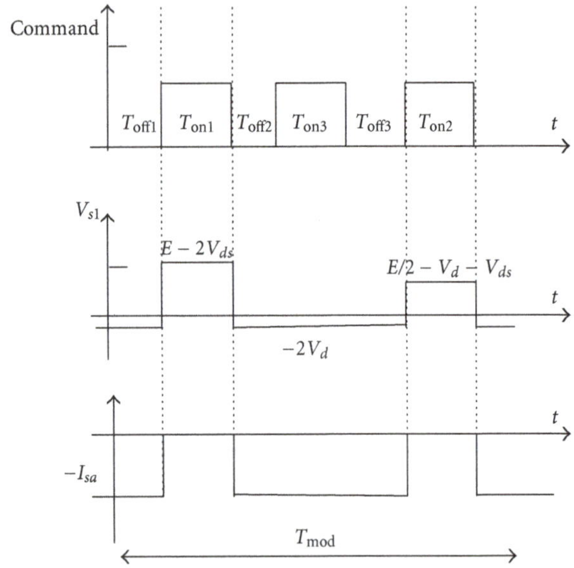

FIGURE 5: Command signals of the controlled devices in the case of $I_{sa} > 0$.

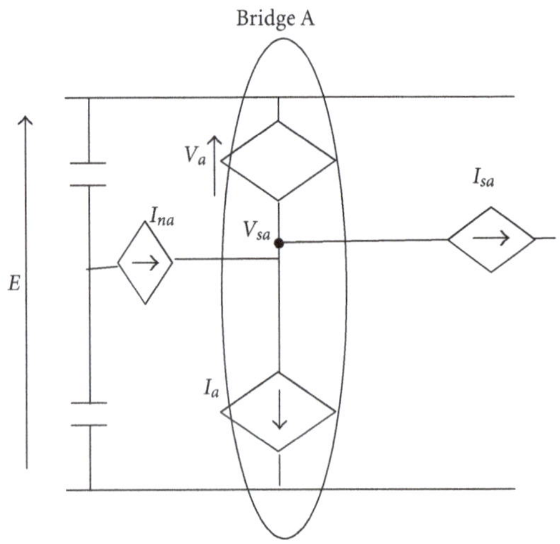

FIGURE 6: A simplified representation of averaged model.

If the current I_{sa} is negative,

$$\langle V_a \rangle = -2Vd\rho_1 + (E - 2Vds)\rho_3 + \left(\frac{E}{2} - Vd - Vds\right)\rho_2 + (2Vd)\rho_M, \tag{7}$$

$$\langle V_a \rangle = -2Vd\rho_1 + (E - 2Vds)\rho_3 + \left(\frac{E}{2} - Vd - Vds\right)\rho_2 + (2Vd)\rho_M, \tag{8}$$

$$\langle I_a \rangle = I_{sa}\rho_3. \tag{9}$$

In order to demonstrate the accuracy of the proposed averaged model, a circuit model of the inverter implemented in Simpower Matlab simulator, using the circuit represented in Figure 1 is used. For the simulation in this section and

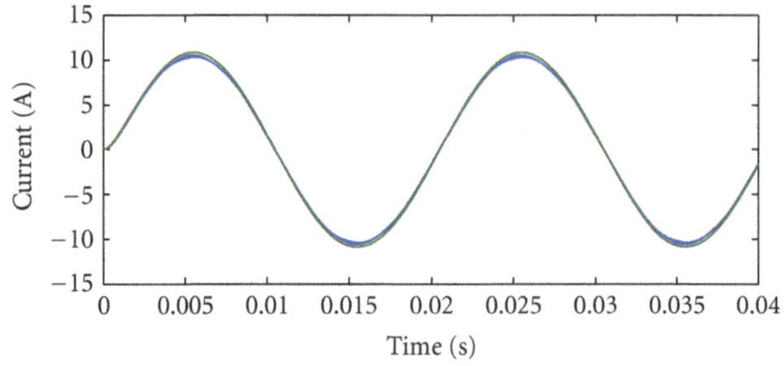

FIGURE 7: The output current obtained with the non-ideal averaged model and the circuit model after filtering (3-levels converter).

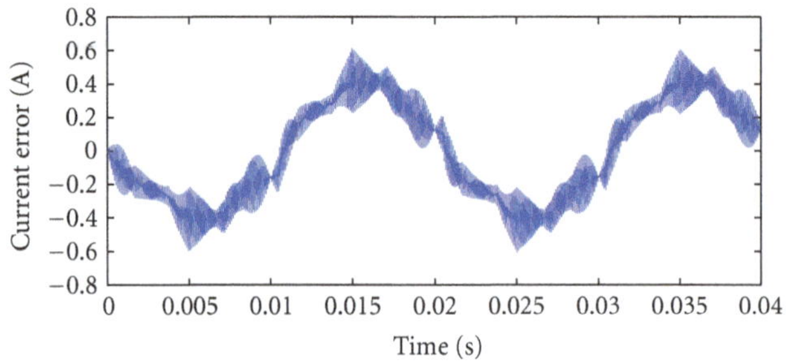

FIGURE 8: The current error.

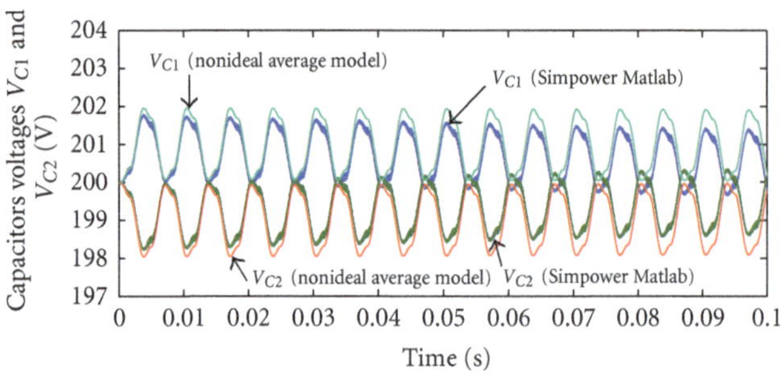

FIGURE 9: Evolution of the capacitors voltages C_1 and C_2 (using Simpower Matlab and the averaged model).

Section 5, we use 2 GHz Core2 Duo processor and 3 GB RAM.

The comparison between the current waveforms (Figure 7) obtained by the Simpower simulator and the current given by the proposed model shows the good accuracy of the averaged model behavior for converter electric analysis. The current error between the two types' models is given in Figure 8.

Figure 9 shows the capacitors voltages (C_1 and C_2) waveforms obtained by the model Simpower simulator and the averaged model. The simulation results given by the proposed averaged model are in good agreement with Simpower simulations.

It is evident that multilevel converter analysis using fine models gives accurate results, but these simulations cost is unaffordable. When long time range simulation is needed, the proposed averaged model can be used, and the obtained results show the accuracy of the proposed model (Table 1).

Table 1: CPU running time (s) necessary to simulate a period (0.02 s).

Type model	Running time (s)
Nonideal averaged model	10
Circuit model (Simpower Matlab)	90

So, we used this model like a reference model in the observer design.

4. Observer Design of Capacitors Voltages

The DC side of the three-level NPC inverter (Figure 1) has a DC source E and two connections capacitors C_1 and C_2. The switching variable γ_k represents the state of the multilevel inverter active switches, S_{kj}, $k \in \{1,2,3\}$, $j \in \{1,2,3,4\}$. The three states of the binary switching γ_k can be defined as

$$\gamma_k = \begin{cases} 1 & (S_{k1} = 1 \wedge S_{k2} = 1) \wedge (S_{k3} = 0 \wedge S_{k4} = 0) \\ 0 & (S_{k1} = 0 \wedge S_{k2} = 1) \wedge (S_{k3} = 1 \wedge S_{k4} = 0) \\ -1 & (S_{k1} = 0 \wedge S_{k2} = 0) \wedge (S_{k3} = 1 \wedge S_{k4} = 1). \end{cases} \quad (10)$$

Applying the Kirchhoff laws to the three-level inverter circuit (Figure 1) and doing some mathematical manipulations, the dynamic equations of the AC currents, I_{sa}, I_{sb}, and I_{sc}, and the capacitor voltages, V_{C1} and V_{C2}, are defined as functions of the circuit parameters and switching variables γ_k:

$$\begin{bmatrix} \frac{dI_{sa}}{dt} \\ \frac{dI_{sb}}{dt} \\ \frac{dI_{sc}}{dt} \\ \frac{dV_{C1}}{dt} \\ \frac{dV_{C2}}{dt} \end{bmatrix} = \begin{bmatrix} -\frac{R}{L} & 0 & 0 & \frac{\psi_{11}}{L} & \frac{\psi_{12}}{L} \\ 0 & -\frac{R}{L} & 0 & \frac{\psi_{21}}{L} & \frac{\psi_{22}}{L} \\ 0 & 0 & -\frac{R}{L} & \frac{\psi_{31}}{L} & \frac{\psi_{31}}{L} \\ -\frac{\Gamma_{11}}{C_1} & -\frac{\Gamma_{12}}{C_1} & -\frac{\Gamma_{13}}{C_1} & 0 & 0 \\ -\frac{\Gamma_{21}}{C_2} & -\frac{\Gamma_{22}}{C_2} & -\frac{\Gamma_{23}}{C_2} & 0 & 0 \end{bmatrix} \times \begin{bmatrix} I_{sa} \\ I_{sb} \\ I_{sc} \\ V_{C1} \\ V_{C2} \end{bmatrix} + \begin{bmatrix} 0 \\ 0 \\ 0 \\ \frac{1}{C_1} \\ \frac{1}{C2} \end{bmatrix} \times idc, \quad (11)$$

$$\Gamma_{1k} = \frac{\gamma_k(\gamma_k + 1)}{2}, \qquad \Gamma_{2k} = \frac{\gamma_k(-\gamma_k + 1)}{2},$$
$$\psi_{ki} = \frac{1}{3}\left(2\Gamma_{ik} - \sum_{j=1, j\neq k}^{3} \Gamma_{ik}\right). \quad (12)$$

In order to obtain a simplified model which contains less equations, we change the switching sequences in the (α, β) coordinates:

$$X_{123} = [T]X_{\alpha\beta}, \quad (13)$$

where T is the Concordia transformation:

$$T = \frac{\sqrt{2}}{\sqrt{3}} \begin{pmatrix} 1 & 0 \\ -\frac{1}{2} & \frac{\sqrt{3}}{2} \\ -\frac{1}{2} & -\frac{\sqrt{3}}{2} \end{pmatrix}. \quad (14)$$

Applying (11) to (8), the new model is represented in (α, β) coordinates by

$$\begin{bmatrix} \frac{dI_{\alpha}}{dt} \\ \frac{dI_{\beta}}{dt} \\ \frac{dV_{C1}}{dt} \\ \frac{dV_{C2}}{dt} \end{bmatrix} = \begin{bmatrix} -\frac{R}{L} & 0 & \frac{\Gamma_{1\alpha}}{L} & \frac{\Gamma_{2\alpha}}{L} \\ 0 & -\frac{R}{L} & \frac{\Gamma_{1\beta}}{L} & \frac{\Gamma_{2\beta}}{L} \\ -\frac{\Gamma_{1\alpha}}{C_1} & -\frac{\Gamma_{1\beta}}{C_1} & 0 & 0 \\ -\frac{\Gamma_{2\alpha}}{C_2} & -\frac{\Gamma_{2\beta}}{C_2} & 0 & 0 \end{bmatrix} \times \begin{bmatrix} I_{\alpha} \\ I_{\beta} \\ V_{C1} \\ V_{C2} \end{bmatrix} + \begin{bmatrix} 0 \\ 0 \\ \frac{1}{C_1} \\ \frac{1}{C_2} \end{bmatrix} \times idc, \quad (15)$$

where

$$\Gamma_{i\alpha} = \frac{\sqrt{2}}{\sqrt{3}}\left(\Gamma_{i1} - \frac{\Gamma_{i2}}{2} - \frac{\Gamma_{i3}}{2}\right), \qquad \Gamma_{i\beta} = \frac{1}{\sqrt{2}}(\Gamma_{i2} - \Gamma_{i3}). \quad (16)$$

This model can be written as

$$\dot{X} = A(S)X + H(S,X),$$
$$y = CX, \quad (17)$$

where $X = (I_\alpha \ \ I_\beta \ \ V_{C1} \ \ V_{C2})^T$, $C = (1 \ \ 1 \ \ 0 \ \ 0)$, $H(S,X) = \begin{bmatrix} 0 \\ 0 \\ 1/C_1 \\ 1/C_2 \end{bmatrix}$.

$$A(S) = \begin{bmatrix} -\frac{R}{L} & 0 & \frac{\Gamma_{1\alpha}}{L} & \frac{\Gamma_{2\alpha}}{L} \\ 0 & -\frac{R}{L} & \frac{\Gamma_{1\beta}}{L} & \frac{\Gamma_{2\beta}}{L} \\ -\frac{\Gamma_{1\alpha}}{C_1} & -\frac{\Gamma_{1\beta}}{C_1} & 0 & 0 \\ -\frac{\Gamma_{2\alpha}}{C_2} & -\frac{\Gamma_{2\beta}}{C_2} & 0 & 0 \end{bmatrix}. \quad (18)$$

For different configurations, the capacitors are not always traversed by the load current. The current can pass by one or two capacitors as it cannot traverse any capacitor if the top switchers of each leg are all closed or all open. Thus, some configurations give observability problem of the studied system. This can be proven by the test of rank, whenever, at least, a capacitor is traversed by the load current:

$$\Theta = \text{rang}\left(\begin{bmatrix} C \\ CA \\ CA^2 \\ CA^3 \end{bmatrix}\right) = 3 \prec 4. \quad (19)$$

It is clear that Θ is not of full rank; thus the system is not observable. So, to design the observer, a convenient representation of the model must be given.

To overcome this difficulty, the idea is to consider two interconnected affine models and to construct the new observer based on those models.

Now, consider that the system (15) can be represented as a set of the interconnected subsystems as follows:

$$\Sigma_1 : \begin{bmatrix} \frac{dI_\alpha}{dt} \\ \frac{dI_\beta}{dt} \\ \frac{dV_{C1}}{dt} \end{bmatrix} = \begin{bmatrix} -\frac{R}{L} & 0 & \frac{\Gamma_{1\alpha}}{L} \\ 0 & -\frac{R}{L} & \frac{\Gamma_{1\beta}}{L} \\ -\frac{\Gamma_{1\alpha}}{C_1} & -\frac{\Gamma_{1\beta}}{C_1} & 0 \end{bmatrix} \begin{bmatrix} I_\alpha \\ I_\beta \\ V_{C1} \end{bmatrix} + \begin{bmatrix} \frac{\Gamma_{2\alpha}}{L} \\ \frac{\Gamma_{2\beta}}{L} \\ 0 \end{bmatrix} V_{C2} + \begin{bmatrix} 0 \\ 0 \\ \frac{1}{C_1} \end{bmatrix} \times idc, \quad (20)$$

$$\Sigma_2 : \begin{bmatrix} \frac{dI_\alpha}{dt} \\ \frac{dI_\beta}{dt} \\ \frac{dV_{C2}}{dt} \end{bmatrix} = \begin{bmatrix} -\frac{R}{L} & 0 & \frac{\Gamma_{2\alpha}}{L} \\ 0 & -\frac{R}{L} & \frac{\Gamma_{2\beta}}{L} \\ -\frac{\Gamma_{2\alpha}}{C_2} & -\frac{\Gamma_{2\beta}}{C_2} & 0 \end{bmatrix} \begin{bmatrix} I_\alpha \\ I_\beta \\ V_{C2} \end{bmatrix} + \begin{bmatrix} \frac{\Gamma_{1\alpha}}{L} \\ \frac{\Gamma_{1\beta}}{L} \\ 0 \end{bmatrix} V_{C1} + \begin{bmatrix} 0 \\ 0 \\ \frac{1}{C_2} \end{bmatrix} \times idc. \quad (21)$$

Each subsystem can be written as

$$\Sigma_j \begin{cases} \dot{\chi}_j = A_j(S_k)\chi_j + \Gamma_i\left(S_k, \chi_j\right), \\ y = C_j\chi_j, \quad j = 1,2 \; k = 1,2,3 \; i = 1,2 \neq j, \end{cases} \quad (22)$$

where $\chi_j = (I_\alpha \; I_\beta \; V_{Cj})^T$ represents the state of the jth subsystem, S_k are the instantaneous inputs applied to the system, and y is the measurable output, and

$$A_j(S_k) = \begin{bmatrix} -\frac{R}{L} & 0 & \frac{\Gamma_{j\alpha}}{L} \\ 0 & -\frac{R}{L} & \frac{\Gamma_{j\beta}}{L} \\ -\frac{\Gamma_{j\alpha}}{C_j} & -\frac{\Gamma_{j\beta}}{C_j} & 0 \end{bmatrix}, \quad (23)$$

$$\Gamma_i\left(S_k, \chi_j\right) = \begin{bmatrix} \frac{\Gamma_{i\alpha}}{L} \\ \frac{\Gamma_{i\beta}}{L} \\ 0 \end{bmatrix} V_{Ci} + \begin{bmatrix} 0 \\ 0 \\ \frac{1}{C_j} \end{bmatrix} \times idc,$$

$$j = 1,2, \quad k = 1,2,3, \quad i = 1,2 \neq j, \quad (24)$$

and $C = (1 \;\; 1 \;\; 0)$.

The function $\Gamma_i(S_k, \chi_j)$ is the interconnection function. It depends on the control S_k and the state vector of each subsystem Σ_j.

From the preceding composition (22), the following system:

$$\varphi_j \begin{cases} Z_j = A_j(S_k)Z_j + \Gamma_i(S_k) - P_j^{-1}C_j(y - \hat{y}), \\ P_j = -\theta_j P_j - A^T(S_k)P_j - P_j A(S_k) + C_j C_j^T, \end{cases} \quad (25)$$

is an observer for the system (16) for $j = 1,2$, where $P_j^{-1}C_j$ is the observer gain which depends on the solution of the Riccati equation (25) for each subsystem Σ_j. The parameter θ_j must be a positive constant and sufficiently greater such that for any positive symmetric matrix $P_j(0)$, the following condition is verified [23]:

$$\forall \theta \geq \theta_0 \quad \exists \gamma \succ 0, \; \delta \succ 0, \; t_0 \succ 0 : \forall t \geq t_0, \\ \gamma I \leq P(0) \leq \delta I. \quad (26)$$

The parameter θ_j determines the convergence rate of the observer. The authors [23] proposed guidelines for the tuning of the parameters of the adaptive observer.

Note that this observer is the deterministic version of the Kalman filter for state affine systems. It is clear that the system observability depends on the applied inputs. Then the convergence of this observer can be proved assuming that the inputs S_j are regularly persistent (see Appendix A); it is a class of admissible inputs that allows to observe the system (for more details, see [19]). This guarantees the works of observer, and the observer gain is well defined, that is, the matrix P_j is nonsingular.

Now, a further result based on regular persistence is introduced.

Lemma 1. *Assume that the input S_j is regularly persistent for system (22) and consider the following Lyapunov differential equation:*

$$\dot{P}_j = -\theta_j P_j - A_j^T(S_k)P_j - P_j A_j(S_k) + C_j C_j^T. \quad (27)$$

With $P_j \succ 0$, then, $\exists\theta_{j0} \succ 0$ such that for any symmetric positive definite matrix $P_j(0)$, $\exists\theta_j \succ \theta_{j0}$, $\exists\alpha_j, \beta_j, t_0 \succ 0$: $\forall t \succ t_0$,

$$\forall t \succ t_0, \quad \alpha_j I \prec P_j(t) \prec \beta_j I, \tag{28}$$

where I is the identity matrix.

The proof of this lemma follows the same steps of Theorem 3.2 in [24].

The system will be represented by a set of interconnected subsystems (Figure 10):

$$\Sigma\begin{cases}\dot{\chi}_1 = A_1(S_k)\chi_1 + \Gamma_2\left(S_k, \chi_j\right)\\ \dot{\chi}_2 = A_2(S_k)\chi_2 + \Gamma_1\left(S_k, \chi_j\right).\end{cases} \tag{29}$$

The idea of this paper is to construct an observer for all the system Σ from the separate observers of the subsystem Σ_j.

If φ_j is an exponential observer for Σ_j, then the following interconnected system:

$$\varphi\begin{cases}\dot{\chi}_1 = A_1(S_k)\chi_1 + \Gamma_1\left(S_k, \chi_j\right) + P_1^{-1}C_1^T(\hat{y} - y)\\ \dot{\chi}_2 = A_2(S_k)\chi_2 + \Gamma_2\left(S_k, \chi_j\right) + P_2^{-1}C_2^T(\hat{y} - y)\\ P_1 = -\theta_1 P_1 - A_1^T(S_k)P_1 - P_1 A_1(S_k) + C_1 C_1^T\\ P_2 = -\theta_2 P_2 - A_2^T(S_k)P_2 - P_2 A_2(S_k) + C_2 C_2^T.\end{cases} \tag{30}$$

is an observer for the interconnected system Σ.

Each observer subsystem φ_j is defined by (25).

Remark 2. The proposed observer functions in the case if the inputs controls are regularly persistent (see Appendix A), which is equivalent to that all subsystem φ_j is observable.

Now, we give the sufficient conditions which ensure the convergence of the interconnected observer φ. For that, we introduce the following assumptions.

Assumption 1. Assume that the input S_j is a regularly persistent input for subsystem Σ_j and admits an exponential observer φ_j for $j = 1, 2$. If an input is regularly persistent, it excites sufficiently the system to obtain information necessary to constructing the not measured variables using the designed observer (see Appendix A). In this case, an observer of the form φ can be designed and the estimation error will be bounded.

Assumption 2. The term $\Gamma_j(S_k, \chi_i)$ does not destroy the observability property of subsystem φ_j under the action of the regularly persistent input S_j. Moreover, $\Gamma_j(S_k, \chi_i)$ is Lipchitz with respect to χ_i and uniformly with respect to S_j for $j = 1, 2$. This condition is verified because these functions are linear.

Then the following result can be established.

Lemma 3. *Consider the interconnected system Σ and the two Assumptions 1 and 2 are verified. Then the system φ (30) is an observer for the system Σ (29).*

The proof of the lemma is given in Appendix B.

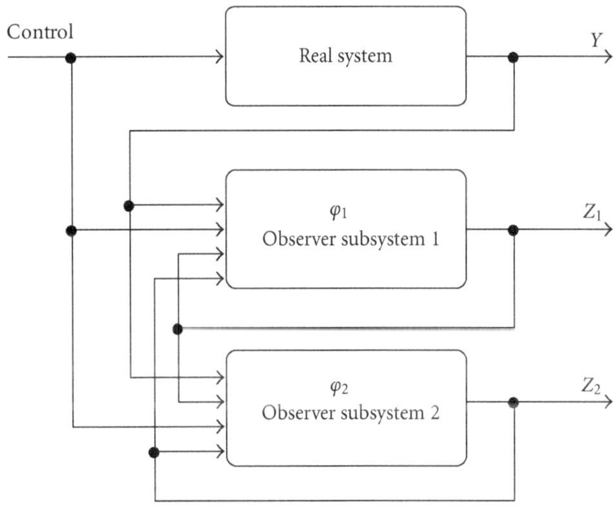

FIGURE 10: Interconnected observer diagram.

5. Simulation Results

In this section, a detailed simulation was carried out to validate the proposed interconnected observer. The general block diagram is shown in Figure 11. In this study the 3-level NPC inverter block was replaced by the averaged model of the inverter to develop the observer. The block of the interconnected observer is given by (30). In order to show the accuracy of the observer a fine model (circuit model) of the inverter (implemented in Matlab Simpower) was used to estimate the capacitors voltages. The estimated capacitors voltages of the two models are compared to the real capacitors voltages. We define the real capacitor voltage by the capacitor voltage given by the voltages sensors of the circuit model implemented in Simpower Matlab.

The different behavior of the interconnected observer was performed. The capacitors values are $C_1 = C_2 = 1000\,\mu$F. The NPC Inverter is connected at three equilibrium phases where each phase composed by an inductor $L = 10$ mHand a resistor $R = 20\,\Omega$. To estimate the NPC capacitor voltages according to (25) the observer parameter gain $\theta_j = 15000$ for $j = 1, 2$ is used. SVPWM was used to control the multilevel inverter described in Section 2. The switching frequency was fixed at 10 kHz. Simulations were carried out using matlab/simulink. Matlab simulator is an important tool where different system models (electrical, mechanical, thermal, etc.) can be developed. In fact, Matlab simulator enables averaged model representation.

Figures 12 and 13 show the comparison between the estimated capacitors voltages C_1 and C_2 (using nonideal average model and circuit model) and the real capacitor voltage.

Figures 14 and 15 show the estimated errors for the estimated capacitors voltages V_{C1} and V_{C2}. The error for each model is calculated by (V_{Ci}-real) – (V_{Ci}-estimated).

In order to study the observer behavior, we changed the load of the first leg ($L = 10$ mH, $R = 20\,\Omega$) to ($L = 10$ mH, $R = 40\,\Omega$) at $t = 0.1$ s (Figures 16 and 17). In other terms,

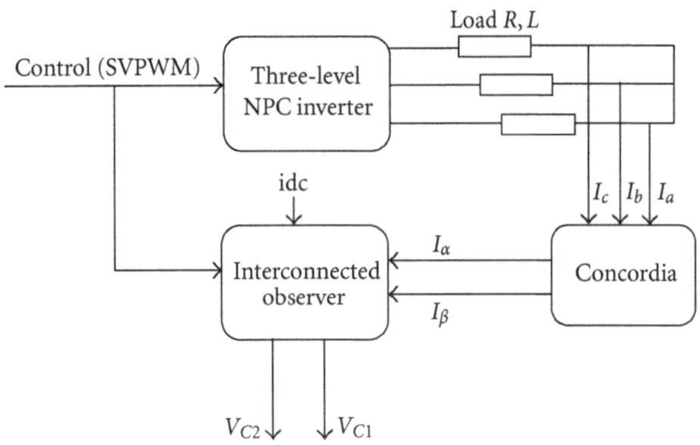

Figure 11: Block diagram of three-level NPC inverter and interconnected observer.

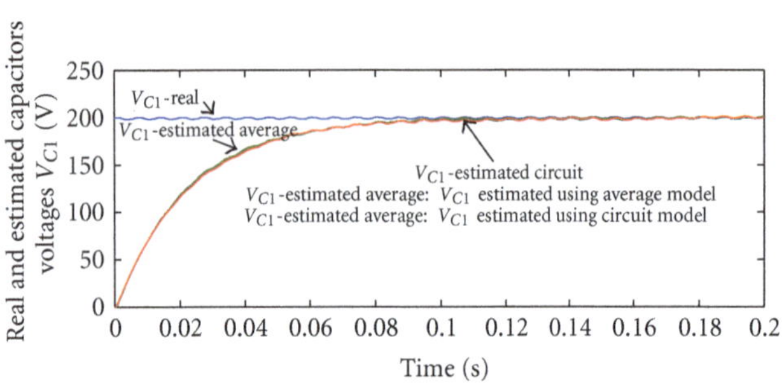

Figure 12: Real and estimated capacitors voltages V_{C1} (V) using nonideal average model and circuit model (Simpower Matlab).

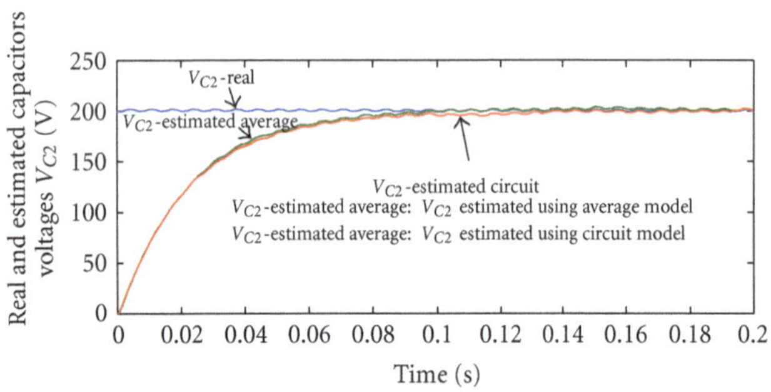

Figure 13: Real and estimated capacitors voltages V_{C2} (V) using nonideal average model and circuit model (Simpower Matlab).

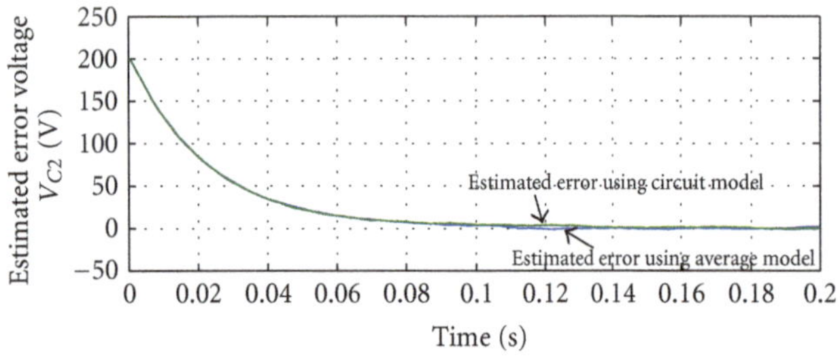

Figure 14: Estimation errors for V_{C1} (V) using nonideal average model and circuit model (Simpower Matlab).

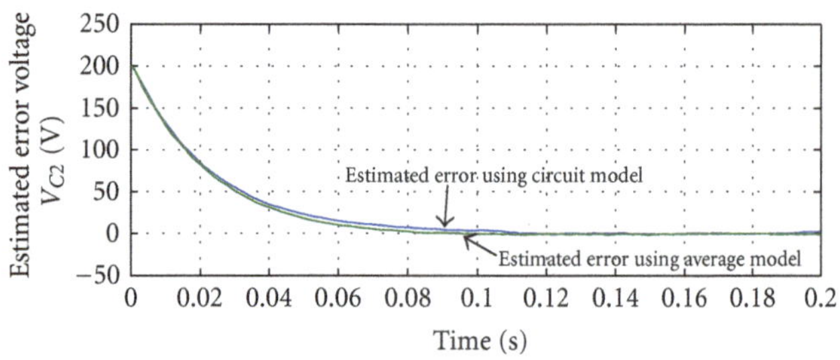

Figure 15: Estimation errors for V_{C2} (V) using nonideal average model and circuit model (Simpower Matlab).

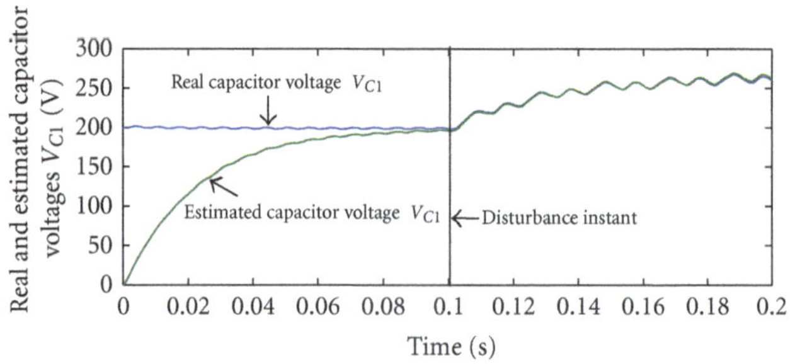

Figure 16: Real and estimated capacitors voltages V_{C1} (V) using nonideal average model with disturbance load.

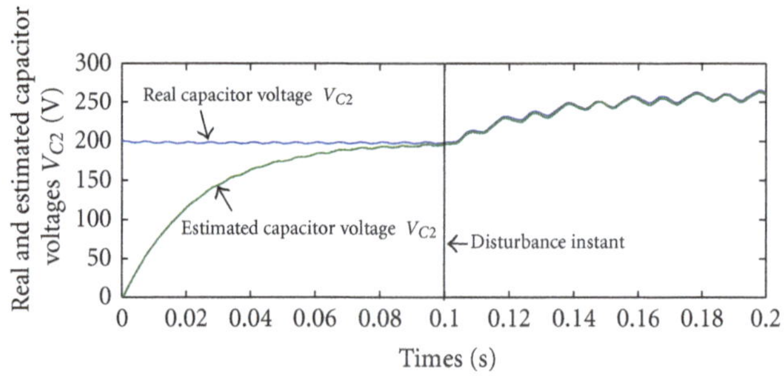

Figure 17: Real and estimated capacitors voltages V_{C2} (V) using nonideal average model with disturbance load.

we have not got equilibrium three phases. Then, Figures 18 and 19 show at $t = 0.1$ s the change of the load of the first leg ($L = 10$ mH, $R = 20\,\Omega$) to ($L = 10$ mH, $R = 40\,\Omega$) and at $t = 0.15$ s we return to the initial load.

Simulations results show the usefulness of this design. In any condition, the observer is able to follow the real capacitor voltages accurately. Moreover, the good agreement between estimated capacitors voltages obtained by the proposed averaged model and the circuit model from Simpower Matlab is illustrated. We notice that simulation time with fine model (circuit model implemented in Simpower Matlab) is eight times (Table 1) greater than the simulation with the averaged model. This is an important reason to use averaged modeling especially in complex converters architecture. Converter states observer and control becomes, more and more easy.

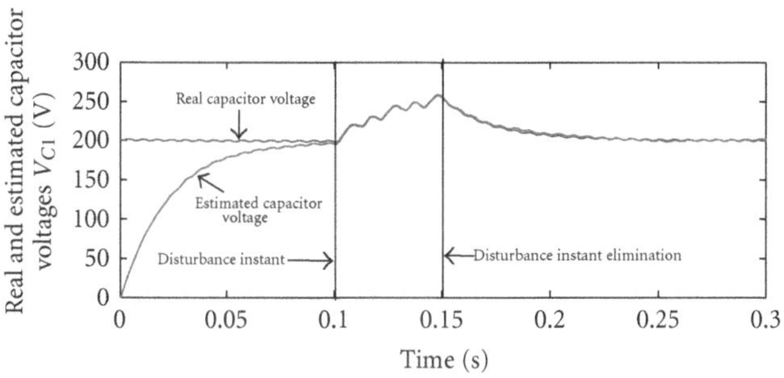

Figure 18: Real and estimated capacitors voltages V_{C1} (V) using nonideal average model with disturbance load and its elimination.

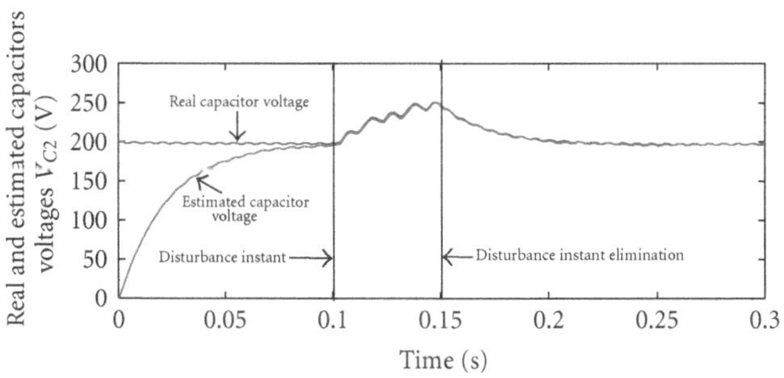

Figure 19: Real and estimated capacitors voltages V_{C2} (V) using nonideal average model with disturbance load and its elimination.

6. Conclusion

The control and monitoring of the capacitors voltages in multilevel NPC converter are essential. Measuring voltages becomes expensive and impractical because of the high voltages and power levels handled in such applications. Thus the advantage of using an observation technique becomes evident.

This paper proposed an adaptive observer to estimate the capacitor voltages in NPC inverter. This observer can reconstruct the capacitor voltages accurately from the nonideal average model inputs and replace the needed capacitor voltage sensors. Simulation results show the good precision of the proposed average model. It can be used for the NPC converters analysis instead of fastidious simulations in circuit type simulators where simulations cost is unaffordable. The obtained results are acceptable, and let us conclude that the observer is well suited for purposes control of the capacitor voltages in multilevel NPC inverter.

Appendices

A. Observability Definitions

We introduce some definitions related with the inputs applied to the system. Consider a state-affine controlled system of the form

$$\Sigma \begin{cases} \dot{x} = A(u)x + B(u), \\ y = Cx, \end{cases} \tag{A.1}$$

where $x \in \mathbb{R}^n$, $u \in \mathbb{R}^m$, $y \in \mathbb{R}^p$ with $A : \mathbb{R}^m \to M(n,m)$, $B : \mathbb{R}^m \to M(n,1)$ continuous, and $C \in M(p,n)$, where $M(k,l)$ denotes the space of $k \times l$ matrices with coefficients in $\mathbb{R}k$ (resp., l) are the number of rows (resp., columns). From now on, we will assume that $B(u) = 0$ without loss of generality.

Notation. let $\varphi_u(\tau,t)$ denotes the transition matrix of

$$\frac{d}{d\tau}\varphi_u(\tau,t) = A(u(\tau))x + \varphi_u(\tau,t), \\ \varphi_u(\tau,t) = I, \tag{A.2}$$

with the classical relation $\varphi_u(t_1,t_2)\varphi_u(t_2,t_3) = \varphi_u(t_1,t_3)$

We then define the following:

(i) The observability gramian

$$\Gamma(t,T,u) = \int_t^{t+T} \varphi_u^T(\tau,t)C^T C\varphi_u(\tau,t)d\tau. \tag{A.3}$$

(ii) The universality index

$$\gamma(t,T,u) = \min_i(\lambda_i(\Gamma(t,T,u))), \tag{A.4}$$

where $\lambda_i(M)$ stands for the eigenvalues of a given matrix M. The input functions are assumed to be measurable and such that $A(u)$ is bounded on the set of admissible inputs of $\mathbb{R}^+$.

Now, we give definitions of regular persistence of input u.

Definition 4 (regular Persistence). A measurable bounded input u is said to be regularly persistent for the state-affine system (Σ) if there exists $T \succ 0$, $\alpha \succ 0$, $t_0 \succ 0$ such that $\gamma(t,T,u) \succ 0$ for every $t \geq t_0$.

B. Regular Persistence

This proof is based on the demonstration in [25].

The dynamics of the estimation error $\varepsilon_j = Z_j - \chi_j$ is given by $\dot{\varepsilon}_j = (A(S_j) - P_j^{-1}C_j^T C_j)\varepsilon_j + \Delta\Gamma_j(S_j,\chi_j,Z_j)$, where $\Delta\Gamma_j(S_j,\chi_j,Z_j) = \Gamma_j(S_j,Z_j) - \Gamma_j(S_j,\chi_j)$ for $j = 1,2$.

Now, let $V = V_1 + V_2$ be a Lyapunov function for the interconnected system Σ, where $V_j(\varepsilon_j) = \varepsilon_j^T P_j \varepsilon_j$ is a Lyapunov function for each subsystem Σ_j.

From the time derivative of $V_j(\varepsilon_j)$, it follows that $\dot{V}_j(\varepsilon_j) \leq -\theta_j V_j(\varepsilon_j) + \varepsilon_j^T P_j \Delta\Gamma_j(S_j,\chi_j,Z_j)$ for $j = 1,2$.

Now, adding and subtracting the term $\Delta\Gamma_j(S_j,\chi_j,Z_j)^T P_j \Delta\Gamma_j(S_j,\chi_j,Z_j)$ we have

$$\dot{V}_j(\varepsilon_j) \leq -\theta_j V_j(\varepsilon_j) + \varepsilon_j^T P_j \Delta\Gamma_j(S_j,\chi_j,Z_j) \\ \pm \Delta\Gamma_j(S_j,\chi_j,Z_j)^T P_j \Delta\Gamma_j(S_j,\chi_j,Z_j). \tag{B.1}$$

Next, regrouping the appropriate terms,

$$\dot{V}_j(\varepsilon_j) \leq -(\theta_j - 1)\|\varepsilon_j\|_{P_j}^2 - \|\varepsilon_j\|_{P_j}^2 + 2\varepsilon_j^T P_j \Delta\Gamma_j(S_j,\chi_j,Z_j) \\ - \|\Delta\Gamma_j(S_j,\chi_j,Z_j)\|_{P_j}^2 + \|\Delta\Gamma_j(S_j,\chi_j,Z_j)\|_{P_j}^2. \tag{B.2}$$

From this inequality, we get $\dot{V}_j(\varepsilon_j) \leq -(\theta_j - 1)\|\varepsilon_j\|^2_{P_j} + \|\Delta\Gamma_j(S_j, \chi_j, Z_j)\|^2_{P_j}$.

Now, from Assumption 2, we have $\|\Delta\Gamma_j(S_k, \chi_j, Z_j)\|^2_{P_k} \leq (\lambda_l \|\varepsilon_l\|^2_{P_j})_{l\in\{1,2\}\neq j}$.

Then, we find $\dot{V}_j(\varepsilon_j) \leq -(\theta_j - 1)\|\varepsilon_j\|^2_{P_j} + (\lambda_l \|\varepsilon_l\|^2_{P_j})_{l\in\{1,2\}\neq j}$.

And replacing this expression in $V = V_1 + V_2$ it follows that:

$$\dot{V}(\varepsilon) \leq \sum_{j=1}^{2} \left(-\left(\theta_j - 1\right) \left\|\varepsilon_j\right\|^2_{P_j} + \left(\lambda_l \|\varepsilon_l\|^2_{P_j}\right)_{l\in\{1,2\}\neq j} \right). \quad \text{(B.3)}$$

Take into account that the inputs are regularly persistent, and then from Lemma 1 the matrix P_j is bounded. Using the lemma on equivalence of norms, that is, there exists a positive constant μ_l such that

$$\|\varepsilon_l\|^2_{P_j} \leq \mu_l \|\varepsilon_l\|^2_{P_l}. \quad \text{(B.4)}$$

Then, we get

$$\dot{V}(\varepsilon) \leq \sum_{j=1}^{2} \left(-\left(\theta_j - 1\right) \left\|\varepsilon_j\right\|^2_{P_j} + \left(\lambda_l \|\varepsilon_l\|^2_{Pl}\right)_{l\in\{1,2\}\neq j} \right). \quad \text{(B.5)}$$

or

$$\dot{V}_j\left(\varepsilon_j\right) \leq -\sum_{j=1}^{2} \left(\left(\theta_j - 1\right) - \mu_l \lambda_l \right) \left\|\varepsilon_j\right\|^2_{P_j}. \quad \text{(B.6)}$$

Finally, we have

$$\dot{V}\left(\varepsilon_j\right) \leq V(\varepsilon(t)) e^{-\gamma(t-t_0)}. \quad \text{(B.7)}$$

For $\gamma = \min(\gamma_1, \gamma_2)$, where $\gamma_j = (\theta_j - 1) - \mu_j \lambda_j$, taking $\varepsilon = \mathrm{col}(\varepsilon_1, \varepsilon_2)$, it is easy to see that

$$\|\varepsilon(t)\| \leq K \|\varepsilon(t_0)\| e^{-\gamma(t-t_0)}. \quad \text{(B.8)}$$

References

[1] G. Baoming and F. Z. Peng, "Speed sensorless vector control induction motor drives fed by cascaded neutral point clamped inverter," in *Proceedings of the 24th Annual IEEE Applied Power Electronics Conference and Exposition (APEC '09)*, pp. 1991–1997, February 2009.

[2] B. Hu, G. Xu, M. Zhang, J. Kang, and L. Xia, "Study on a novel clamped topology of multilevel converters," in *Proceedings of the IEEE International Electric Machines and Drives Conference (IEMDC '09)*, pp. 379–384, May 2009.

[3] T. A. Lipo, M. D. Manjrekar, and P. Steimer, "Hybrid multilevel power conversion system: a competitive solution for high power applications," in *Proceedings of the 34th IEEE-IAS Annual Meeting*, pp. 1520–1527, October 1999.

[4] O. Vodyakho and C. C. Mi, "Three-level inverter-based shunt active power filter in three-phase three-wire and four-wire systems," *IEEE Transactions on Power Electronics*, vol. 24, no. 5, pp. 1350–1363, 20089.

[5] F. J. C. Padilha, W. I. Suemitsu, M. D. Bellar, and P. M. Lourenco, "Low-cost gate drive circuit for three-level neutral-point-clamped voltage-source inverter," *IEEE Transactions on Industrial Electronics*, vol. 56, no. 4, pp. 1196–1204, 2009.

[6] T. Wanjekeche, D. V. Nicolae, and A. A. Jimoh, "A cascaded NPC/H-bridge inverter with simplified control strategy and minimum component count," in *Proceedings of the IEEE Africon*, vol. 4, pp. 769–777, September 2009.

[7] R. P. Aguilera and D. E. Quevedo, "Capacitor voltage estimation for predictive control algorithm of flying capacitor converters," in *Proceedings of the IEEE International Conference on Industrial Technology (ICIT '09)*, pp. 1–6, February 2009.

[8] T. A. Meynard, H. Foch, P. Thomas, J. Courault, R. Jakob, and M. Nahrstaedt, "Multicell converters: basic concepts and industry applications," *IEEE Transactions on Industrial Electronics*, vol. 49, no. 5, pp. 955–964, 2002.

[9] E. Babaei, "A cascade multilevel converter topology with reduced number of switches," *IEEE Transactions on Power Electronics*, vol. 23, no. 6, pp. 2657–2664, 2008.

[10] A. Nabae, I. Takahashi, and H. Akagi, "A neutral-point clamped PWM inverter," in *Proceedings of the IEEE-IAS*, pp. 761–766, October 1981.

[11] H. Zhang, S. J. Finney, A. Massoud, and B. W. Williams, "An SVM algorithm to balance the capacitor voltages of the three-level NPC active power filter," *IEEE Transactions on Power Electronics*, vol. 23, no. 6, pp. 2694–2702, 2008.

[12] J. A. Ulrich and A. R. Bendre, "Floating capacitor voltage regulation in diode clamped hybrid multilevel converters," in *Proceedings of the IEEE Electric Ship Technologies Symposium (ESTS '09)*, pp. 197–202, April 2009.

[13] Y. S. Lai, Y. K. Chou, and S. Y. Pai, "Simple PWM technique of capacitor voltage balance for three-level inverter with DC-link voltage sensor only," in *Proceedings of the 33rd Annual Conference of the IEEE Industrial Electronics Society (IECON '07)*, pp. 1749–1754, November 2007.

[14] A. Birouche, *Contribution sur la synthèse d'observateurs pour les systèmes dynamiques hybrides [Thèse de doctorat]*, Institut National de Polytechnique de Lorraine, Loraine, France, 2006.

[15] S. Bonnabel and P. Rouchon, "Control and observer design for nonlinear finite and infinite dimensional systems," in *Lecture Notes in Control and Information Sciences*, vol. 322, pp. 53–65, Springer, 2005.

[16] M. Morari, A. Bemporad, and G. Ferrari-Trecate, "Observability and controllability of piecewise affine and hybrid systems," *IEEE Transactions on Automatic Control*, vol. 45, no. 10, pp. 1864–1876, 2000.

[17] H. Nademi, A. Das, and L. Norum, "Nonlinear observer-based capacitor voltage estimation for sliding mode current controller in NPC multilevel converters," in *Proceedings of the IEEE Trondheim PowerTech*, pp. 1–7, 2011.

[18] A. Gopinath, A. A. S. Mohamed, and M. R. Baiju, "Fractal based space vector PWM for multilevel inverters—a novel approach," *IEEE Transactions on Industrial Electronics*, vol. 56, no. 4, pp. 1230–1237, 2009.

[19] M. A. Trabelsi, "Modélisation et Commande des Systèmes Physiques à Topologie Variable: Application au Convertisseur Multicellulaire," L'Institut National des Sciences Appliquées de Lyon, 2009.

[20] J. Chen, R. Erickson, and D. Maksimović, "Averaged switch modeling of boundary conduction mode dc-to-dc converters," in *Proceedings of the 27th Annual Conference of the IEEE Industrial Electronics Society (IECON '2001)*, pp. 844–849, December 2001.

[21] B. Urmila and D. Subba Rayudu, "Optimum space vector PWM algorithm for three-level inverter," *ARPN Journal of Engineering and Applied Sciences*, vol. 6, no. 9, pp. 24–36, 2011.

[22] K. Ammous, E. Haouas, and S. Abid, "Averaged modelling of multilevel converters," *The International Journal for*

Computation and Mathematics in Electrical and Electronic Engineering, vol. 29, no. 3, pp. 626–646, 2010.

[23] V. Andrieu, L. Praly, and A. Astolfi, "High gain observers with updated gain and homogeneous correction terms," *Automatica*, vol. 45, no. 2, pp. 422–428, 2009.

[24] H. Khalil, *Nonlinear Systems*, Prentice-Hall, 2nd edition, 1996.

[25] J. De León Morales, M. F. Escalante, and M. T. Mata-Jiménez, "Observer for DC voltages in a cascaded H-bridge multilevel STATCOM," *IET Electric Power Applications*, vol. 1, no. 6, pp. 879–889, 2007.

2

Analysis of a Microgrid under Transient Conditions Using Voltage and Frequency Controller

Monika Jain,[1] Sushma Gupta,[1] Deepika Masand,[2] and Gayatri Agnihotri[1]

[1] *Electrical Engineering Department, Maulana Azad National Institute (M.A.N.I.T), Deemed University, Bhopal (M.P.) 462052, India*
[2] *Electrical and Electronic Department of Engineering, Oriental college of science and technology (O.I.S.T), R.G.P.V University, Bhopal (M.P.) 462021, India*

Correspondence should be addressed to Monika Jain, chahatfeb2035@yahoo.co.in

Academic Editor: Pavol Bauer

This paper presents an investigation of voltage-and-frequency-(VF-) based battery energy storage system (BESS) controller used in micro grid for analyzing the optimum capability of plant. Microgrid is formed by using three hydropower plants feeding three-phase four-wire load. The proposed controller is used for load balancing, harmonic elimination, load leveling, and neutral current compensation. The proposed BESS controller permits the selection of an optimum voltage level of battery and allows independent current control of each phase. The main emphasis is given on maintaining constant voltage and frequency within the micro grid during transient conditions. Micro grid with power plant and its controller is modeled in MATLAB/Simulink using Power System Blockset (PSB) toolboxes.

1. Introduction

Existing and new generating plants cause depletion of fossil fuels, global warming, and pollution. This diverts researchers' attention towards renewable energy resources. The renewable energy sources such as wind, small-hydro, biomass, geothermal are inexhaustible in nature and easily available in our country. Also remotely located villages, islands, hills, military equipment, and so forth are some of the areas which are mainly isolated from the power system grid and in these areas wind and hydro-energy are available abundantly. Thus for supplying energy in such areas, isolated system, micro-or minigrid system, is an emerging concept and draws the attention of many researchers. Microgrid is the small scale energy system equipped with distributed energy resources, electronic equipments, loads, ancillary facilities, and so forth. In micro grid system, micro-generators are operated parallel to the loads, usually through some voltage-source-inverter-(VSI-) based or current-source-inverter-based system that can operate in controllable systems. The concept of micro grid was first introduced in 2001 by Lasseter [1]. A number of new technologies and innovative ideas are proposed [2–10] till now. Uno et al. [2] describe the supply and demand stability in micro grid for different connections. Later on the scheme based on power electronics devices with distributed generation (DG) is used to form micro grid [3]. The micro grid thus formed can work in grid connected mode as well as in islanded mode with the help of power electronic converter. A new concept was introduced in which two inverters are used, one in shunt and another in series with individual DG formsing micro grid [4]. The former is used for balancing the voltage and the latter is used for balanced the line current. Small-signal stability for conventional power systems is well established, but for inverter-based micro grids there is need to analysis, how the circuit and control features give rise to particular oscillatory modes and which of them have poor damping. Thus the author develops the state-space modeling for such inverters based micro-grid [5]. Moving further micro grid not only deals with power sharing or transient stability problems but also deals with over voltage, under voltage, oscillations, offset voltage, and so on. Hence Wu et al. describe the existing problems and the possible solution in micro grid [6]. When micro grid is working in islanded mode proper synchronization between generators is required, if not so large transient current will flow. A concept of soft connection scheme based on classical zero crossing detector was introduced for synchronization of generators forming micro grid [7]. In spite of synchronization there is another problem of proper load sharing in autonomous micro grid. Thus a new concept was developed by Majumder et al. [8], in which high gain angle droop control scheme is used for proper load sharing in micro grid. A supplementary loop is proposed for each DG to stabilize the system using high angle droop gain. Further a mathematical model that predicts the effect of voltage offset on reactive power sharing between inverters in micro grid was introduced [9]. Active synchronizing control scheme was introduced by author, which is required when micro grid consists of multiple DGs. Further the author concluded that the scheme is useful for islanded and reconnected mode [10].

Past research indicates that a self excited induction generator (SEIG) has emerged as a suitable candidate for isolated system [11–14] because of its many advantages such as its small size and weight, robust construction, reduced unit size and maintenance cost, brushless rotor construction, absence of a separate excitation source, and self protection against severe over loads and short circuits. In order to utilize the full potential of the energy source, several such generators may have to be operated in parallel. There is no problem of synchronization and hunting in SEIG thus here it is used as generator in forming micro grid.

In this paper micro grid is formed using hydro power plants located at some distance far from each power plant. Hydro-electric plants can provide cheap, clean, and reliable electricity for villages. Power generated by these plants is almost constant for 24 h a day. If light load condition arises the extra power will be stored in battery and if load increases then deficit power will be supplied by the battery. Hydro-electric plants convert the kinetic energy of a waterfall into electric energy [15]. In this paper main emphasis is given on the transient condition within the micro grid and optimum utilization of BESS controller during this condition.

Focus of the Paper. So many authors work on the power sharing between generators during grid connected mode and islanded mode. In case of grid connected mode, the generators of micro grid are connected to the load as well as to the main grid. The loads are shared by the main grid and generator of micro grid. Hence the generators of micro grid are not fully loaded. Now if a fault occurs in the main grid it gets isolated from the rest of the system (forming micro grid), this mode is known as islanded mode. The generators of micro grid (which are not fully loaded) now shared power within themselves and maintain constant voltage and frequency at load ends.

But this paper deals with islanded mode only and the generators which are forming micro grid are fully loaded during healthy condition. If the fault occurs in any of the generator and it gets isolated then also the load of the faulty generator is shared by the rest of the generators in micro grid and thus maintains voltage and frequency constant at the load ends.

In addition, this paper also describes the concept of micro grid in which SEIGs are connected in parallel forming micro grid with their individual load. The hydroturbines are used as prime mover for SEIGs which are used to deliver power. The system consists of squirrel cage induction generator with capacitor bank. The capacitor bank is used to cater reactive power requirement. A controller is connected in parallel to the load to improve the stability of the system under transients. In spite of the above advantage SEIG is superior compared to other electric generator because there is no need of synchronization between two asynchronous generators.

2. System Description and Working

The schematic diagram of micro grid is shown in Figure 1. The system consists of three small hydro-turbines which are used as prime movers for SEIGs. Single hydro power plant consists of asynchronous generator (SEIG) with the excitation capacitor, consumer load and a voltage source converter with a battery energy storage system (BESS) based voltage and frequency controller. The self-excited induction generator has been found to be an attractive competitor to the synchronous generator due to several advantages as discussed in above section. A major drawback of SEIG's is their poor voltage regulation under varying load conditions even for fixed prime movers speeds. The star connected 3-phase capacitor bank is used for generator excitation and value of excitation capacitor is selected to generate the rated voltage at no load condition while additional demand of power during full load and faulty condition is met by controller because of having capability of bidirectional flow of active and reactive power.

The reactive power requirement can be compensated by converter based compensators like DSTATCOM and DVR. This paper deals with BESS as a controller. The BESS controller consists of DSTATCOM and a battery connected parallel to DC link of DSTATCOM. Basic principle of active and reactive power sharing starts with the active and reactive power flows between two ac sources through line impedance $Z(= R + jX)$ as shown in Figure 2. The active power P and reactive power Q are given by:

$$P = \frac{V_1}{(R^2 + X^2)[R(V_1 - V_2\cos\delta) + V_2\sin\delta]},$$
$$Q = \frac{V_1}{(R^2 + X^2)[-RV_2\sin\delta + X(V_1 - V_2\cos\delta)]}. \quad (1)$$

If the line is mainly inductive, where the resistance can be neglected, the typical phase angle between two voltages is

$$P = \frac{V_1 V_2}{X(\sin\delta)},$$
$$Q = \frac{V_1}{X(V_1 - V_2)}. \quad (2)$$

These equations imply that the active power is proportional to phase angle and reactive power is proportional to voltage differences between the two systems, respectively. For this

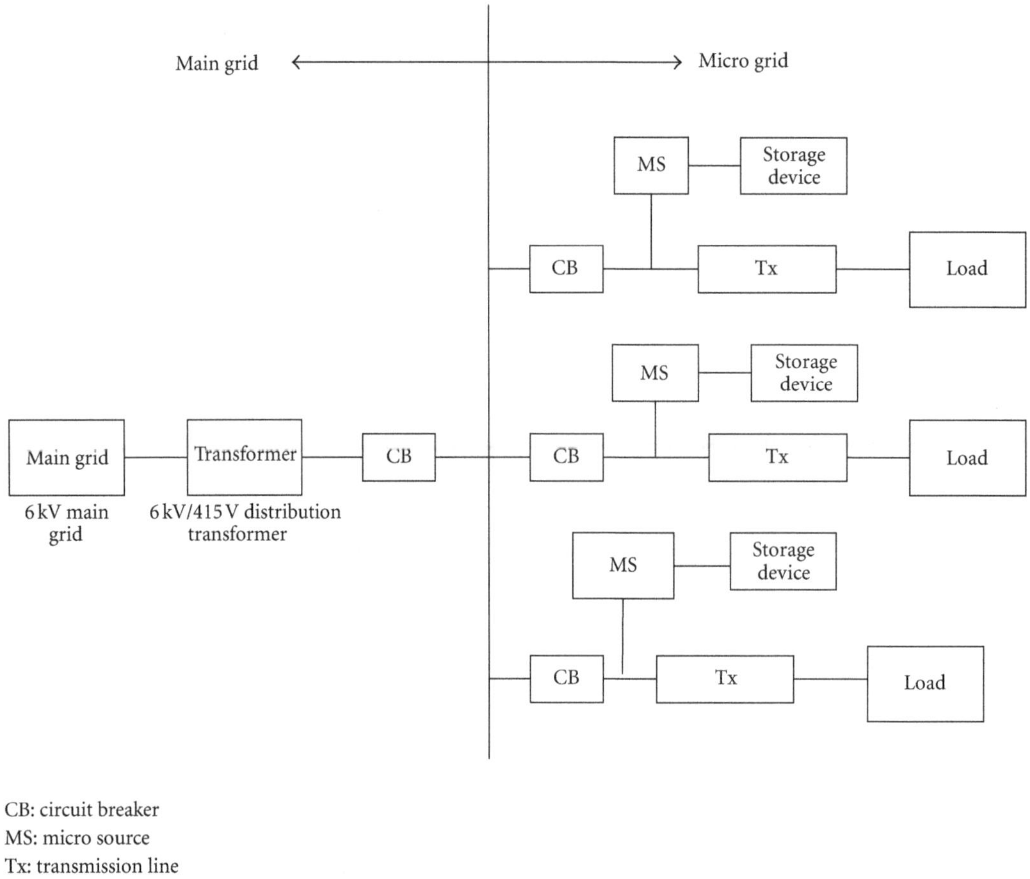

FIGURE 1: The micro grid with micro source, load, and controller.

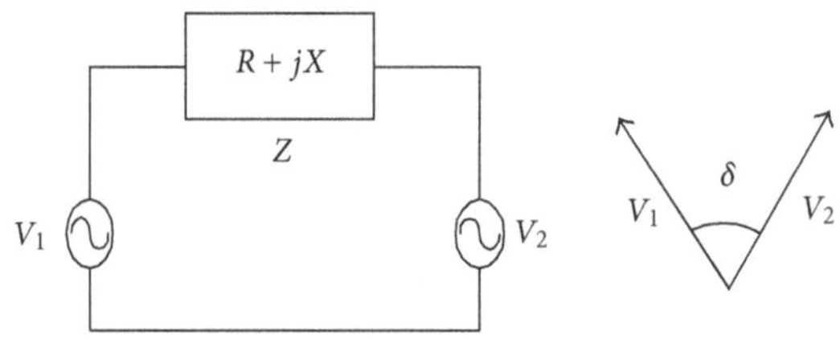

FIGURE 2: Two generators connected by line impedance.

reason, in BESS controller two PI controllers are there to control active and reactive powers by varying its output frequency (that generates the phase-angle difference as a result) and voltage levels. As a result, the *P-F* and *Q-V* droop strategies provide a stable active and reactive power sharing between the generators. Thus constant power requirement of SEIGs and load is met by IGBT based current controlled voltage source converter (CC-VSC) along with battery energy storage system (BESS) at its DC link. The output of VSC is connected through the AC filtering inductor at the point of common coupling between every micro source and load.

The basic principle of regulating the frequency by the controller is that it maintains the constant output power at the generator terminal voltage because input power from the turbine is constant so the frequency at the terminal remains constant [11]. In given configuration if no load condition arises the extra power is used to charge the battery and during heavy load condition CC-VSC is able to regulate the reactive power and frequency. If one of the SEIG is shut down due to faulty condition, the terminal voltage and frequency are still regulated.

3. Control Strategy

The proposed system consists of three hydro power plants which are operating parallel forming micro grid. Each plant consists of self excited induction generator, with load and its controller. The control strategy for the other two plants is also the same. Figure 3 shows the control strategy of the proposed voltage and frequency controller which is based on the generation of reference source currents [13–15]. Three-phase reference source currents are having two components, one is in-phase component or active power

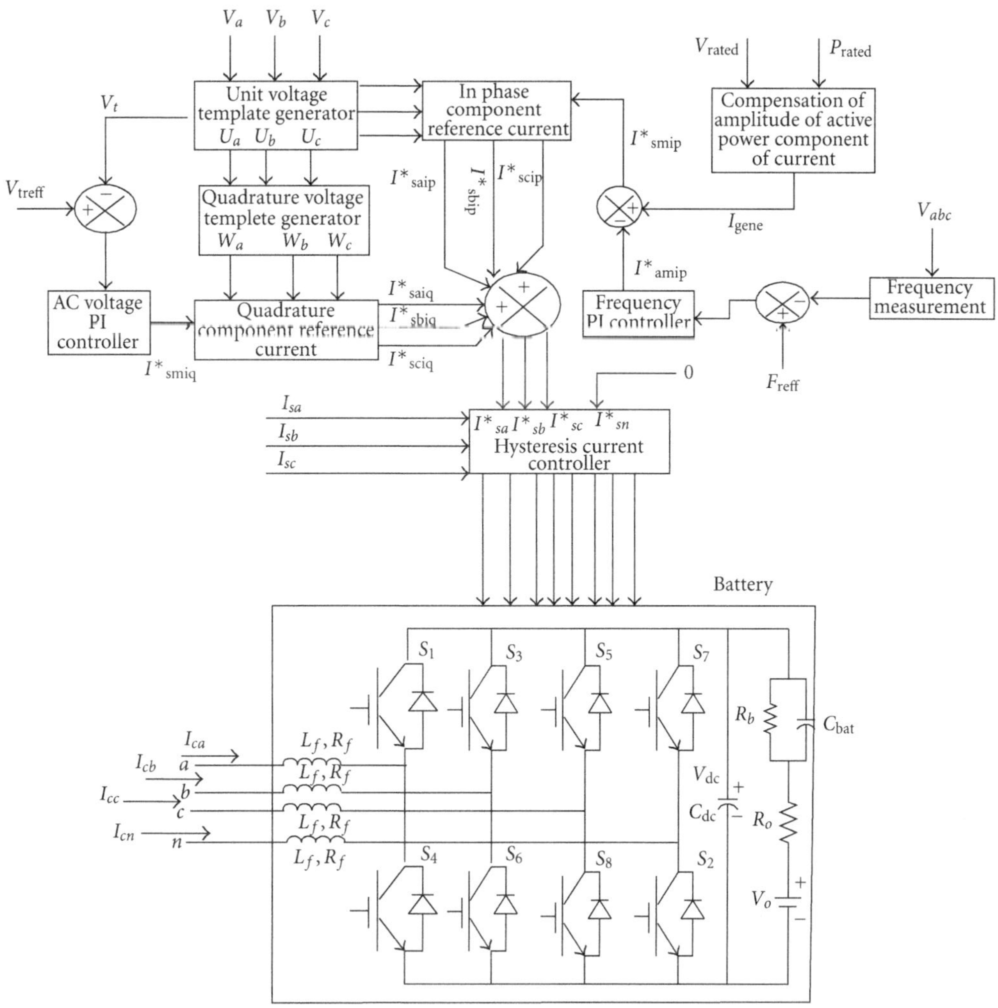

FIGURE 3: Schematic diagram of control scheme for SEIG –STATCOM system.

component (i^*_{saip}, i^*_{sbip}, i^*_{scip}) while the other one is in-quadrature component or reactive components (i^*_{saiq}, i^*_{sbiq}, i^*_{sciq}).

In-phase component of reference source current is estimated by taking the difference of amplitude of rated generated current (I_{gene}) and output of frequency PI controller (I_{smip}) [13].

The amplitude of rated current of the generator is calculated as

$$I_{gene} = \frac{\sqrt{2}(P_{rated})}{\sqrt{3}(V_{rated})}, \quad (3)$$

where P_{rated} is the rated generated power (P_{gene}) and V_{rated} is rms voltage of the asynchronous generator.

The frequency error is given as

$$f_{error(q)} = f_{reff(q)} - f_{(q)}, \quad (4)$$

where f_{reff} is reference frequency (50 Hz in present system) and f is the frequency of the voltage of an asynchronous generator. The instantaneous value of f is estimated using phase locked loop (PLL).

At the qth sampling instant, the output of PI frequency controller (P_c) is as:

$$P_{c(q)} = P_{c(q-1)} + K_{pf}\left\{f_{error(q)} - f_{error(q-1)}\right\} + K_{if} f_{error(q)}, \quad (5)$$

where K_{pf} and K_{if} are the proportional and integral gains of frequency PI controller.

From (1) and (2) at the qth sampling instant, the amplitude of active component of current is

$$I^*_{smip} = I_{gene(q)} - I_{smip(q)}. \quad (6)$$

The instantaneous values of in-phase components of reference source currents are estimated as

$$i^*_{saip} = I^*_{smip} U_a; \qquad i^*_{sbip} = I^*_{smip} U_b; \qquad i^*_{scip} = I^*_{smip} U_c. \quad (7)$$

Multiplication of I^*_{smip} with in-phase unit amplitude templates (U_a, U_b, U_c) yields the in-phase component of reference source currents. These templates are three-phase sinusoidal functions which are derived by dividing the ac voltages

V_a, V_b, and V_c by their amplitude V_t. The instantaneous line voltages at the IAGs terminals (V_a, V_b, and V_c) are considered close to sinusoidal and their amplitude is computed as

$$V_t = \left\{\left(\frac{2}{3}\right)\left(V_a^2 + V_b^2 + V_c^2\right)\right\}^{1/2}. \tag{8}$$

The unity amplitude templates having instantaneous value in phase with instantaneous voltage (V_a, V_b, and V_c), which are derived as [3, 8, 9]:

$$U_a = \frac{V_a}{V_t}; \qquad U_b = \frac{V_b}{V_t}; \qquad U_c = \frac{V_c}{V_t}. \tag{9}$$

To generate the quadrature component of reference source current, another set of sinusoidal quadrature unity amplitude templates (W_a, W_b, W_c) is obtained from in-phase unit vectors (U_a, U_b and U_c). The multiplication of these components with output of PI terminal voltage controller (I^*_{smiq}) gives the quadrature or reactive component of reference source currents.

The ac voltage error V_{error} at the qth sampling instant is

$$V_{\text{error}(q)} = V_{\text{treff}(q)} - V_{t(q)}, \tag{10}$$

where $V_{\text{treff}(q)}$ is the amplitude of reference ac terminal voltage and $V_{t(q)}$ is the amplitude of the sensed three-phase ac voltage at the IAG terminals at qth instant.

The output of the PI controller ($I^*_{\text{smiq}(q)}$) for maintaining constant ac terminal voltage at the qth sampling instant is expressed as

$$I^*_{\text{smiq}(q)} = I^*_{\text{smiq}(q-1)} + K_{pa}\left\{V_{\text{error}(q)} - V_{\text{error}(q-1)}\right\} + K_{ia}V_{\text{error}(q)}, \tag{11}$$

where K_{pa} and K_{ia} are the proportional and integral gain constants of the proportional integral (PI) controller. $V_{\text{error}(q)}$ and $V_{\text{error}(q-1)}$ are the voltage errors in qth and $(q-1)$th instant. The output of PI controller ($I^*_{\text{smiq}(q-1)}$) is considered the amplitude of quadrature component of the reference source current at $(q-1)$th instant.

The instantaneous quadrature components of the reference source currents are estimated as

$$i^*_{\text{saiq}} = I^*_{\text{smiq}}\, W_a; \qquad i^*_{\text{sbiq}} = I^*_{\text{smiq}}\, W_b; \qquad i^*_{\text{sciq}} = I^*_{\text{smiq}}\, W_c, \tag{12}$$

where W_a, W_b and W_c are another set of unit vectors having a phase shift of 90° leading with respect to the corresponding unit vectors U_a, U_b, and U_c which are estimated as

$$\begin{aligned} W_a &= \frac{-U_b}{\sqrt{3}} + \frac{U_c}{\sqrt{3}}, \\ W_b &= \frac{(\sqrt{3}U_a)}{\sqrt{2}} + \frac{(U_b - U_c)}{2\sqrt{3}}, \\ W_c &= \frac{-(\sqrt{3}U_a)}{\sqrt{2}} + \frac{(U_b - U_c)}{2\sqrt{3}}. \end{aligned} \tag{13}$$

The total reference source currents are the sum of in-phase and quadrature components of the reference source current as:

$$\begin{aligned} i^*_{\text{sa}} &= i^*_{\text{saiq}} + i^*_{\text{saip}}, \\ i^*_{\text{sb}} &= i^*_{\text{sbiq}} + i^*_{\text{sbip}}, \\ i^*_{\text{sc}} &= i^*_{\text{sciq}} + i^*_{\text{scip}}. \end{aligned} \tag{14}$$

The reference source currents (i^*_{sa}, i^*_{sb}, and i^*_{sc}) are compared with the sensed source currents (i_{sa}, i_{sb}, and i_{sc}). The current errors are computed as

$$\begin{aligned} i_{\text{sa error}} &= i^*_{\text{sa}} - i_{\text{sa}}, \\ i_{\text{sb error}} &= i^*_{\text{sb}} - i_{\text{sb}}, \\ i_{\text{sc error}} &= i^*_{\text{sc}} - i_{\text{sc}}. \end{aligned} \tag{15}$$

These error signals are then given to ON/OFF switching patterns of the gate drive (IGBT) signals to generated pulse from the hysteresis current controller.

3.1. Design of BESS Controller. Battery Energy Storage System (BESS) consist of lead-acid batteries, power conversion unit, and control system. The latter two are already discussed in the above section. The batteries are mainly used by power generating utilities, power distribution utilities, and power consumers. Main economic benefits of using BESS by power distributor and consumers is its ability to shave peak demands of loads by mean of discharging stored battery energy.

Lead-acid batteries are electrochemical devices that can be charge and discharge many times. Each cell of this battery consists of two electrodes. The positive electrode is of lead oxide (PbO_2) and negative electrode is of sponge lead (Pb) [16]. When these electrodes are immersed in sulfuric acid electrolyte, a nominal open-circuit potential of 2V/cell is created. When circuit between two electrodes is closed, battery discharges its stored energy.

Thevenin equivalent circuit of the battery-based model [17, 18] is shown in Figure 4.

The terminal voltage of the equivalent battery is obtained as

$$V_{\text{battery}} = \left(\frac{2\sqrt{2}}{\sqrt{3}}\right)V, \tag{16}$$

where V is the line rms voltage ($V = 415$ V).

At no load, voltage across the terminals of SEIG is rated voltage of 415 V. The battery voltage must be more than the peak of line voltage for satisfactory operation of Hysteresis controller. A slightly higher round-off value of 750 V is considered.

Since the battery is an energy storage unit, its energy is represented in kWh when a battery is used to model the

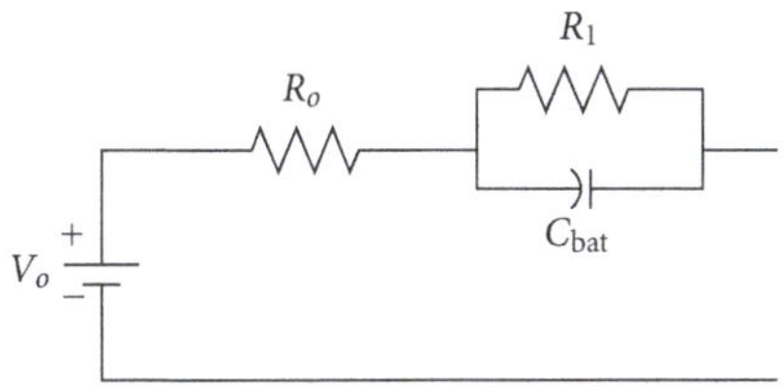

FIGURE 4: Thevenin equivalent circuit of battery model.

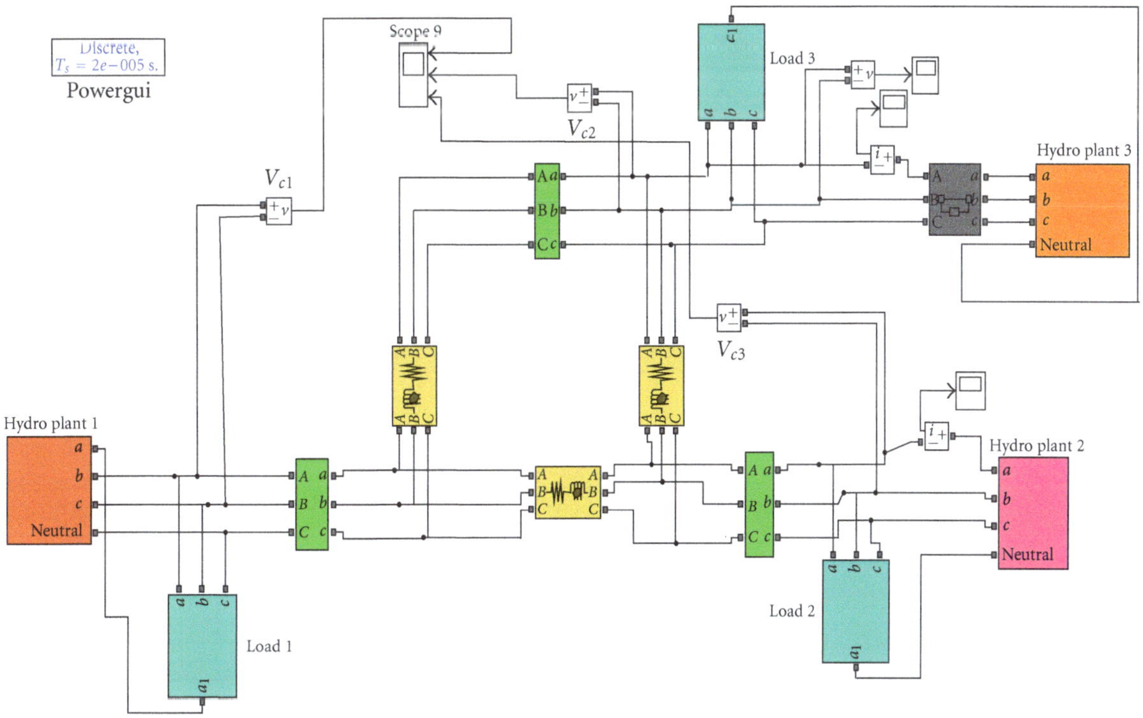

FIGURE 5: MATLAB/Simulink model of microgrid.

battery unit, the equivalent capacitance can be determined from:

$$C_{bat} = \frac{(\text{kwh} * 3600 * 10^3)}{0.5(V_{\text{ocmax}}^2 - V_{\text{ocmax}}^2)}, \tag{17}$$

$$C_{bat} = \frac{(6 * 4 * 3600 * 10^3)}{0.5[(760^2) - (740^2)]}, \tag{18}$$

$$C_{bat} = 5760F.$$

In the given Thevenins equivalent model R_o is the equivalent resistance (external + internal) of a parallel/series combination of the battery, which is usually a small value. The parallel circuit of R_b and C_{bat} is used to describe the energy and voltage during charging or discharging. R_b in parallel with C_{bat} represents self discharging of the battery, since the self discharging current of a battery is small, the value of resistance R_b is large. From these equations, different parameters of a battery (R_b, C_{bat}) are selected and given in the appendix. Here it is considered that the battery is having energy storage of 6 kW for 4 h peaking capacity, and its variation in voltage of 740–760 V.

The greatest potential for BES is in multifunction applications. These include

(i) Peak shaving of customer load.

(ii) Load leveling—also benefits the electric utility.

(iii) Energy/power for uninterruptable loads.

(iv) Generation reserve requirement.

(v) Automatic generation control (area control error (ACE) requirement).

(vi) Deferment of new transmission construction.

(vii) Reduction of magnetic field effects.

4. MATlAB Simulink Model

Figure 5 shows the MATLAB/Simulink based model of micro grid in which micro sources connected in parallel operate along its controller, respectively. Each micro source has its

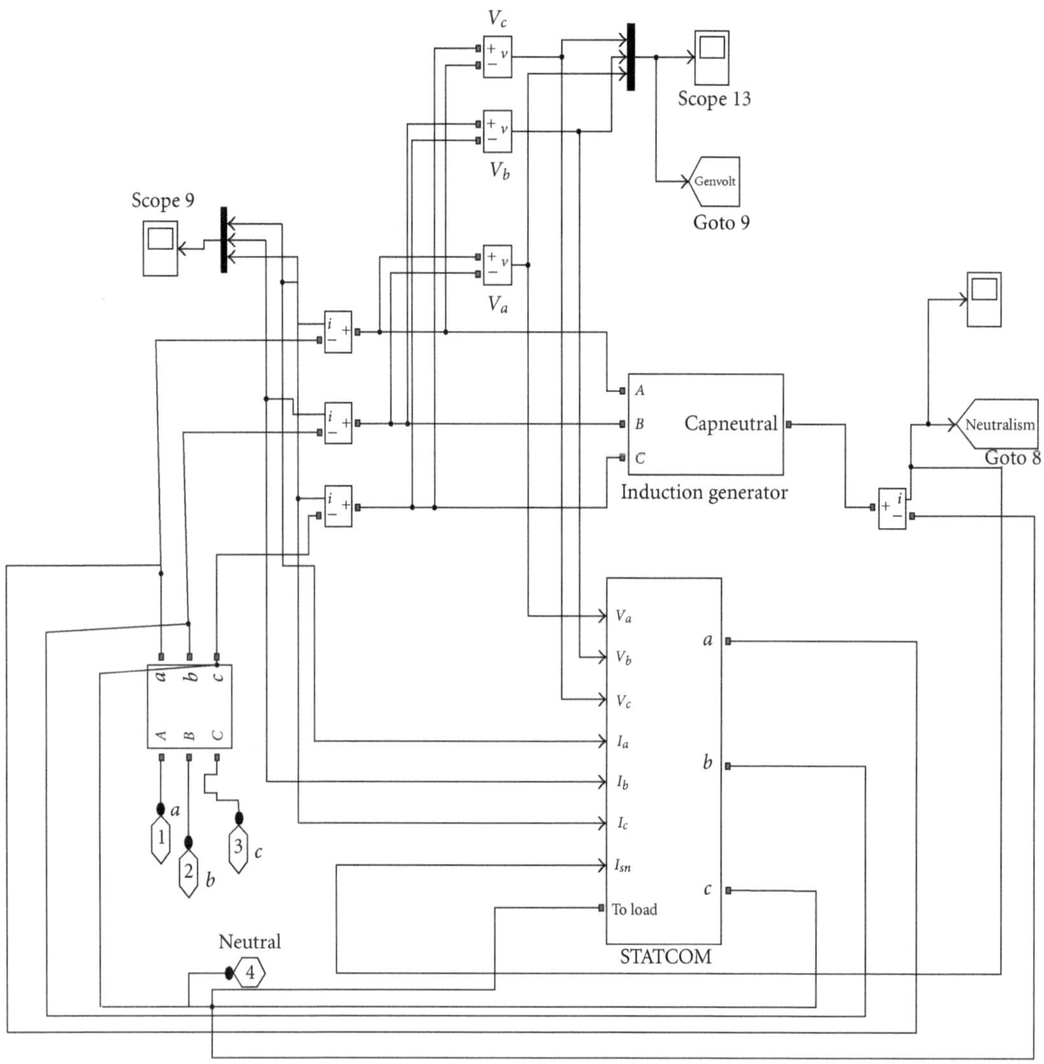

FIGURE 6: Sub system of MATLAB/Simulink model of hydropower plant using VF based BESS Controller.

own load and is connected to the other micro source through short transmission line. Line resistance and reactance is also considered. The model consists of three hydro power plants each having induction generator with its prime movers, each of 7.5 kW, 415 V, 50 Hz, 4-pole and each having star connected capacitor bank to provide reactive power at no load condition. The additional requirement of power during transient condition is provided by voltage frequency based BESS controller. The sub models of DSTATCOM controllers are shown in Figures 6 and 7 which are formed by all equations which are describe in Section 3. The proposed system is modeled using simulink and PSB toolboxes in MATLAB version of 7.8, in discrete mode at 5e-6 sec step size with ode 23tb (stiff/TR-BDF-2) solver along with the saturation characteristics of the machine as determined by synchronous speed test. Available universal bridge is used to model the VSC. Thevenin's equivalent battery energy storage system is modeled as in [13, 14, 16, 18] using passive components and a dc battery. The resistive/inductive balanced/unbalanced load is considered as linear load.

5. Results and Discussion

The performance of the proposed controller in micro grid is demonstrated with balanced/unbalanced linear, nonlinear and dynamic loads. Simulation waveforms for the entire three micro sources are presented in Figures 8–11. The subscript 1, 2, 3 are used for plant 1, plant 2, and plant 3. All the waveforms are represented using these subscripts for all type of loads. The waveforms consist of generator voltage (V_{gabc}), generator current (I_{gabc}), load current (I_{Labc}), controller current (I_{cabc}), load neutral current (I_{Ln}), load neutral current controller (I_{Lncc}), neutral current (I_n), terminal voltage (V_t), frequency (f), and battery current (I_{bat}) with respect to time (sec.), presented in the figures. Generator "3" is disconnected at 2.7 sec and reconnected at 2.9 sec. The voltage and current

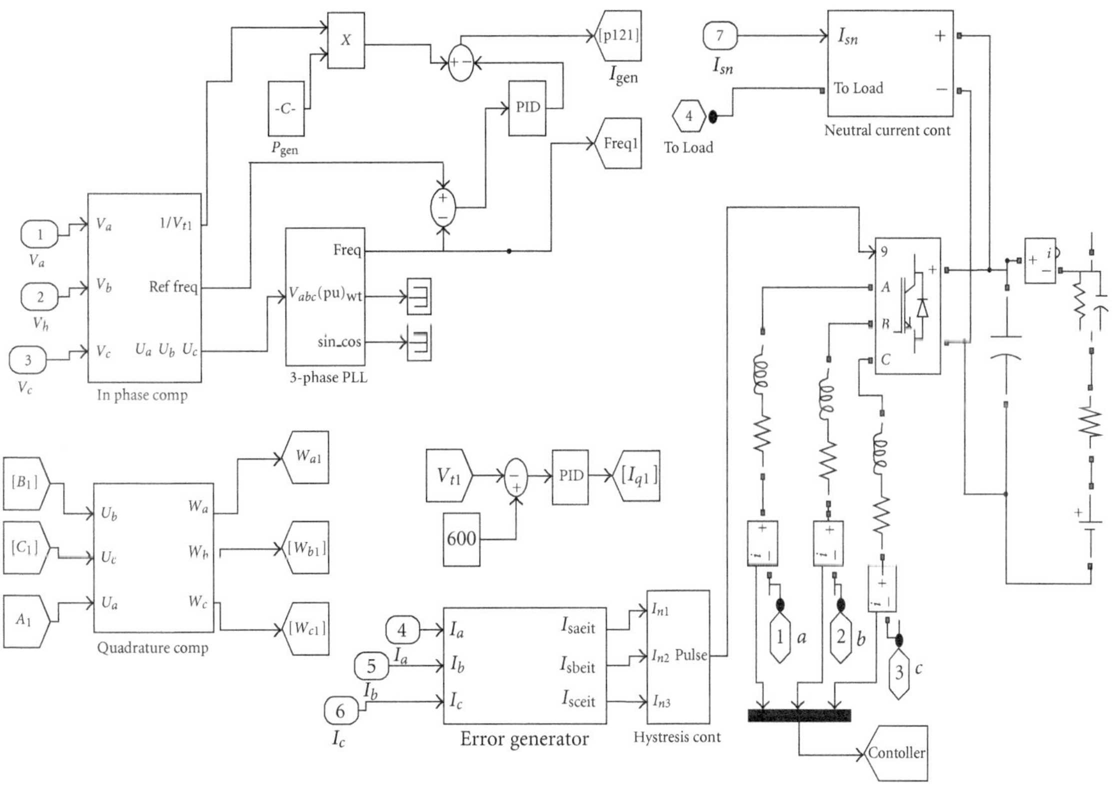

Figure 7: Sub system of MATLAB/Simulink model of hydropower plant using VF-based BESS Controller.

of the disconnected generator remain same after connection to the micro grid. In this paper the following four cases are discussed.

5.1. Performance of Microgrid with BESS Feeding Balanced/Unbalanced Resistive Load. The performance of BESS based VF controller for micro grid with balanced/unbalanced resistive load is shown in Figures 8(a) and 8(b). Before the application of consumer load, the battery consumes all the generated active power. Star connected three single phase load each of 2.5 KW is applied at 2.2 sec. At 2.3 sec one phase is open and later on at 2.4 sec another phase is open hence load becomes unbalanced. Results include charging and discharging of battery and maintaining load balancing. Later on one by one phases are reconnected to the system. Hence the balanced, unbalanced, and no load condition voltage and frequency of the system remain constant.

5.2. Performance of Microgrid with BESS Controller after Removing One Micro-Source. The performance of BESS based VF controller is shown in Figures 8(a) and 8(b). At 2.6 sec balanced resistive loads is applied to the system and at 2.7 sec fault is occur between hydro plant "3" and its load "3" so it get disconnected from the rest of the micro source. The plant remains disconnected from "2.7 to 2.9 sec" and again it is connected to the system at "2.9 sec." During the disconnected interval 2.7 sec to 2.9 sec there is discharging of battery "1" and "2" as they have to share the extra load of micro source "3." As is seen in waveform of Figure 8, terminal voltage and frequency of the system remain constant even after removal of one micro source.

5.3. Performance of Microgrid with BESS Feeding Balanced/Unbalanced Reactive Load and Removal of One Micro-Source. Similarly, the performance of BESS based VF controller for micro grid with balanced/unbalanced reactive load is shown in Figures 9(a) and 9(b). Star connected three single phase load of each 2.5 KW and 1.87 KVAR is applied at 2.2 sec. At 2.3 sec one phase is open and later on at 2.4 sec another phase is open hence load becomes unbalanced and results in charging and discharging of battery, which shows load balancing of BESS fed VF controller. Again at 2.6 sec balanced load is applied to the system. Now at 2.7 sec a 3-phase fault occurs between load and hydroplant "3" so this plant is out of order and unable to support its load from 2.7 sec to 2.9 sec. It is seen from the waveform even if one plant is off the BESS based VF controller works and supplies the power during faulty condition. In this way, BESS keeps the overall generated power constant at the generator terminal which in turn regulates the system frequency.

5.4. Performance of Microgrid with BESS Feeding Balanced/Unbalanced Nonlinear Load and Removal off One Micro Source. Similarly, the performance of BESS based VF controller for micro grid with balanced/unbalanced non-linear load is shown in Figures 10(a) and 10(b). Star connected three single phase diode rectifier based non-linear load is applied at 2.2 sec. At 2.3 sec one phase is open and later on

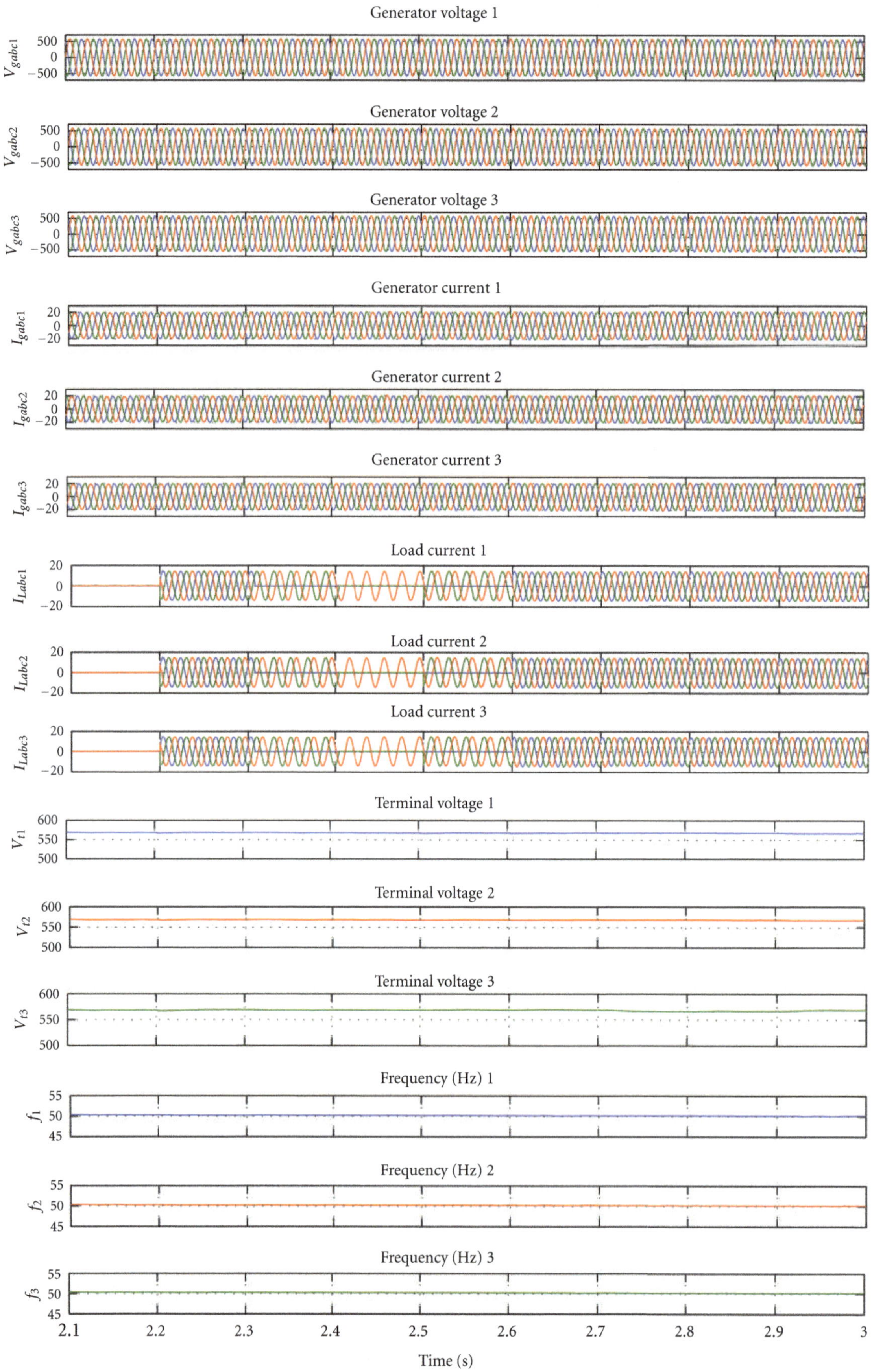

(a)

FIGURE 8: Continued.

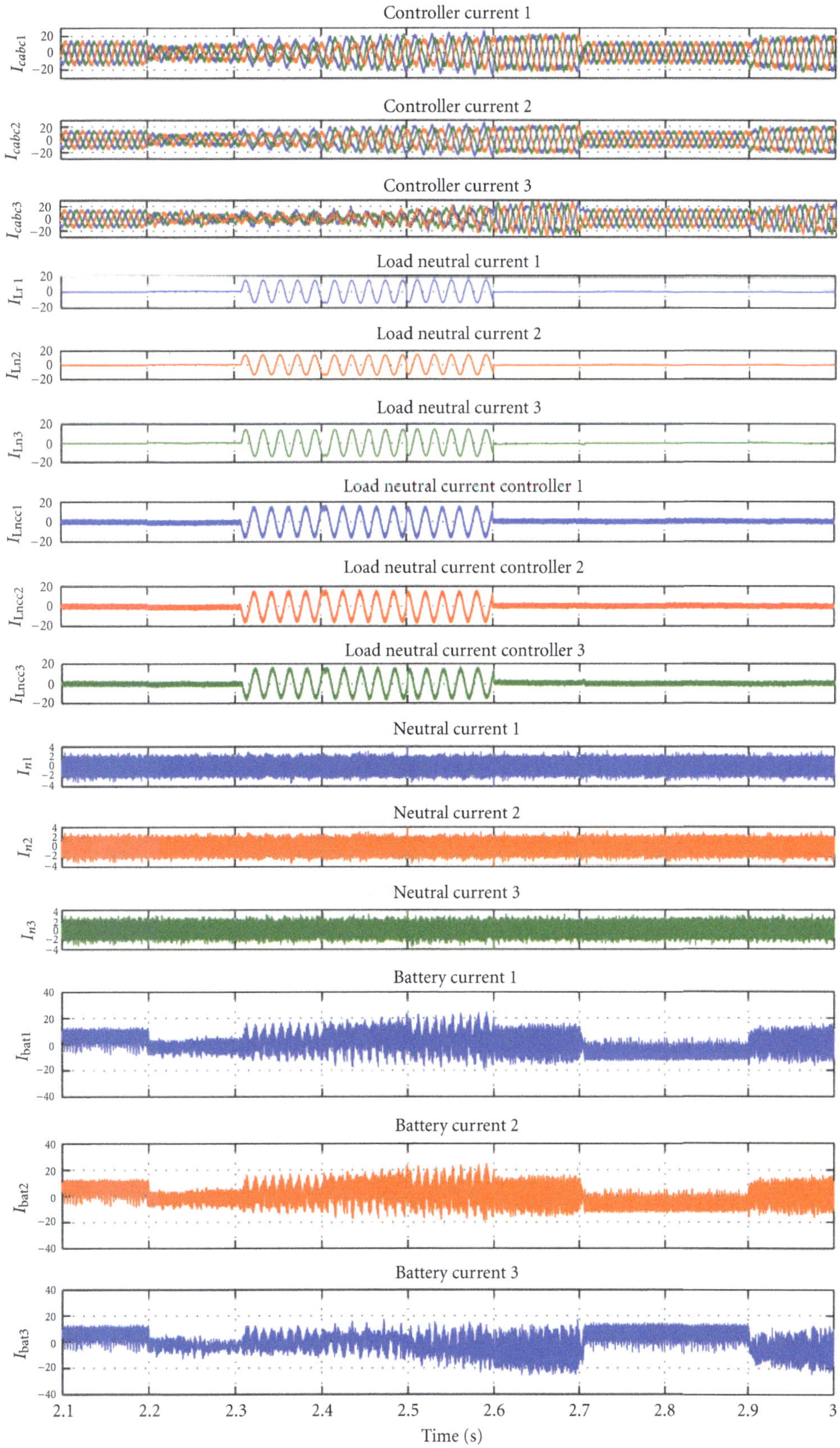

(b)

Figure 8: (a) Transient waveform of microgrid with BESS based VFC feeding balanced/unbalanced resistive load and shunting of one plant. (b) Transient waveform of microgrid with BESS based VFC feeding balanced/unbalanced resistive load and shunting of one plant.

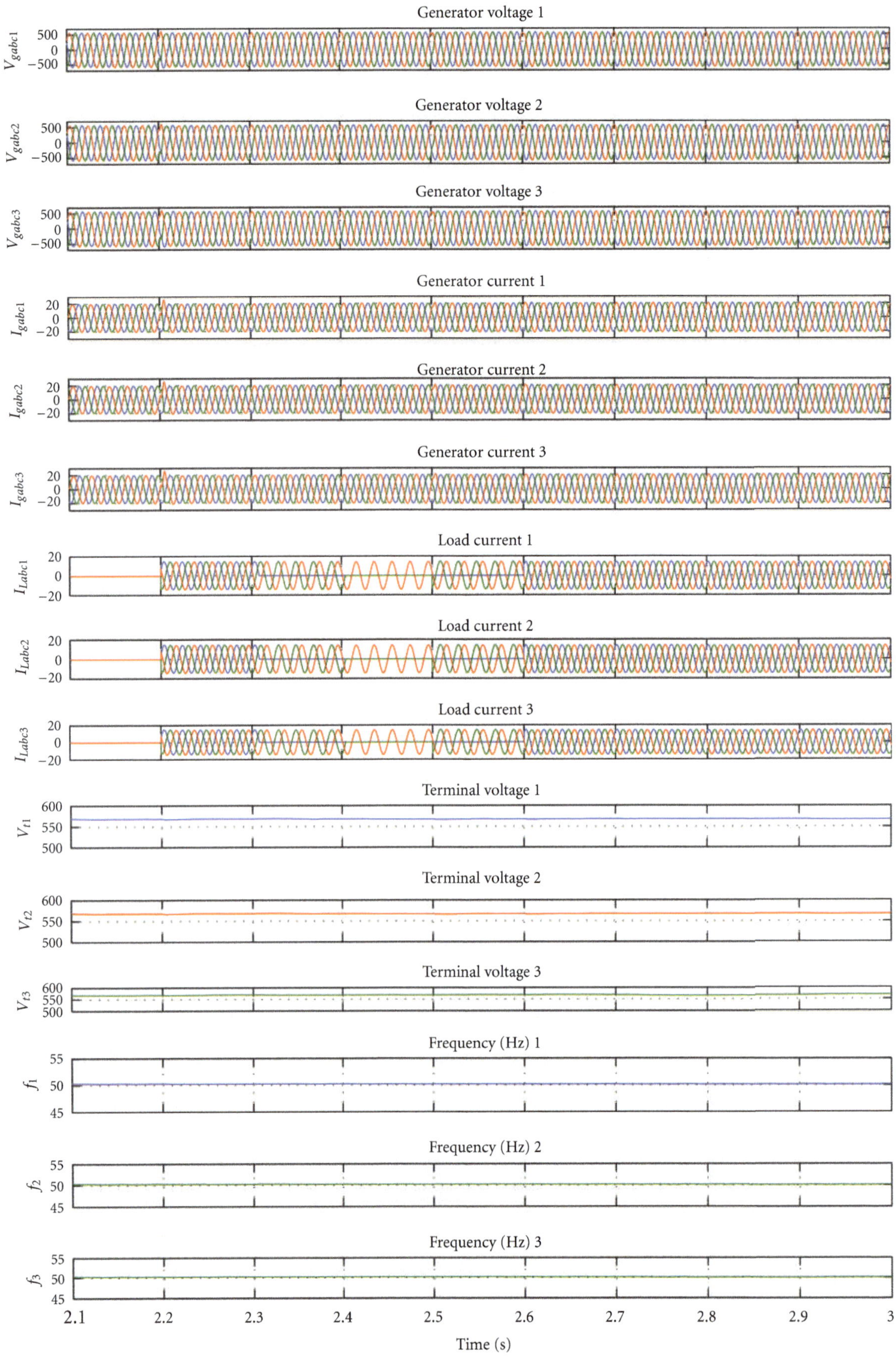

(a)

FIGURE 9: Continued.

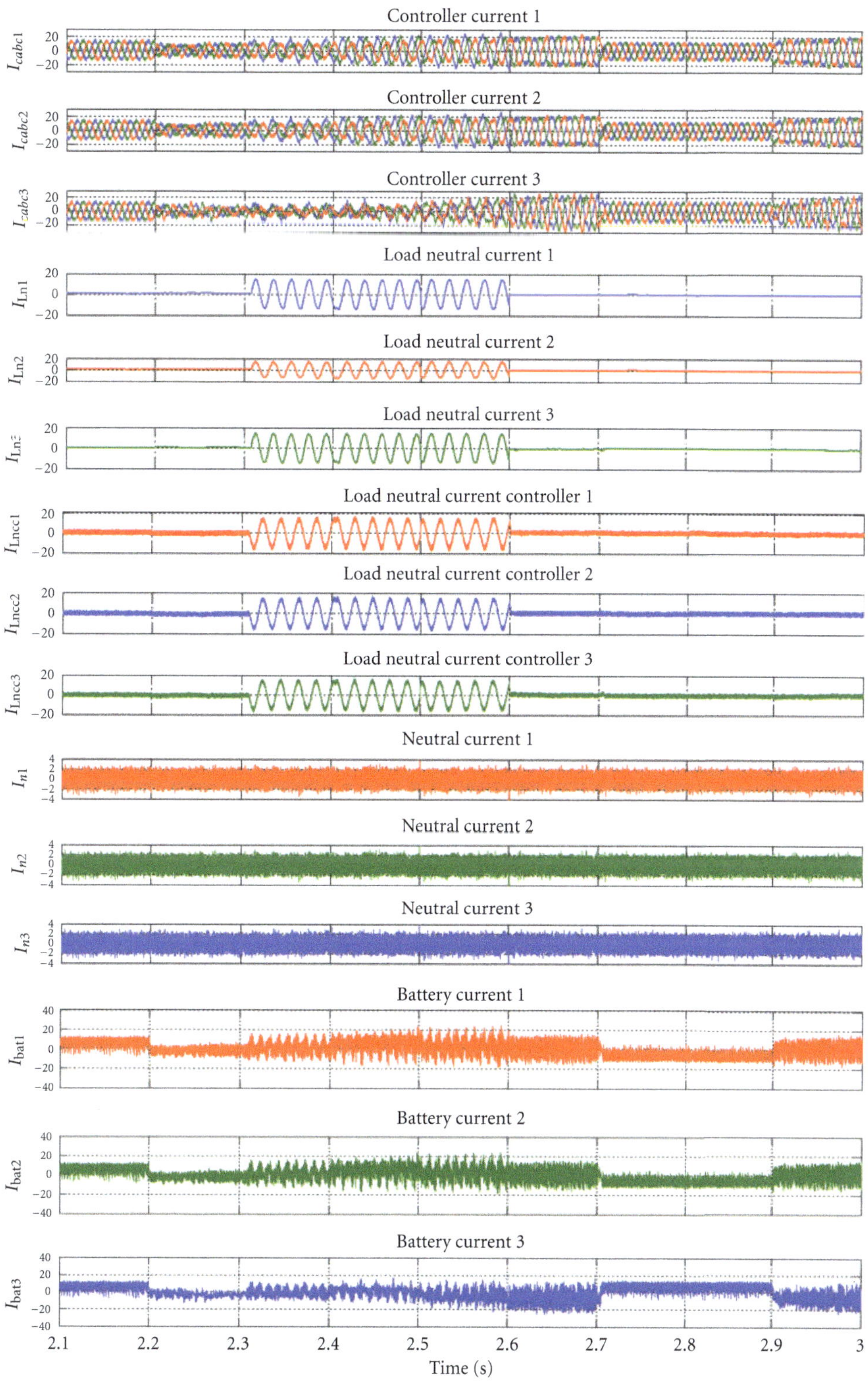

(b)

FIGURE 9: (a) Transient waveform of microgrid with BESS based VFC feeding balanced/unbalanced reactive load and removal of one plant. (b) Transient waveform of microgrid with BESS based VFC feeding balanced/unbalanced reactive load and removal of one plant.

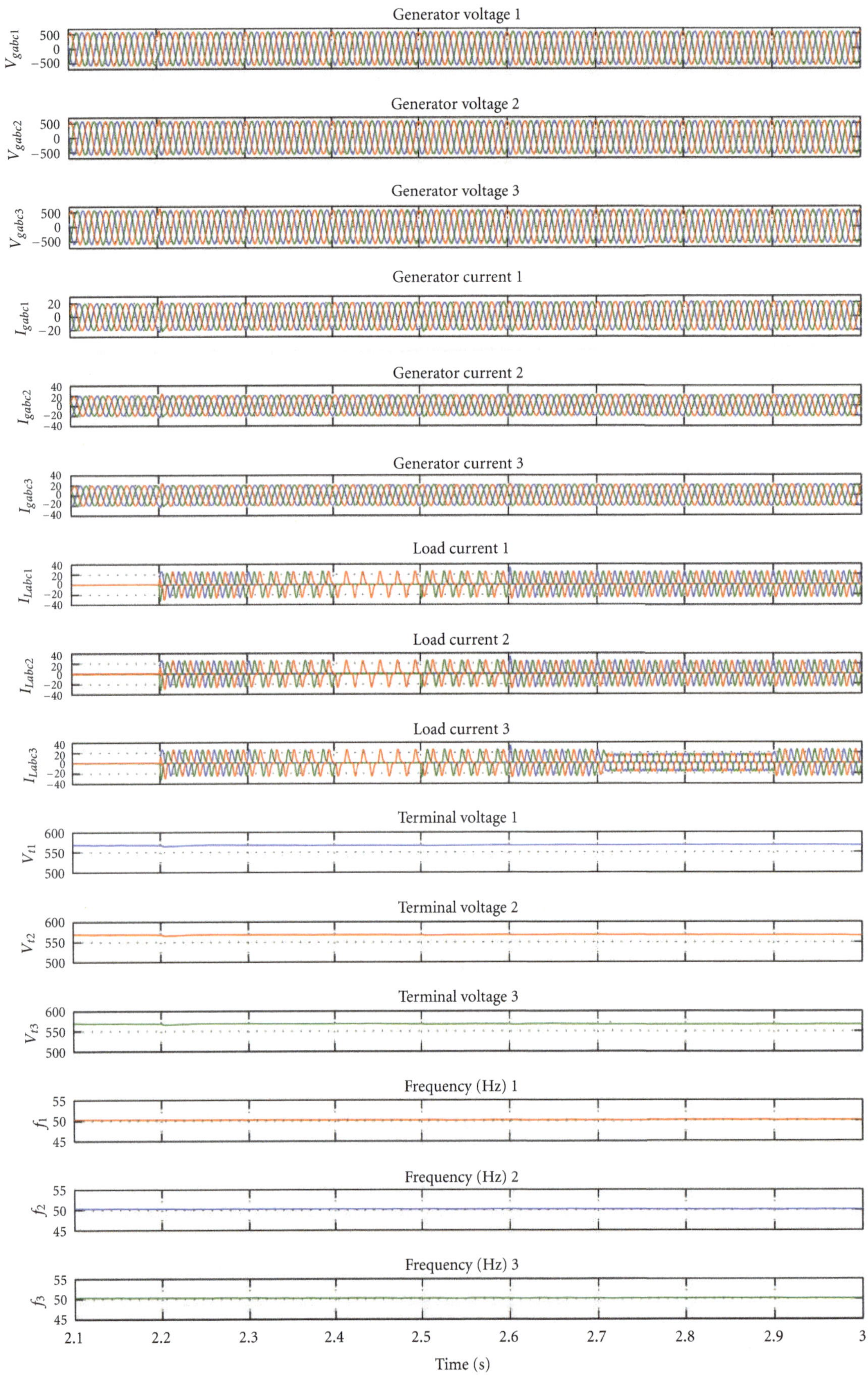

(a)

Figure 10: Continued.

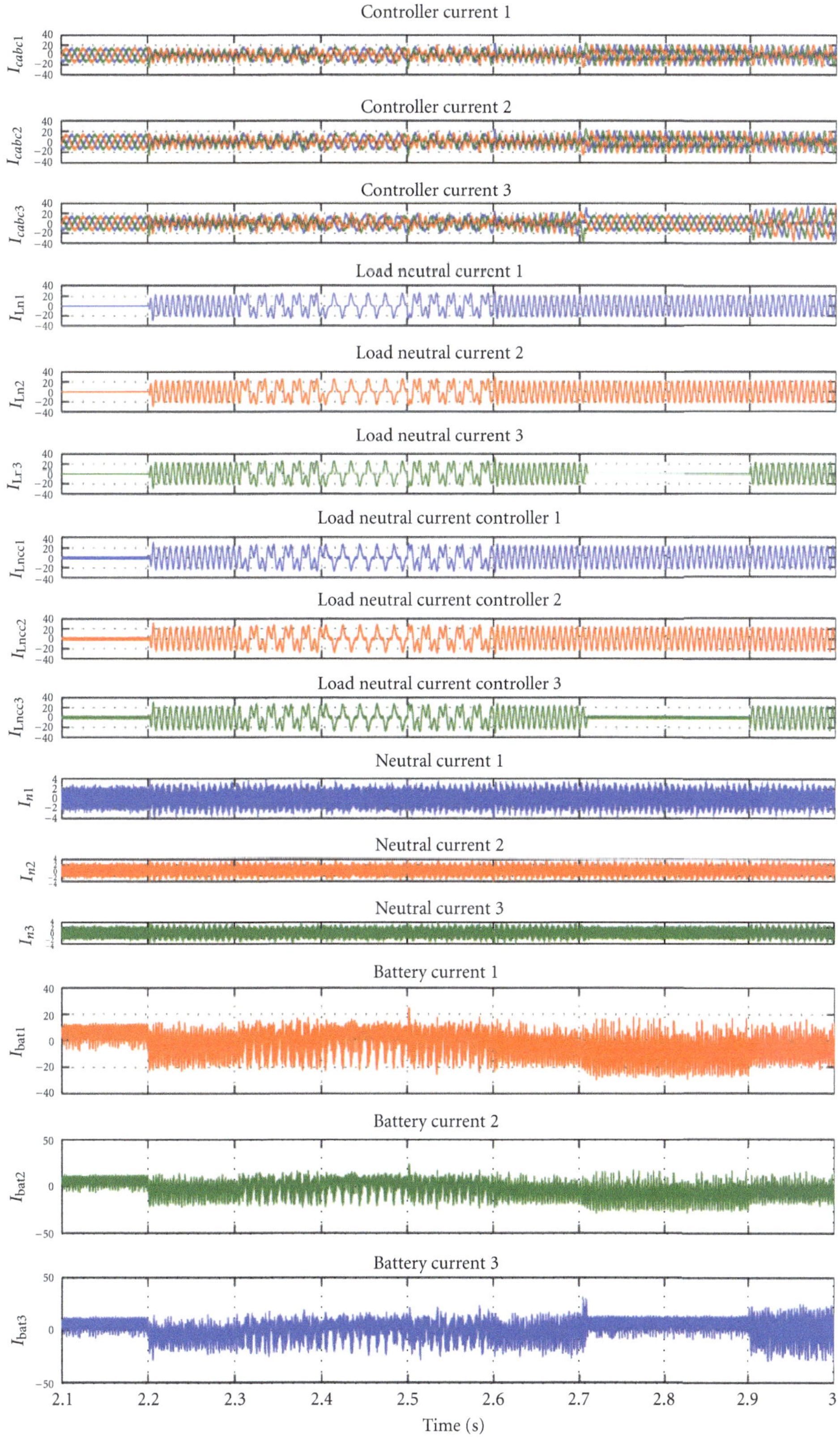

(b)

FIGURE 10: (a) Transient waveform of microgrid with BESS based VFC feeding balanced/unbalanced nonlinear load and removal of one plant. (b) Transient waveform of microgrid with BESS based VFC feeding balanced/unbalanced nonlinear load and removal of one plant.

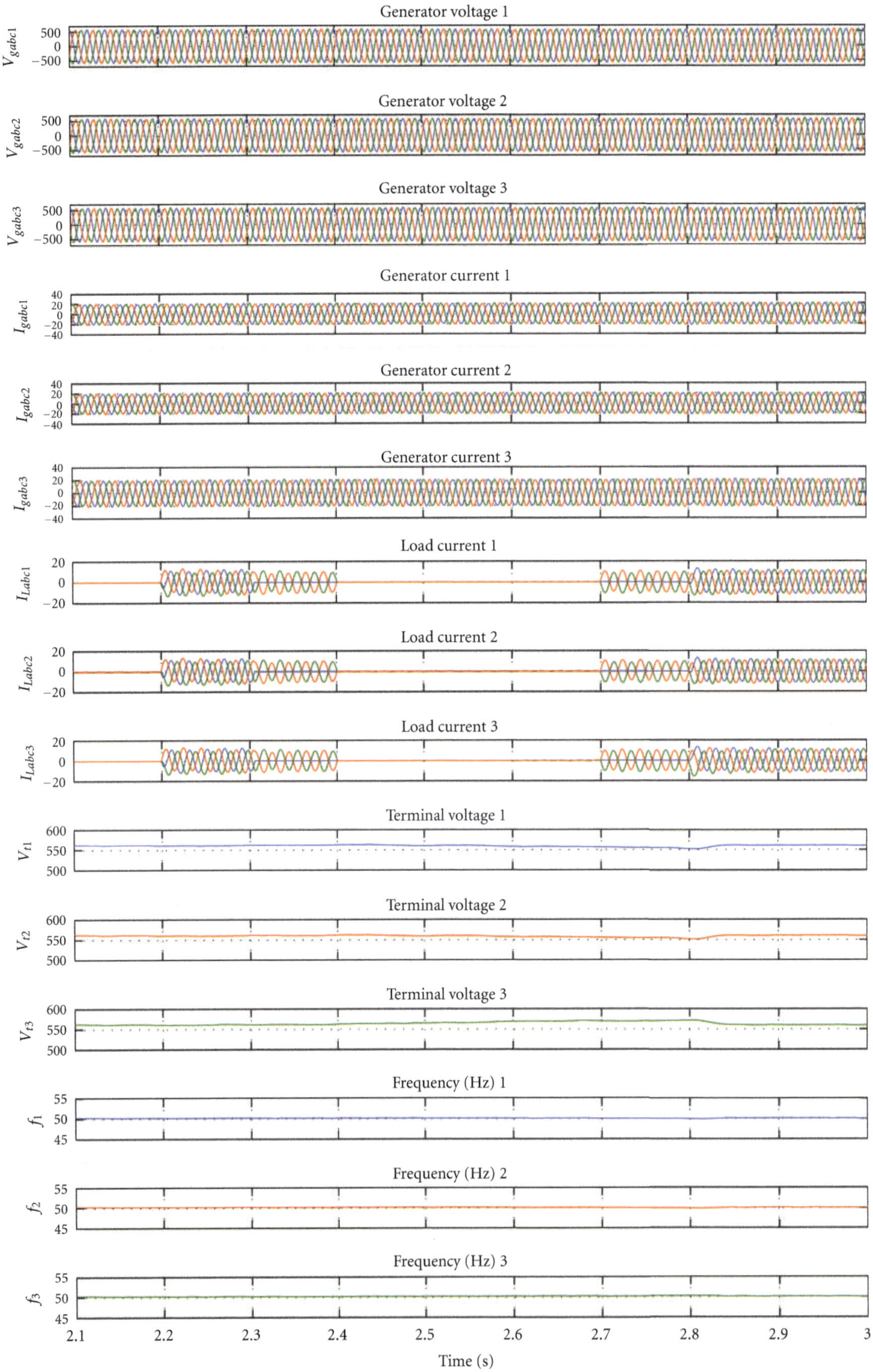

(a)

FIGURE 11: Continued.

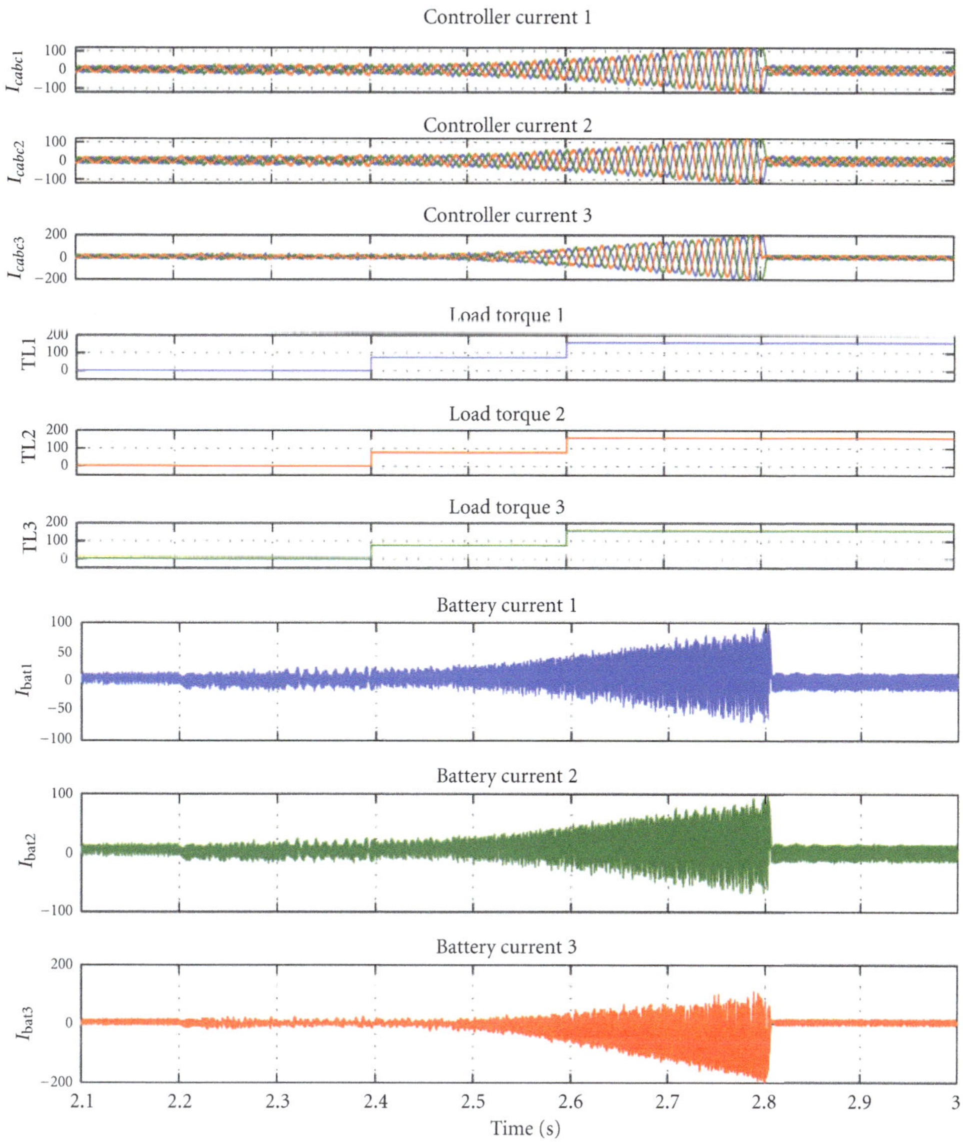

(b)

FIGURE 11: (a) Transient waveform of microgrid with BESS based VFC feeding balanced dynamic load and removal of one plant. (b) Transient waveform of microgrid with BESS based VFC feeding balanced dynamic load and removal of one plant.

at 2.4 sec another phase is open hence load becomes unbalanced and results in charging and discharging of battery, which shows load balancing of BESS fed VF controller. Again at 2.6 sec balanced load is applied to the system. Now at 2.7 sec a 3-phase fault occurs between load and hydroplant "3" so this plant is out of order and unable to support its load from 2.7 sec to 2.9 sec. As is seen from the waveform even if one plant is off the BESS based VF controller works and supplies the power during faulty condition. In this way, BESS keeps the overall generated power constant at the generator terminal which in turn regulates the system frequency.

5.5. Performance of Microgrid with BESS Feeding Balance Dynamic Load and Removal of One Micro-Source. The performance of BESS controller for micro grid with balanced dynamic load is shown in Figures 11(a) and 11(b). At 2.2 sec direct on line starting of induction motor of 2 kW, 50 Hz asynchronous motor is applied and sudden changes are observed in load current, due to high starting current. A minor discharge of battery is observed to regulate the frequency but after 4-5 cycles when sudden starting current is controlled, voltage and current at the generator terminal remain constant. At 2.4 sec load torque is 75 Nm and at 2.4 it is increased to 157 Nm; if it is applied on the motor shaft then the motor load current is increased but the controller performance is desirable. Now at 2.7 sec a 3-phase fault occurs between load and hydro plant "3" so this plant is out of order and unable to support its load from 2.7 sec to 2.9 sec. As is seen from the waveform even if one plant is off the BESS based VF controller works and supplies the power to load during faulty condition. In this way, BESS keeps the overall generated power constant at the generator terminal which in turn regulates the system frequency.

6. Conclusion

A detailed analysis of a micro grid with three micro sources under different load conditions is dealt with in this paper. The proposed BESS controller based voltage and frequency with each micro source regulates the terminal voltage and frequency of the source and hence stabilizes the micro grid. The mathematical model for BESS controller is analyzed for transient and dynamic condition. The controller provides desirable feature of synchronous condenser operating in capacitive and inductive modes. The capability of BESS based VF controller is demonstrated for load balancing, load leveling, and harmonic eliminations in linear, nonlinear, and dynamic loads. The BESS controller also works satisfactorily during removal of one micro source but as the whole system is connected in parallel other controller shares the load of that removed power generating plant and hence maintains the voltage and frequency of the whole system constant. Though due to battery energy storage system the size and the cost of the controller is increased but reliability of system with its VF control, outweighs this added cost. The implementation of this technology reinforces the use of such system where the grid connection has not reached yet and also in remote areas where renewable sources are available in abundance.

Appendix

(I) Machine parameters: 7.5 kW, 415 V, 50 Hz, Y-connected, 4-pole, $R_s = 1\,\Omega$, $R_r = 0.77\,\Omega$, $X_{\mathrm{lr}} = X_{\mathrm{ls}} = 1.5\,\Omega$, $J = 0.1384$ kg-m^2.

(II) Controller parameters:

$$K_{\mathrm{pf}} = 2,\ K_{\mathrm{if}} = 10,$$

$$K_{\mathrm{pd}} = 0.02,\ K_{\mathrm{id}} = 0.0025.$$

(I) Battery parameters:

$L_f = 4\,\mathrm{mH}$, $R_f = 0.1\,\Omega$ and $C_{\mathrm{dc}} = 4000\,\mu\mathrm{f}$, $R_1 = 10\,\mathrm{K}$,

$R_o = 0.01\,\Omega$, $C_{\mathrm{bat}} = 5760$ F.

(IV) Consumer Loads

(a) Resistive load = 2.5 kW single phase loads.

(b) Reactive load = 2.5 kW, 1.875 KVAR 0.8 PF lagging single phase loads.

(c) Nonlinear load = 2.5 kW with 100 μF capacitor and 8 mH inductor at DC end of single phase diode rectifier.

(d) Dynamic load = induction motor 2 kW, 50 Hz, 4 pole, 415 V.

References

[1] B. Lasseter, "Microgrids (distributed power generation)," in *Proceedings of IEEE Power Engineering Society Winter Meeting*, vol. 1, pp. 146–149, Columbus, Ohio, USA, February 2001.

[2] Y. Uno, G. Fujita, R. Yokoyama, M. Matubara, T. Toyoshima, and T. Tsukui, "Evaluation of micro-grid supply and demand stability for different interconnections," in *Proceedings of the IEEE International Power and Energy Conference (PECon '06)*, pp. 612–617, November 2006.

[3] F. Katiraei, M. R. Iravani, and P. W. Lehn, "Micro-grid autonomous operation during and subsequent to islanding process," *IEEE Transactions on Power Delivery*, vol. 20, no. 1, pp. 248–257, 2005.

[4] Y. Wei Li, D. M. Vilathgamuwa, and P. Chiang Loh, "A grid-interfacing power quality compensator for three-phase three-wire microgrid applications," *IEEE Transactions on Power Electronics*, vol. 21, no. 4, pp. 1021–1031, 2006.

[5] N. Pogaku, M. Prodanović, and T. C. Green, "Modeling, analysis and testing of autonomous operation of an inverter-based microgrid," *IEEE Transactions on Power Electronics*, vol. 22, no. 2, pp. 613–625, 2007.

[6] C. X. Wu, F. S. Wen, and Y. L. Lou, "The existed problems and possible solutions of micro-grid based on distributed generation," in *Proceedings of the 3rd International Conference on Deregulation and Restructuring and Power Technologies (DRPT '08)*, pp. 2763–2768, April 2008.

[7] B. Adhikary, B. Ghimire, and P. Karki, "Interconnection of two micro hydro units forming a mini-grid system using soft connection," in *Proceedings of IEEE Region 10 Conference (TENCON '09)*, November 2009.

[8] R. Majumder, B. Chaudhuri, A. Ghosh, R. Majumder, G. Ledwich, and F. Zare, "Improvement of stability and load sharing in an autonomous microgrid using supplementary droop control loop," *IEEE Transactions on Power Systems*, vol. 25, no. 2, pp. 796–808, 2010.

[9] S. V. Iyer, M. N. Belur, and M. C. Chandorkar, "Analysis and mitigation of voltage offsets in multi-inverter microgrids," *IEEE Transactions on Energy Conversion*, vol. 26, no. 1, pp. 354–363, 2011.

[10] C. Cho, J.-H. Jeon, J.-Y. Kim, S. Kwon, K. Park, and S. Kim, "Active synchronizing control of a microgrid," *IEEE Transactions on Power Electronics*, vol. 26, no. 12, pp. 3707–3719, 2011.

[11] C. V. Nayar, J. Perahia, F. Thomas, S. J. Phillips, T. Pryor, and W. L. James, "Investigation of capacitor-excited induction generators and permanent magnet alternators for small scale wind power generation," *Renewable Energy*, vol. 1, no. 3-4, pp. 381–388, 1991.

[12] A. Chandra, B. Singh, B. N. Singh, and K. A. Haddad, "An improved control algorithm of shunt active filter for voltage regulation, harmonic elimination, power-factor correction, and balancing of nonlinear loads," *IEEE Transactions on Power Electronics*, vol. 15, no. 3, pp. 495–507, 2000.

[13] B. Singh, G. Kasal, A. Chandra, and K. Haddad, "Battery based voltage and frequency controller for parallel operated isolated asynchronous generators," in *Proceedings of IEEE International Symposium on Industrial Electronics (ISIE '07)*, pp. 883–888, June 2007.

[14] B. Singh and G. K. Kasal, "Voltage and frequency controller for a three-phase four-wire autonomous wind energy conversion system," *IEEE Transactions on Energy Conversion*, vol. 23, no. 2, pp. 509–518, 2008.

[15] P. K. Goel, B. Singh, S. S. Murthy, and N. Kishore, "Isolated wind-hydro hybrid system using cage generators and battery

storage," *IEEE Transactions on Industrial Electronics*, vol. 58, no. 4, pp. 1141–1153, 2011.

[16] M. D. Aderson and D. S. Carr, "Battery energy storage technology," *Proceedings of the IEEE*, vol. 18, no. 3, pp. 475–479, 1993.

[17] M. Ceraolo, "New dynamical models of lead-acid batteries," *IEEE Transactions on Power Systems*, vol. 15, no. 4, pp. 1184–1190, 2000.

[18] N. Jantharamin and L. Zhang, "A new dynamic model for lead-acid batteries," in *Proceedings of the 4th IET International Conference on Power Electronics, Machines and Drives (PEMD '08)*, pp. 86–90, April 2008.

3

A Review on the Faults of Electric Machines Used in Electric Ships

Dionysios V. Spyropoulos and Epaminondas D. Mitronikas

Electromechanical Energy Conversion Laboratory, Department of Electrical and Computer Engineering, University of Patras, 26504 Rio, Patras, Greece

Correspondence should be addressed to Dionysios V. Spyropoulos; dionspyrop@ece.upatras.gr

Academic Editor: George Antonopoulos

Electric propulsion systems are today widely applied in modern ships, including transport ships and warships. The ship of the future will be fully electric, and not only its propulsion system but also all the other services will depend on electric power. The robust and reliable operation of the ship's power system is essential. In this work, a review on the mechanical and electrical faults of electric machines that are used in electric ships is presented.

1. Introduction

The idea to use electric power for the propulsion of ships is not new, as from the beginning of the twentieth century electric ship propulsion systems had shown advantages compared to the traditional ones using internal combustion engines. Since the 1980s, there has been an explosion in the number and variety of electrical propelled ships built around the world, from cruise liners to warships. Since the beginning of the last century, both AC and DC propulsion systems have been used. Moreover, the great evolution in the field of power electronics offered the means for efficient AC machines control, even in the case of high-power machines like those used in ship propulsion systems [1]. The ship of the future will be fully electric, as not only concerning its propulsion system but also all the other vital functions like navigation, and ship services will depend on electric power. To achieve this goal, the integrated power and propulsion system has to be designed properly in order to be robust, to achieve uninterrupted operation, to be fault tolerant, and able to continue its operation in case of failure through reconfiguration. Thus, timely hardware and electric propulsion machine diagnostics play a crucial role in maintaining the trouble-free operation of the ship.

The "heart" of the ship's integrated power system is the electric machine. The widespread application of electric machines for power generation, propulsion, thermal management systems, and so forth, makes their robust and reliable operation a crucial aspect. Usually, induction motors and permanent magnet motors are used in ship propulsion systems with high power ratings.

In this paper, the most common electrical and mechanical faults of these machines are presented.

2. Modern Electric Ship Propulsion System

The vast majority of ships are propelled directly from fossil fuel powered engines using mechanical transmission and gearboxes, as shown in Figure 1. The electrical part of such a system achieves 95% efficiency in nominal operation. Nowadays, an increasing number of ship propulsion systems manufacturers consider developing electric propulsion systems for commercial and cargo ships, while respective programs are under development for adopting electric propulsion in warships [2–4]. In such propulsion systems, generators convert prime mover rotational power into electric power, which is transmitted in electric propulsion motors in order to produce the necessary rotational power, as shown in Figure 2. Electric propulsion systems offer many advantages in terms of fuel economy, maneuverability, performance, and reduced manpower [5, 6].

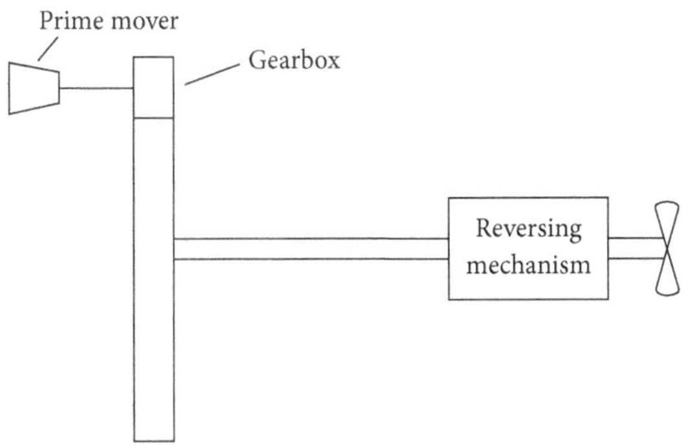

FIGURE 1: Propulsion system with mechanical transmission [5].

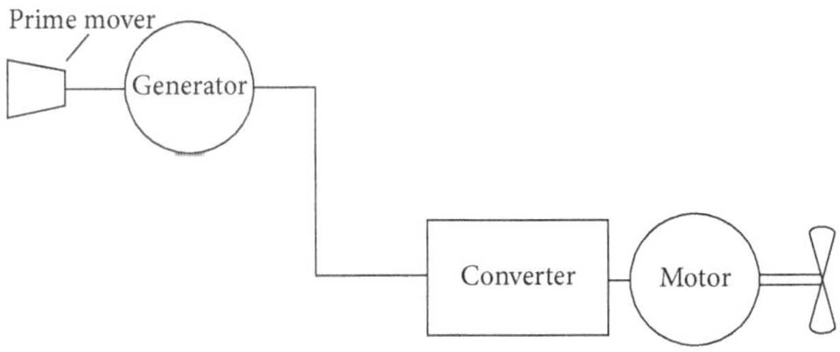

FIGURE 2: Propulsion system in an electric ship [5].

As it has been already noted, the ship of the future will be fully electric (All Electric Ship). The integrated system of this kind of ship will use powerful electric machines driven by advanced power electronic devices which will be controlled by sophisticated software [6]. The total needs for propulsion, navigation, and other services like ventilation systems, refrigeration plants, pumps, cranes, hoists, winches, and so forth, have to be covered by the produced electric power. Such a system is shown in Figure 3. Moreover, in the case of warships, the presence of more electric power demanding equipment (e.g., weapon systems) means that more installed power and reliability are required.

Taking into consideration that the fully electric ship is rather a novel technology approach, it has to be designed taking advantage of the already mature industrial technology concerning drives, control, and hardware diagnostics.

More specifically, hardware and electric machine diagnostics will play an important role in maintaining the continuous and reliable operation of the ship.

From the viewpoint of the engineer, the accuracy of the embedded diagnostic systems is essential in order to implement reconfigurable control algorithms in case of fault, thus achieving the ship's uninterrupted operation and survivability. As far as maintenance is concerned, high-level diagnostic accuracy is important for the isolation and quick repair of the fault and moreover for the more effective scheduling of preventive maintenance [8]. The most important part of the ship's integrated power system is undoubtedly the electric machines that are used. Particularly for the high-power electrical machines, the replacement procedure in case of fault can reach up to very high costs, while onboard replacement of a faulty unit is impossible. Consequently, their robust design and continuous condition monitoring are crucial in order to timely diagnose and repair any fault before it may lead to the complete failure of the system.

3. Electric Machines and Fault Diagnosis

3.1. Overview. The types of electric propulsion and supplemental motors usually used nowadays in ships are induction (asynchronous) motors, synchronous motors, and permanently excited synchronous motors (permanent magnet motors).

Induction motors (IMs) are currently the most widespread motors. This is due to their low cost, simple manufacturing, reliability, and robustness, while their efficiency is up to 96%.

Permanent magnet (PM) motors have similar stator construction as the induction machines, but the conventional rotor is replaced by an arrangement of permanent magnets, thus achieving higher torque density and relatively improved efficiency (up to 98%).

Finally, synchronous motors achieve fairly high efficiency (about 97%), requiring, though, a DC excitation source (in the rotor).

As already noted, electrical machines and drives have very low maintenance requirements; however, the ship's challenging operating environment makes them more vulnerable to failures. For this reason, many machines used in ship installations include specific resistors for reducing the wetness whereas for preventing problems caused by ship vibrations these are mounted on special bases. In any case, if a severe fault occurs it is extremely difficult to repair it onboard while any loss of electric propulsion power onboard would have devastating consequences. Thus, it is essential to diagnose the fault at its very early stages so that the maintenance procedure can be programmed before the fault leads to catastrophic events.

The main faults in the aforementioned electric machines can be classified as [9–15]:

(i) stator faults resulting in the open or short circuits on one or more stator windings;

(ii) broken rotor bar or end-ring faults on induction motors;

(iii) static or/and dynamic air-gap irregularities (rotor eccentricity);

(iv) bearing and gearbox failures;

(v) short circuit in the rotor field windings;

(vi) defects of the permanent magnets of PM motors.

Each fault disturbs the motor's normal operation producing several symptoms, like unbalanced line currents and air-gap voltages, torque and speed pulsations, decreased efficiency and average torque, excessive heating, and increased losses.

Through the industrial experience several surveys of faults of electrical machines have been conducted. In Table 1 [11], the results of the surveys of IEEE-IAS, ERPI, and Allianz are compared concerning Induction machines.

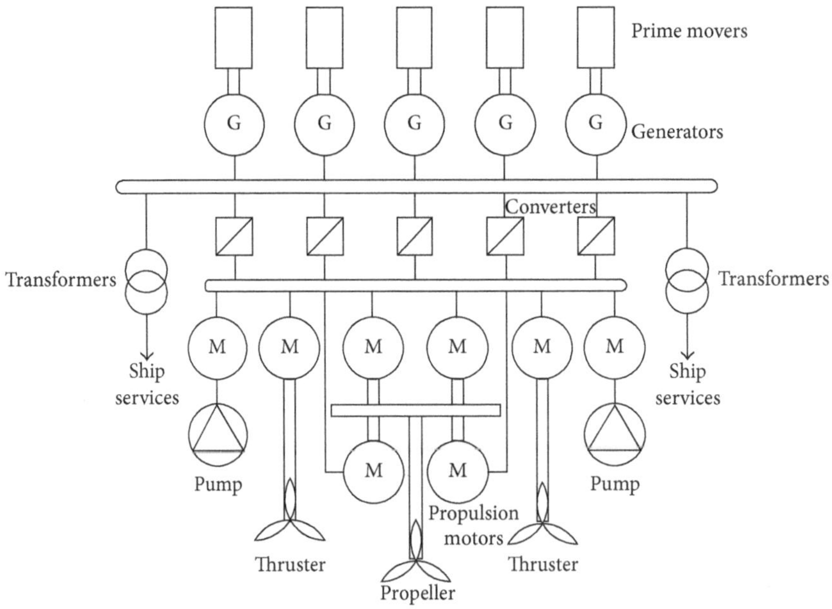

FIGURE 3: "All Electric Ship" power system topology [7].

TABLE 1: Comparison of IEEE-IAS, ERPI, and Allianz surveys.

Major components	IEEE-IAS % of failures	ERPI % of failures	Allianz % of failures
Bearing related	44	41	13
Stator related	26	36	66
Rotor related	8	9	13
Other	22	14	8

The Allianz survey focuses on medium- to high-voltage large induction machines like these that are used in electric ships and it is clear that stator-related faults are the most prominent for high-power machines. Moreover, in [16] a survey that focuses on electric motors used in offshore applications shows that the main faults are also due to bearing and stator winding defects. In [17], several real cases of faults in high-power electric machines are presented, analyzing thoroughly the causes (e.g., pages 45, 51, and 58.).

Traditionally, monitoring the line currents and analyzing them using several signal processing methods (FFT, Wavelets, Park's vector approach, etc.) [10, 15], one can reliably diagnose the existence and severity of the fault. The motor current signature analysis (MCSA) is broadly used, as it is a reliable and nonintrusive method, while for the monitoring, current sensors already incorporated in the drive system can be effectively used. In MCSA, the spectrum of the machine's line currents is investigated, in order to detect a fault. Several characteristic frequencies of the current identify the presence or not of each fault, while their amplitudes determine the fault's severity. Alternatively, instantaneous power and torque monitoring, temperature monitoring, acoustic methods, vibration signal analysis, and zero or negative sequence voltage monitoring have also been proposed; however, while the MCSA is the less intrusive for the system method, it is the most used one providing reliable results [9–14].

In the following paragraphs a more detailed approach for each electrical and mechanical fault in IM and PM machines is presented while the characteristic current frequencies which indicate each fault using the MCSA technique are given.

3.2. Stator Faults in AC Machines. The stator windings of medium- and high-voltage AC machines are normally form-wound coils. Induction and synchronous permanent magnet machines have the same stator structure. Their insulation is based on advanced materials compared to small induction machines and its constant monitoring is essential, given that any fault will lead to catastrophic results. Usually insulation defects are the result of long-term aging or contamination mainly because of the very high operating temperature. While the major failure contributors are abnormal moisture, poor ventilation, high vibration, abnormal voltage, and persistent overloading [16], an insulation failure, typically, takes less than 30 s to expand and finally leads to the destruction of the stator core [9–14, 18–21].

The existence of a stator fault could be detected using the line currents of the machine. Under ideal conditions, the current Park vector components (I_d, I_q) are calculated in terms of mains phase variables (I_a, I_b, I_c) as [19–21]:

$$I_d = \sqrt{\frac{2}{3}}I_a - \sqrt{\frac{1}{6}}I_b - \sqrt{\frac{1}{6}}I_c,$$
$$I_q = \sqrt{\frac{1}{2}}I_b - \sqrt{\frac{1}{2}}I_c. \quad (1)$$

Under ideal conditions, the 3-phase currents lead to a Park's vector with the following components:

$$I_d = \frac{\sqrt{6}}{2} I_M \sin(\omega t),$$
$$I_q = \frac{\sqrt{6}}{2} I_M \sin\left(\omega t - \frac{\pi}{2}\right), \tag{2}$$

where I_M is the peak value of motor phase current and ω is the angular frequency. The space vector magnitude is given as follows:

$$Mg = \left|\sqrt{I_d^2 + I_q^2}\right|. \tag{3}$$

The corresponding representation of (3) is a circular locus centered at the origin of the coordinates, as shown in Figure 4.

Under a fault condition, the three-phase currents will carry additional components leading to a different representation of the Park's vector, as shown in Figure 5. The change of the pattern of the Park's vector, in addition to the presence of an AC component at twice the supply frequency $2f_s$, is an indicator of the stator fault.

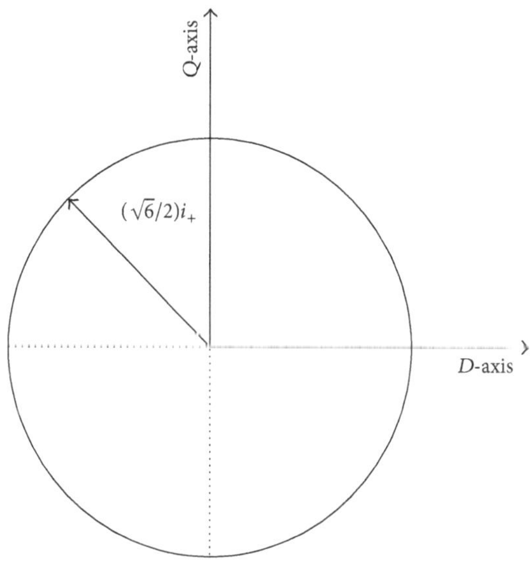

FIGURE 4: The current Park vector representation under ideal conditions.

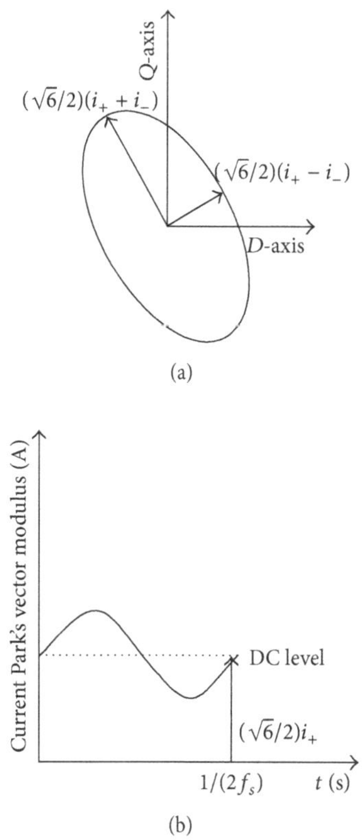

FIGURE 5: (a) Current Park's vector representation under abnormal conditions and (b) current Park's vector modulus under abnormal conditions.

3.3. Broken Rotor Bar/End-Ring Failures of Asynchronous Motors. Asynchronous Cage Motors are divided into two categories, depending on the construction of the cage, cast and fabricated. The material used for cast cage is aluminum, whereas for fabricated is copper. For high-power motors, fabricated cage rotors are used. Broken bar faults are caused by a variety of reasons: thermal, mechanical, magnetic, residual, dynamic, and environmental stresses [9–14]. These stresses originate from human, manufacturing, and functioning errors causing asymmetry to the geometry of the machine and consequently inducing asymmetrical spatial strain. The currents in a healthy rotor are presented in Figure 6. Under faulty conditions (broken bar/end-ring fault), the amplitude of the current changes in the area of the fault, as shown in Figure 7 [22].

Upon the occurrence of this type of fault, the characteristic frequency of the supply current that occurs is [9–14, 22–25]:

$$f_{\text{brb}} = (1 \pm 2ks) f_s, \quad k = 1, 2, 3, \ldots, \tag{4}$$

where f_s is the supply frequency, k is an integer order number, and s is the slip. As the fault severity increases, the amplitudes also increase considerably.

3.4. Bearing Faults. Bearing faults are very common in both low- and high-voltage electric motors and special attention is required. Almost 40%–50% of all motors' failures are bearing related. For high-power motors, both rolling ball and sleeve bearing are used. The major causes for bearing failures are the thermal, mechanical, and electrical stresses, contamination and corrosion, improper installation, and improper lubrication. It is generally believed that bearing failures develop from general roughness and gradually lead to defect of the bearing components [9–13, 16, 26, 27]. Bearings faults condition monitoring is important for asynchronous and permanent magnet machines that are used in the electric ship, while any bearing defection will cause speed and torque pulsations, thus reducing the efficiency of the machine.

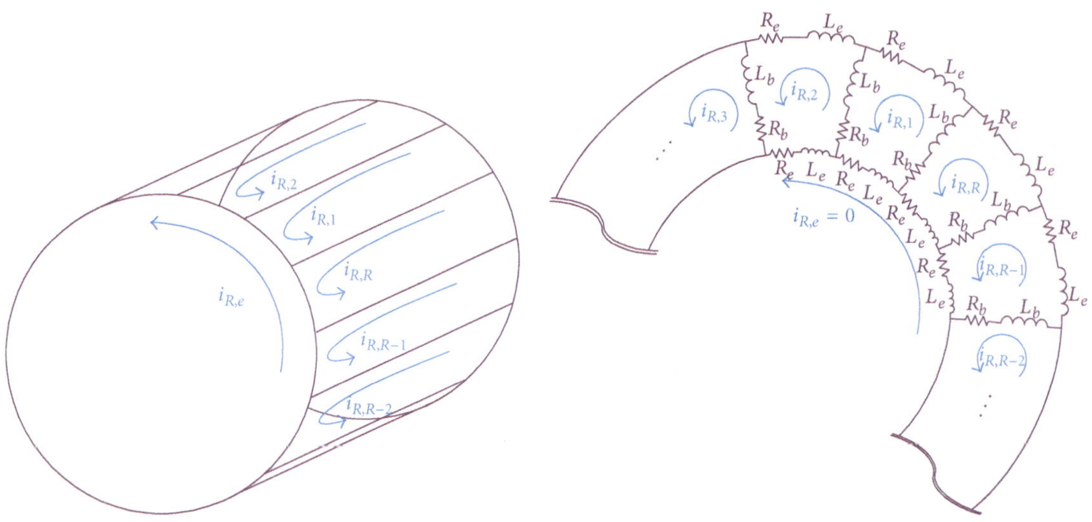

(a) (b)

FIGURE 6: (a) Sketch of the rotor of a healthy squirrel cage asynchronous machine. (b) Equivalent circuit of the rotor of a healthy squirrel cage asynchronous machine [22].

The diagnosis of the bearing failure can also be done through the examination of the current spectrum, the characteristic frequencies are given as follows [9–13, 27]:

$$f_{\text{bng}} = \left| f_s \pm k f_u \right|, \tag{5}$$

where f_s is the supply frequency, k is an integer order number, and f_u is a characteristic vibration frequency strongly connected to the geometry of the bearings as shown in [9].

3.5. Eccentricity-Related Faults. Machine eccentricity is the condition where unequal air gap exists between the rotor and the stator. If the degree of the eccentricity becomes large, the resulting unbalanced radial forces may cause rub between the stator and the rotor leading to the destruction of the machine. There are two kinds of eccentricity: static eccentricity and dynamic eccentricity, as presented in Figure 8, where the arrows indicate the misalignment between the rotor center and the rotational center at each case. In the case of the static air gap eccentricity, the position of the minimum radial air gap length is fixed in space. On the other hand, in the case of the dynamic eccentricity the center of the rotor does not coincide with the center of rotation, thus the position of the minimum air gap rotates with the rotor. The main causes of static eccentricity are the ovality of the stator core or the incorrect positioning of the stator or the rotor at the commissioning stage. However, dynamic eccentricity may be caused by a bent rotor shaft, misalignment or wear of the bearings, mechanical resonance at a critical speed, and so forth.

It should be noted that a degree of eccentricity at around 10% is generally admissible. However, machine manufacturers keep the total level of eccentricity even lower in order to reduce vibration and noise [9].

Investigating the current spectrum, eccentricity-fault-related harmonics are given as follows [9–13]:

$$f_{\text{ecc}} = f_s \left[\left(nR \pm n_d \right) \left(\frac{(1-s)}{p} \right) \pm n_{\text{ws}} \right], \tag{6}$$

where $n_d = 0$ in the case of static eccentricity, f_s is the supply frequency, R is the number of rotor slots, s is the slip, p is the number of pole pairs, n is a positive integer, and n_{ws} is the order of the stator time harmonics that are present in the power supply driving the motor.

Usually, in practical cases there is a combination of static and dynamic eccentricity, and the frequency harmonics that indicate the fault are given as follows:

$$f_{\text{ecc}} = f_s \left(1 \pm \frac{(1-s)}{p} \right). \tag{7}$$

Moreover, in [28] the diagnosis of dynamic eccentricity fault in PM motors is achieved investigating the current harmonics given by

$$f_{\text{ecc}} = f_s \left(1 \pm \frac{(2k-1)}{p} \right), \tag{8}$$

where k is an integer.

3.6. Rotor Magnets Defects in Permanent Magnet Motors. Monitoring the magnet status of PM motors is very important since local or uniform demagnetization caused by operating stresses lead to irreversible degradation of motor efficiency. Defects leading to demagnetization or disintegration of permanent magnets are caused by thermal, mechanical, or environmental stresses [5, 29].

Investigating the current spectrum, permanent magnet-fault-related harmonics are given as follows [29]:

$$f_{\text{dm}} = f_s \left(1 \pm \frac{k}{p} \right), \tag{9}$$

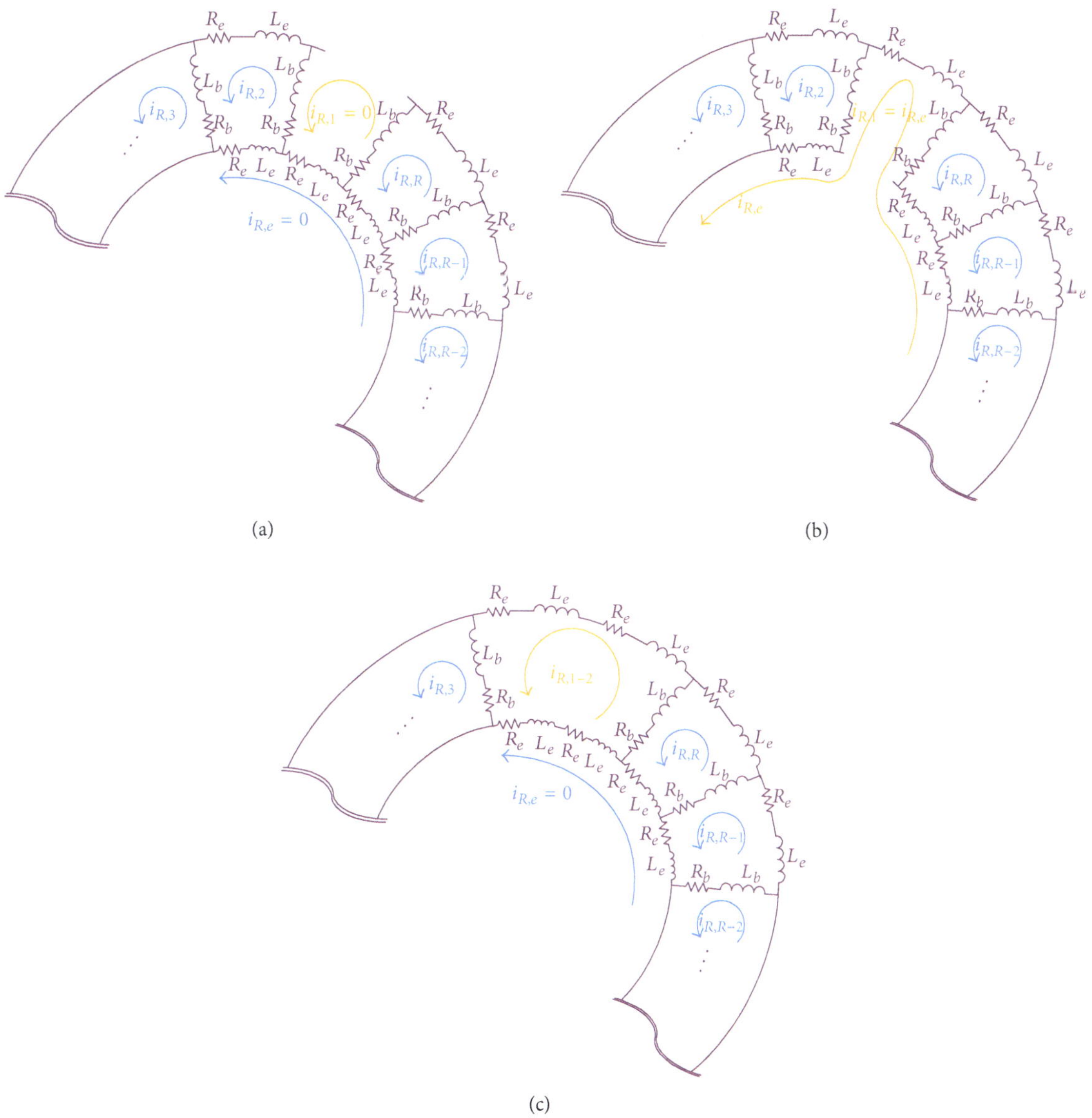

FIGURE 7: (a), (b) Equivalent circuit of the rotor of a squirrel cage asynchronous machine under end-ring fault. (c) Equivalent circuit of the rotor of a squirrel cage asynchronous machine under broken bar fault [22].

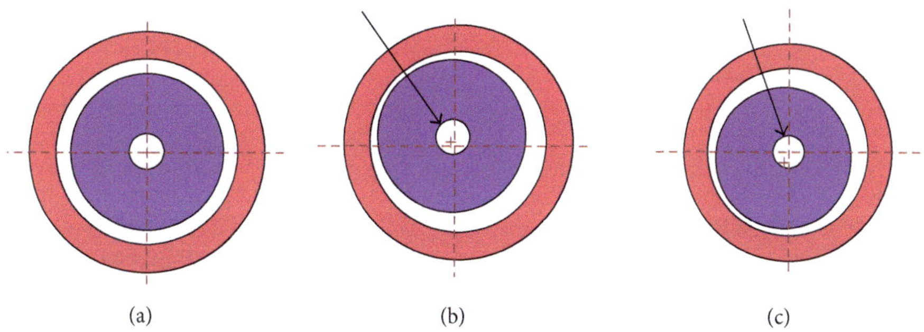

FIGURE 8: Sketch of the rotor of an induction machine: (a) healthy case, (b) under static eccentricity, and (c) under dynamic eccentricity [22].

where f_s is the supply frequency, p the number of pole pairs, and k is a positive integer.

4. Conclusions

In this paper, a review of electrical and mechanical faults of asynchronous and permanents magnet machines is presented. Moreover, for each type of fault, the characteristic frequencies of the current spectrum that indicate the presence and severity of the failure are given. In the literature, it has also been proposed that other signals can be used for the fault diagnosis, like the torque, the power, the supply voltages, and the motor speed. Using a combination of these signals and the appropriate signal processing techniques, the engineer can extract accurate information about the condition of the machine. The field of condition monitoring, fault diagnosis, and fault prognosis is of great research interest; the last years and novel frequency ranges are investigated as fault indicators while state-of-the-art signal processing techniques are applied for the online condition monitoring of the machines under varying operating conditions.

References

[1] T. J. McCoy, "Trends in ship electric propulsion," in *Proceedings of the IEEE Power Engineering Society Summer Meeting*, pp. 343–346, July 2002.

[2] The Frontrunner, "A new era in ship propulsion: The first large size (154K) LNG carrier with electric propulsion is now in operation," ABB Marine, March 2007.

[3] Rolls Royce, "New electric hybrid propulsion system," http://www.rolls-royce.com/Images/hsg_brochure_tcm92-26884.pdf.

[4] VEM-GROUP, http://www.vem-group.com/fileadmin/content/pdf/Produkte_Komponente/Mittel_und_Hochspannung/Schiffbau/schiffbau_0908_e.pdf.

[5] C. G. Hodge and D. J. Mattick, "The electric warship," *Transactions of Institute of Marine Engineers*, vol. 108, pp. 109–125.

[6] E. D. Mitronikas and E. C. Tatakis, "Migrating the experience of industrial systems to electric ships: propulsion motors and fault detection," in *Proceedings of the 1st Marinlive International Workshop on Electric Machines and Power Converters*, 2012.

[7] S. A. Gertsos, J. M. Prousalidis, and C. A. Frangopoulos, "Electric Propulsion: from Infancy to adolensense," in *Proceedings of the 8th International Marine Design Conference*, Athens, Greece, May 2003.

[8] K. P. Logan, "Intelligent diagnostic requirements of future all-electric ship integrated power system," *IEEE Transactions on Industry Applications*, vol. 43, no. 1, pp. 139–149, 2007.

[9] S. Nandi, H. A. Toliyat, and X. Li, "Condition monitoring and fault diagnosis of electrical motors—a review," *IEEE Transactions on Energy Conversion*, vol. 20, no. 4, pp. 719–729, 2005.

[10] M. E. H. Benbouzid and G. B. Kliman, "What stator current processing-based technique to use for induction motor rotor faults diagnosis?" *IEEE Transactions on Energy Conversion*, vol. 18, no. 2, pp. 238–244, 2003.

[11] P. Zhang, Y. Du, T. G. Habetler, and B. Lu, "A survey of condition monitoring and protection methods for medium-voltage induction motors," *IEEE Transactions on Industry Applications*, vol. 47, no. 1, pp. 34–46, 2011.

[12] M. E. H. Benbouzid, "A review of induction motors signature analysis as a medium for faults detection," *IEEE Transactions on Industrial Electronics*, vol. 47, no. 5, pp. 984–993, 2000.

[13] A. M. Trzynadlowski and E. Ritchie, "Comparative investigation of diagnostic media for induction motors: a case of rotor cage faults," *IEEE Transactions on Industrial Electronics*, vol. 47, no. 5, pp. 1092–1099, 2000.

[14] G. S. Khalaf and A. F. Mohamed Haider, "Diagnosis and fault tolerant control of the Induction motors techniques a review," *Australian Journal of Basic and Applied Sciences*, vol. 4, no. 2, pp. 227–246, 2010.

[15] S. H. Kia, H. Henao, and G. A. Capolino, "Digital signal processing for induction machines diagnosis—a review," in *Proceedings of the 33rd Annual Conference of the IEEE Industrial Electronics Society (IECON '07)*, pp. 1155–1162, November 2007.

[16] O. V. Thorsen and M. Dalva, "Survey of faults on induction motors in offshore oil industry, petrochemical industry, gas terminals, and oil refineries," *IEEE Transactions on Industry Applications*, vol. 31, no. 5, pp. 1186–1196, 1995.

[17] P. Tavner, L. Ran, J. Penman, and H. Sedding, *Condition Monitoring of Rotating Electrical Machines*, IET Power and Energy Series, IET, 2008.

[18] A. Stavrou, H. Sedding, and J. Penman, "Current monitoring for detecting inter-turn short circuits in induction motors," in *Proceedings of the IEEE International Electric Machines and Drives Conference*, pp. 345–347, 1999.

[19] S. M. A. Cruz and A. J. Marques Cardoso, "Stator winding fault diagnosis in three-phase synchronous and asynchronous motors, by the extended park's vector approach," *IEEE Transactions on Industry Applications*, vol. 37, no. 5, pp. 1227–1233, 2001.

[20] I. Jaksch and P. Fuchs, "Using space transformation for exact diagnostic of induction motor stator faults," in *Proceedings of the International Symposium on Diagnostics of Electric Machines, Power Electronics, and Drives (SDEMPED '05)*, Vienna, Austria, September 2005.

[21] D. V. Spyropoulos and E. D. Mitronikas, "Induction motor stator fault diagnosis technique using Park vector approach and complex wavelets," in *Proceedings of the 20th International Conference on Electrical Machines (ICEM '12)*, pp. 1730–1734, 2012.

[22] I. P. Georgakopoulos, *Development of fault diagnosis methods in controlled motor systems composed by electronic power converters and asynchronous machine [Ph.D. thesis]*, University of Patras, Patras, Greece, 2010.

[23] J. Milimonfared, H. M. Kelk, S. Nandi, A. D. Minassians, and H. A. Toliyat, "A novel approach for broken-rotor-bar detection in cage induction motors," *IEEE Transactions on Industry Applications*, vol. 35, no. 5, pp. 1000–1006, 1999.

[24] I. P. Tsoumas, G. Georgoulas, E. D. Mitronikas, and A. N. Safacas, "Asynchronous machine rotor fault diagnosis technique using complex wavelets," *IEEE Transactions on Energy Conversion*, vol. 23, no. 2, pp. 444–459, 2008.

[25] H. Henao, H. Razik, and G. Capolino, "Analytical approach of the stator current frequency harmonics computation for detection of induction machine rotor faults," *IEEE Transactions on Industry Applications*, vol. 41, no. 3, pp. 801–807, 2005.

[26] W. Zhou, T. G. Habetler, and R. G. Harley, "Bearing condition monitoring methods for electric machines: a general review," in *Proceedings of the IEEE International Symposium on Diagnostics for Electric Machines, Power Electronics and Drives (SDEMPED '07)*, pp. 3–6, September 2007.

[27] M. J. Devaney and L. Eren, "Detecting motor bearing faults: monitoring an induction motor's current and detecting bearing

failure," *IEEE Instrumentation and Measurement Magazine*, vol. 7, no. 4, pp. 30–50, 2004.

[28] B. M. Ebrahimi, J. Faiz, and M. J. Roshtkhari, "Static-, dynamic-, and mixed-eccentricity fault diagnoses in permanent-magnet synchronous motors," *IEEE Transactions on Industrial Electronics*, vol. 56, no. 11, pp. 4727–4739, 2009.

[29] D. Casadei, F. Filippetti, C. Rossi, and A. Stefani, "Magnets faults characterization for permanent magnet synchronous motors," in *Proceedings of the IEEE International Symposium on Diagnostics for Electric Machines, Power Electronics and Drives (SDEMPED '09)*, September 2009.

4

ANN Current Controller Based on PI-Fuzzy Adaptive System for Shunt Power Active Filter

Moulay Tahar Lamchich

Department of Physics, Faculty of Sciences Semlalia, Cadi Ayyad University, 40000 Marrakech, Morocco

Correspondence should be addressed to M. T. Lamchich, lamchich@ucam.ac.ma

Academic Editor: Jose Pomilio

This paper deals with the use of triphase shunt active filter which is able to compensate current harmonics, reactive power, and current unbalance produced by nonlinear loads. To perform the identification of disturbing currents, a very simple control method is introduced. It's formed by a DC voltage regulator and a balance between the average power of load and the active power supplied by the grid. The output current of the voltage source inverter (VSI) must track the reference current. This is done by a neural controller based on a PI-Fuzzy adaptive system as reference corrector. Also to regulate the DC link capacitor voltage a fuzzy logic adaptive PI controller is used.

1. Introduction

Due to the increased use of nonlinear electrical loads, such as power electronics supplies, harmonics currents are generated in the level of these loads and injected back to the grid causing its voltage distortion at harmonics currents' frequencies. The distorted voltage generates distorted currents which will be absorbed by sensitive loads and causing losses in the lines. Also, the most consuming electrical power loads are almost inductive, and then they contribute to the degradation of grid power factor at the point of common connection (PCC).

Conventionally, passive filters were adopted for the reduction of harmonics and to improve power factor; therefore, they have the disadvantages such as fixed frequency compensation, resonance, and large size. These limitations were avoided by the use of active filters which utilized a switch-mode power electronic converter to supply harmonic currents equal to those in the load currents.

The main objective of the power active filter (PAF) is to compensate reactive power, current harmonics, neutral current, and unbalancing of nonlinear loads by injecting compensating currents.

Almost, the method-based instantaneous active and reactive power (pq method) [1] is currently performed for disturbing current identification. The major disadvantages of a bloc identification-based pq method are essentially as follows:

(i) it is not effective under distorted and unbalanced mains voltages conditions;

(ii) the time delays introduced by pass filters, which are used to separate the average and oscillating parts of powers, degrades the dynamic performance of active filter;

(iii) this method requires more computational calculation.

The controller of the PAF is comprised of an inner current loop which actively shapes the line currents and an outer voltage control loop which regulates the magnitude of the line currents.

This paper presents the application of a fuzzy inference system to determine the parameters of a PI controller used for the DC voltage regulation.

The current loop uses a technique-based artificial neural networks (ANN). The PI-Fuzzy adaptive system is adopted, as reference corrector, for the process learning.

The simulation results, in power system blockset (PSB) Matlab/Simulink environment, will show the effectiveness of the proposed approach for the control of the power active filter.

2. General Structure

The main circuit of the shunt active filter control, as used in the Matlab/Simulink environment, is shown by Figure 1. It is composed of

(i) a three wires power network and the output filter;

(ii) the nonlinear load based on a rectifier;

(iii) a reference current identification bloc based on a balance between powers;

(iv) two regulation blocs, of capacitor voltage and the inverter, based on a neurofuzzy logic technique;

(v) the active filter consisting of a voltage source inverter (VSI) with a capacitor in its DC side.

3. Disturbing Current Identification

The compensation of the current harmonics and reactive power, generated by the nonlinear load, has an object to impose that the AC source currents must be sinusoidal and in phase with source voltages.

So the power active filter (PAF), used as a controlled current source, is controlled by determining its reference currents to inject the compensating currents.

The purpose of it is to produce output currents for tracking the reference current in order to reduce the harmonic components and compensate the reactive power.

We can proceed as follows to determine the reference currents for the control of the PAF:

(i) we start by determining the reference grid currents (noted i_{sj}^*; $j = a, b, c$) witch realized these two conditions (sinusoidal form with unit power factor).

(ii) the inverter reference currents (noted i_{oj}^*; $j = a, b, c$), which permit the PAF control, are deduced from the following equation:

$$i_{oj}^* = i_{cj} - i_{sj}^*; \quad j = a, b, c, \tag{1}$$

where i_{cj}; $j = a, b, c$ are the load currents.

(iii) as show in Figure 1, we evaluate the maximal value (noted IREF) of the reference grid currents, and we use a PLL circuit to generate the sinusoidal forms from the source voltages.

(iv) the maximal value IREF can be deduced from a balance between active power supplied by the AC source and fundamental active power component absorbed by nonlinear load plus losses of inverter for regulating the DC capacitor voltage.

Therefore, The peak of source reference current (IREF) has two components; the first one (I_{sA}^*) is corresponding to the average load active power, and the second one (I_{sD}^*) is obtained from the DC capacitor voltage regulation as follows [2]:

$$\mathrm{IREF} = I_{\mathrm{sA}}^* + I_{\mathrm{sD}}^* \tag{2}$$

The average power of load is obtained using a low-pass filter as shown by Figure 2.

By considering the unity power factor for grid currents, the component I_{sA}^* is defined as follows:

$$I_{\mathrm{sA}}^* = \frac{2}{3}\frac{P_{\mathrm{Lav}}}{V_{\mathrm{sm}}}, \tag{3}$$

where V_{sm} is the peak voltage of AC source.

The second component of AC source current I_{sD}^* is obtained from the DC capacitor voltage regulation as shown by Figure 3.

We can also justify this procedure, for the determination of the reference current, by calculation of instantaneous power components of the p-q theory in a stationary $\alpha\beta$-frame [3].

The load absorbed the instantaneous active power as follows:

$$p_l = v_1 i_1 + v_2 i_2 + v_3 i_3 = v_\alpha i_\alpha + v_\beta i_\beta. \tag{4}$$

This power may be decomposed into oscillating component called harmonic power ($\tilde{p}_l$) and average component called fundamental ($\overline{p}_l$).

For generation of the reference current, a balance between instantaneous powers supplied by the source (noted p_s) and the PAF (p_f) and drained by the load is to be computed.

For obtaining inphase current with supply voltage, the harmonic term of instantaneous active power and all terms of instantaneous reactive power should be removed. In this case, balance of power yields as follows:

$$p_s = \overline{p}_l \qquad p_f = p_l - p_s = \tilde{p}_l. \tag{5}$$

In normal operating condition, the PAF should be able to supply $p_f = \tilde{p}_l$, and consequently voltage level of capacitor remains constant for duration of steady state while it varies during period of transient. In order to control capacitor voltage level on the DC side of the inverter is essential to adjust previous equation.

4. DC-Bus Voltage Control

The DC link voltage control is performed to define the second component of AC source current I_{sD}^* (or harmonic power generated by the PAF) as described below.

In this paper, a PI controller, which its parameters are adjusted by a fuzzy inference system, is used to regulate the DC link voltage [4]. This technique, which consists of gains variation during the system control, takes the PI controller adaptable to nonlinear system and its parameters variation.

The principle scheme of this technique is illustrated by Figure 3.

The PI parameters (K_p, K_i) are calculated using the intermediate values (K_p', K_i'), given by the fuzzy controller, as follows [5]:

$$\begin{aligned} K_p &= \left(K_{p_{\max}} - K_{p_{\min}}\right) K_p' + K_{p_{\min}}, \\ K_i &= \left(K_{i_{\max}} - K_{i_{\min}}\right) K_i' + K_{i_{\min}}, \end{aligned} \tag{6}$$

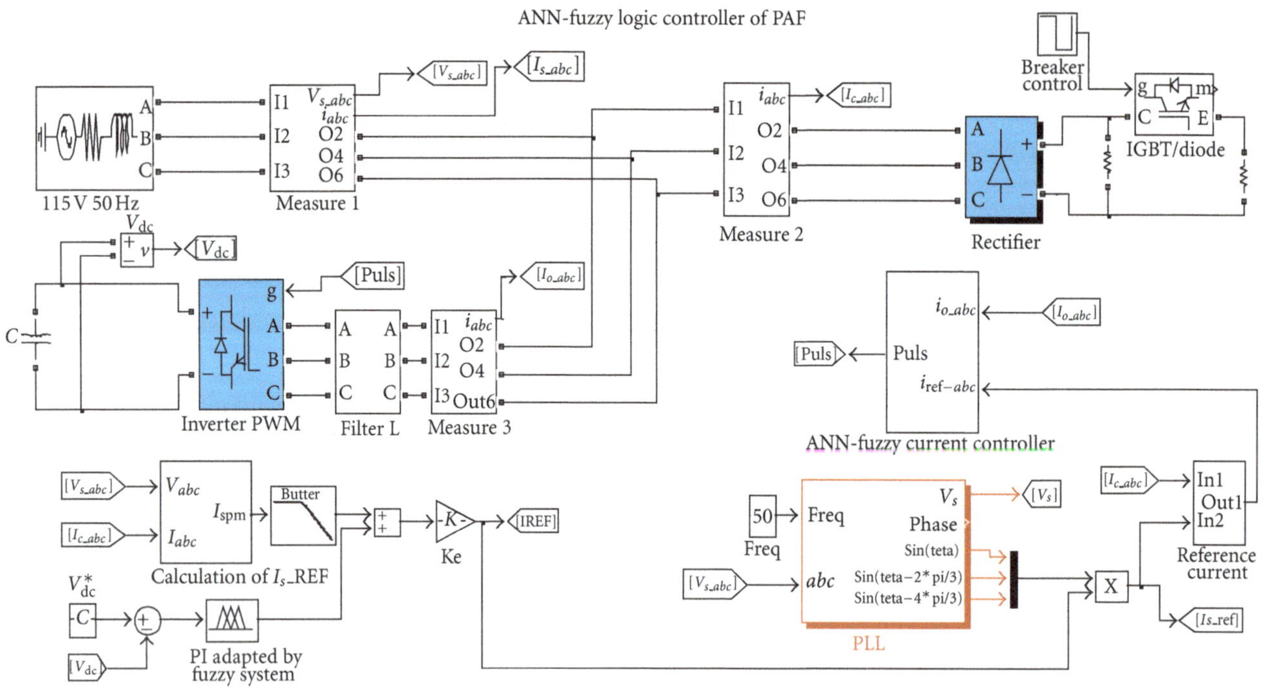

FIGURE 1: General structure of the PFA control.

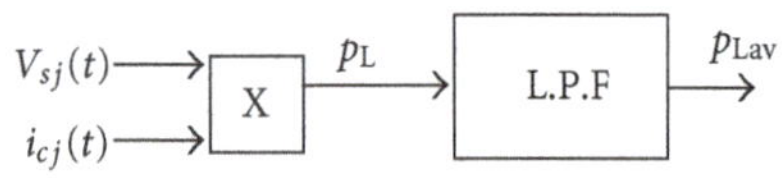

FIGURE 2: Calculation of the average power of load.

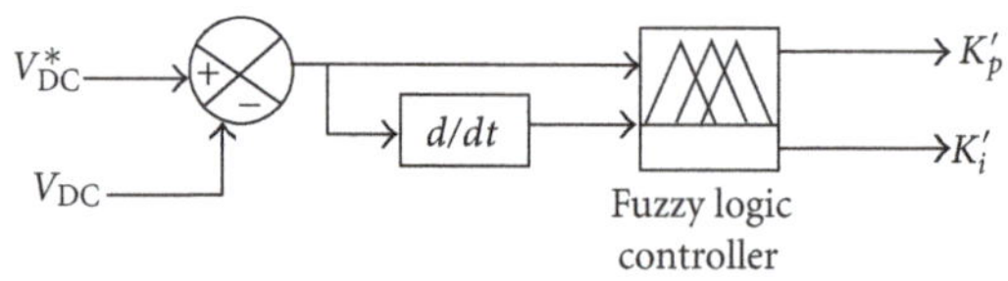

FIGURE 4: K_p' and K_i' evaluated by Fuzzy system.

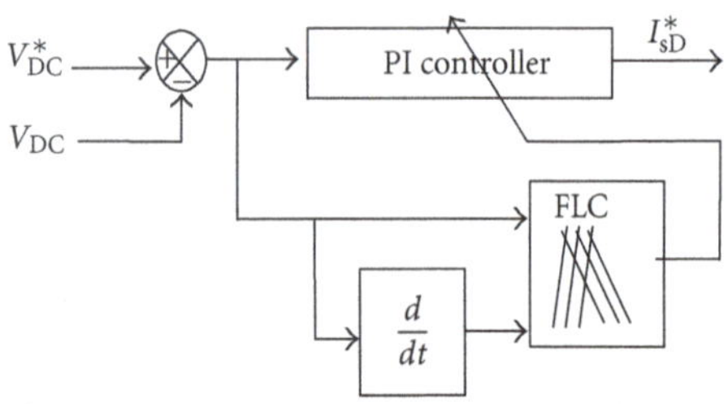

FIGURE 3: VDC regulation with PI adapted by Fuzzy inference system.

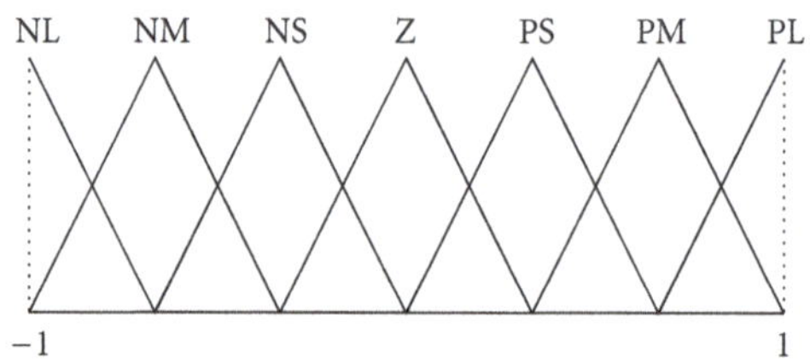

FIGURE 5: Membership function for the error and slope.

where the extremes values of the gains are defined from the parameters given by Ziegler-Nichols method.

Both quantities (K_p', K_i'), corresponding to the output of the fuzzy system (Figure 4), are normalized in the range [0 1].

The membership functions are chosen the same, respectively, for the error and the slope and for the parameters K_p' and K_i' (Figures 5 and 6).

We have chosen to characterize this fuzzy controller by seven sets, respectively, for the inputs and only two sets for the outputs.

The linguistic rules, for the fuzzy logic output variables K_p' and K_i' [6], are resumed in the Table 1.

5. ANN Current Control

To obtain the desired switching signals according to either output inverter currents or grid currents to follow the reference ones, a technique-based artificial neural network has been introduced.

Neural networks, which have the potential to provide an improved method of deriving nonlinear models, have self adapting capabilities which make them well suitable to handle nonlinearities, uncertainties, and parameter variations.

In the last years, a big interest concerned the use of the artificial intelligence techniques (neural networks, fuzzy logic, genetic algorithms) in identification and nonlinear

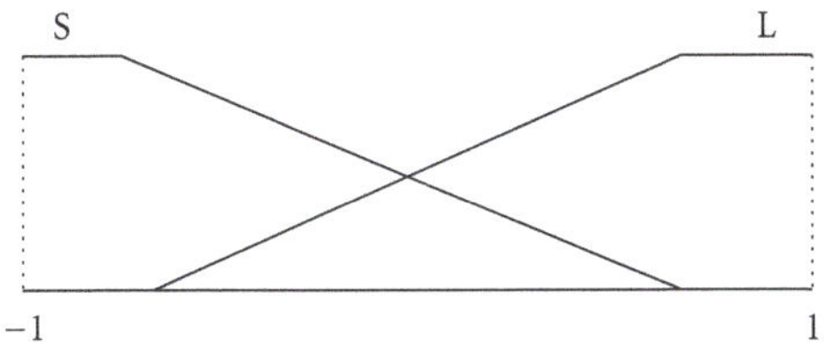

FIGURE 6: Membership function for the parameters K_p' and K_i'.

TABLE 1: Fuzzy rules for the outputs K_p' and K_i'.

Slope	Error						
	NL	NM	NS	Z	PS	PM	PL
NL	L/L	L/L	L/L	L/L	L/L	L/L	L/L
NM	S/L	L/L	L/S	L/S	L/S	L/L	S/L
NS	S/L	S/L	L/L	L/S	L/L	S/L	S/L
Z	S/L	S/L	S/L	L/S	S/L	S/L	S/L
PS	S/L	S/L	L/L	L/S	L/L	S/L	S/L
PM	S/L	L/L	L/S	L/S	L/S	L/L	S/L
PL	L/L	L/L	L/L	L/L	L/L	L/L	L/L

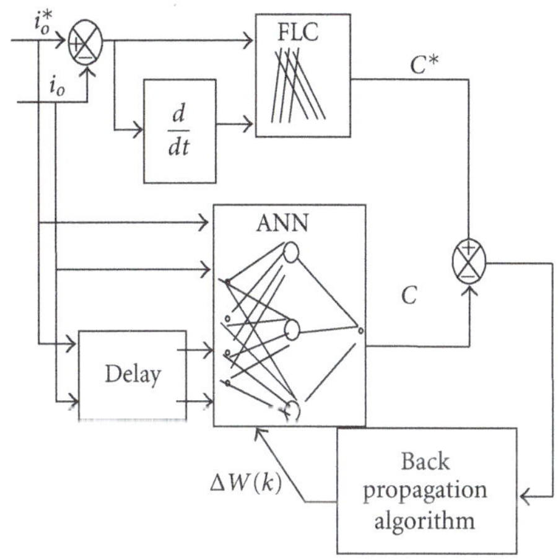

FIGURE 7: Learning scheme of the ANN for generating the inverter switching control.

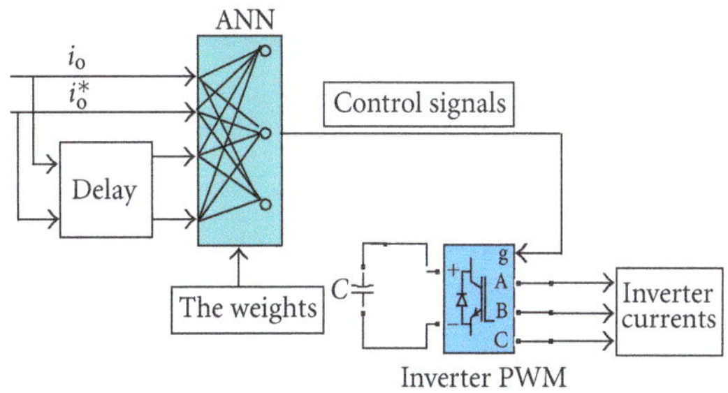

FIGURE 8: ANN corrector developed for the PAF control.

control systems, this is mainly due to their capacities of learning and generalization [7–9].

In this object, the present work consists, partially, in developing a neural corrector for the control of the active power filter currents.

In this control system, we developed a neural controller by learning the behavior of a reference corrector. To do this, we have chosen, as reference corrector, a PI-Fuzzy adaptive system, which has good performances and could be able to adapt to changes in load.

The learning structure of this PI-Fuzzy corrector is given by Figure 7.

The ANN is a three-layer network: the first layer has four nodes, the hidden layer has three neurons, and the output layer has only one neuron corresponding to the inverter switching signal.

As mentioned below, the PI-Fuzzy reference corrector is used for the phase of learning, allowing to determine the parameters of this neural corrector and particularly its weights.

The learning procedure used was the Levenberg Marquardt backpropagation algorithm. The weights are generated until the error between the calculated pattern and the desired pattern, corresponding to the output of the fuzzy system as reference model, is very small and acceptable.

After this stage, the weights final values, corresponding to the steady state of the PAF control, are saved and applied, in the second phase, steadily to the neural controller (Figure 8).

Approximations of ANN weights values, for one phase of inverter, are given as follows

$$\begin{bmatrix} W_{11} = 4.20 \\ W_{12} = 2.45 \\ W_{13} = -0.3 \\ W_{14} = 1.85 \\ W_{k1} = -0.28 \end{bmatrix} \begin{bmatrix} W_{21} = 7.10 \\ W_{22} = -3.55 \\ W_{23} = 0.07 \\ W_{24} = -3.3 \\ W_{k2} = 0.55 \end{bmatrix} \begin{bmatrix} W_{31} = 4.20 \\ W_{32} = 2.45 \\ W_{33} = -0.3 \\ W_{34} = 1.85 \\ W_{k3} = -0.28 \end{bmatrix}. \quad (7)$$

These weights are applied to the neural controller without the fuzzy system used only as reference corrector for the learning state.

6. Simulation Results

Simulations were performed to show the effectiveness of the PAF regulation by means of PI-Fuzzy adaptive corrector for the DC link voltage control and by the use of ANN corrector for the PAF current control.

A nonlinear load, consisting of a three-phase diode-based rectifier and a resistive load in its DC side, is fed by sinusoidal and symmetrical mains phase voltages (127 Vrms, 50 Hz).

Figures 9 and 10 show the load current without compensation (only one phase current is represented for the clearness) and its harmonics spectrum, respectively. It is clearly shown that the total harmonic distortion (THD) is relatively high (the calculated value is approximately 22%).

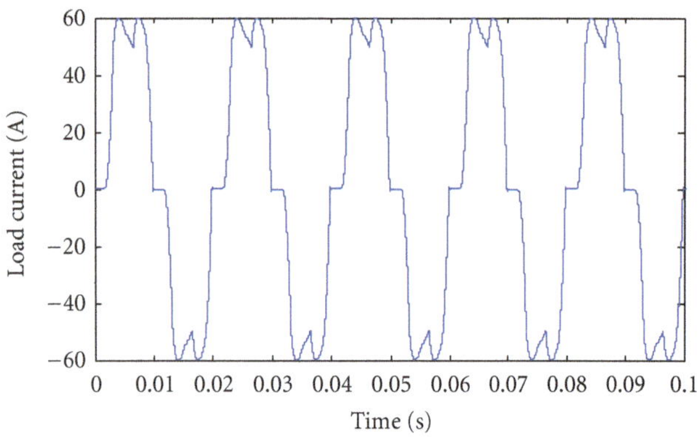

Figure 9: Load current before compensation.

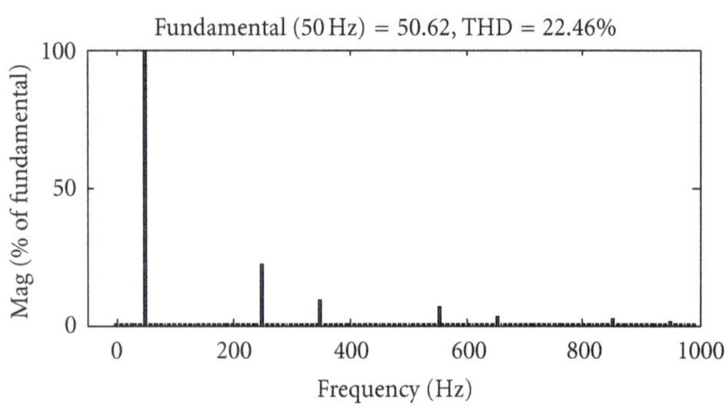

Figure 10: Load current spectrum before compensation.

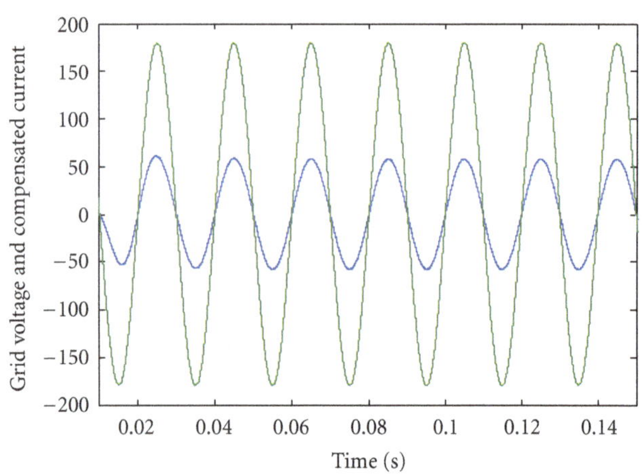

Figure 11: Source voltage and compensated current.

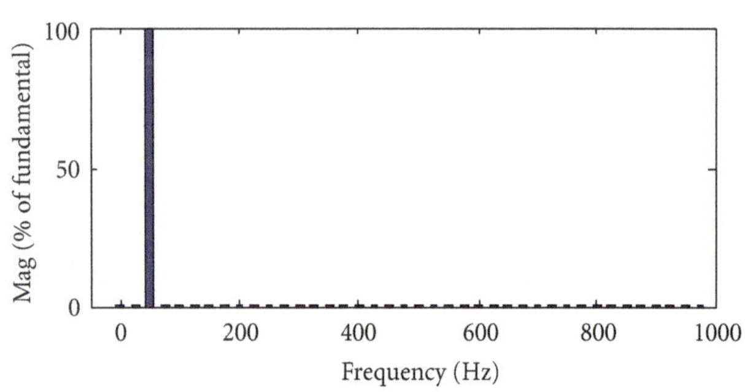

Figure 12: Grid current spectrum after compensation.

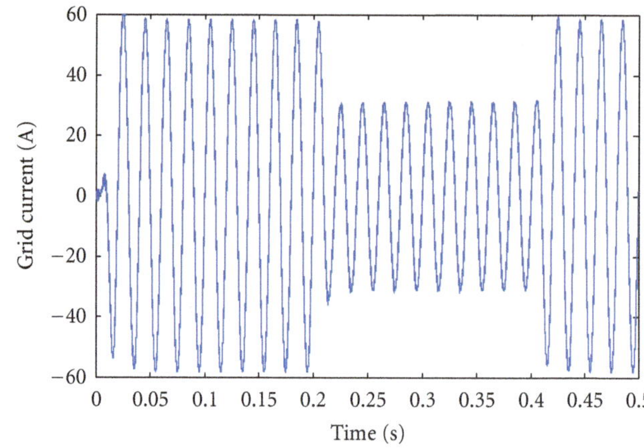

Figure 13: Grid current transient response.

The simulation results of the proposed control algorithm are presented in the next figures. Thus, Figure 11 shows that the source current becomes closely sinusoidal and inphase with source voltage. The source power factor is near unity, and the THD is enormously reduced (the harmonics spectrum, as represented in Figure 12, is reduced to only the fundamental and the calculated value of THD which is approximately 1.2%).

To show the dynamics of the proposed control algorithm, a step change is made in the load. The result is presented in Figure 13. It is obviously shown that the transient response time is very short, and so the dynamic of the controlled system is high.

7. Conclusion

The shunt power active filter (PAF) is basically used to compensate harmonics, produced in the case of a nonlinear load, and reactive power. In this paper, we have presented the simulation results, using the Matlab/Simulink PSB Blocs, for the PAF control.

The reference current's identification is accomplished by an approach based on a powers balance between the active power provided by AC source and that absorbed by the load.

The DC voltage regulation is done by using a PI-Fuzzy adaptive controller. The output current is then used for the currents reference's identification.

To control the PAF, a neurofuzzy controller is adopted, with a PI-Fuzzy system as a reference corrector for the learning stage, and which imposes the generated current to track closely the reference current.

The first investigations, presented here, of the control algorithm prove its effectiveness and its high dynamics. It will be completed in a future work by using other artificial techniques and by considering others PAF configurations.

References

[1] H. Akagi, Y. Kanazawa, and A. Nabae, "Instantaneous reactive power compensators comprising switching devices without energy storage components," *IEEE Transactions on Industry Applications*, vol. IA-20, no. 3, pp. 625–630, 1984.

[2] A. Jami and S. H. Hosseini, "Implementation of a novel control strategy for shunt active filter," *The ECTI Transactions on Electrical Engineering, Electronics, and Communications*, vol. 4, no. 1, pp. 40–46, 2006.
[3] H. Akagi, Y. Kanazawa, and A. Nabae, "Generalized theory of the instantaneous reactive power in three-phase circuits," pp. 1375–1386, International Power Electronics Conference (IPEC' 83), Tokyo, Japan, 1983.
[4] A. Dell'Aquila and A. Lecci, "A current control for three-phase four-wire shunt active filters," *Automatika*, vol. 44, no. 3-4, pp. 129–135, 2003.
[5] Z. Y. Zhao, M. Tomizuka, and S. Isaka, "Fuzzy gain scheduling of PID controllers," *IEEE Transactions on Systems, Man and Cybernetics*, vol. 23, no. 5, pp. 1392–1398, 1993.
[6] P. Viljamaa, *Fuzzy gain scheduling and tuning of multivariable fuzzy control methods of fuzzy computing in control systems [Ph.D. thesis]*, Temper University of Technology, 2000.
[7] G. Dreyfus, J. Martinez, M. Samuelides et al., *Réseaux De Neurons : Méthodologie Et Applications*, Editions Eyrolles, 2002.
[8] D. Ould Abdeslam, P. Wira, J. Mercklé, Y. A. Chapuis, and D. Flieller, "Stratégie neuromimétique d'identification et de commande d'un filtre actif parallèle," , *Revue Des Systèmes*, vol. 9, no. 1, pp. 35–64, 2006.
[9] M. Chakphed and P. Suttichai, "Active power filter for three-phase four-wire electric systems using neural networks," *Electric Power Systems Research*, vol. 60, no. 3, pp. 179–192, 2002.

5

Transformer Model in Wide Frequency Bandwidth for Power Electronics Systems

Carlos Gonzalez-Garcia and Jorge Pleite

Electronic Technology Department, Carlos III University of Madrid, Avenida de la Universidad 30, Leganés, 28911 Madrid, Spain

Correspondence should be addressed to Carlos Gonzalez-Garcia; cggphd@gmail.com

Academic Editor: Francesco Profumo

The development of the smart grids leads to new challenges on the power electronics equipment and power transformers. The use of power electronic transformer presents several advantages, but new problems related with the application of high frequency voltage and current components come across. Thus, an accurate knowledge of the transformer behavior in a wide frequency range is mandatory. A novel modeling procedure to relate the transformer physical behavior and its frequency response by means of electrical parameters is presented. Its usability is demonstrated by an example where a power transformer is used as filter and voltage reducer in an AC-DC-AC converter.

1. Introduction

The future trends on the power delivery are focused on the development of a smart grid. This concept implies new challenges as interconnection or treatment of renewable energy resources and the application of technologies such as telecommunication or power electronics whose products and solutions should play an important role in the future of the medium voltage distribution.

The power transformer is widely spread in electrical systems providing basic functionalities such as voltage insulation and voltage adaptation. As an essential element on the energy distribution, it must be involved in the new smart grid perspective, and therefore new roles regarding its manufacturing, maintenance, use, and operation must be taken into account. One of these roles consists in the analysis of its behavior in a wide frequency bandwidth, not limited to the 50 or 60 Hz power frequency.

Considering the transformer maintenance program, most of routine tests to assess the transformer condition are based on power frequency (50/60 Hz) measurements (load and no load losses, capacitance, and Tan-Delta). However, the FRA test measures the impedance between two terminals of the transformer in the 20 Hz to 20 MHz bandwidth. It has been extensively proven that this procedure allows the detection of failures that are not visible for other techniques [1].

Considering the power delivery, the conventional line-frequency transformers are not able to deal with power quality problems (e.g., sags, swells, flicker, and harmonics). To solve this task, there have been some attempts (although still under development) for the installation of additional equipment as power electronics converter that operates at higher switching frequencies in the medium voltage grid. It is also applied in the LV grid due to the advances in semiconductor technology (faster switching actions, higher blocking voltages, and higher power densities) and the development of new magnetic materials with low loss densities at higher operating frequencies [2, 3]. The application of PETs is extensible to traction applications [4] or Plug-in Electrical Vehicle (PEV) charging stations [5].

This solution implies replacing the typical power frequency distribution transformer by a high frequency switched one, commonly named as PET or Intelligent Solid-State Transformer (ISST) consisting in a Medium Frequency Transformer (MFT) and one or several power electronic converters. As a result, fast switching actions are applied directly to transformer terminals, so that overall magnetic volume is reduced and more compact converter designs are

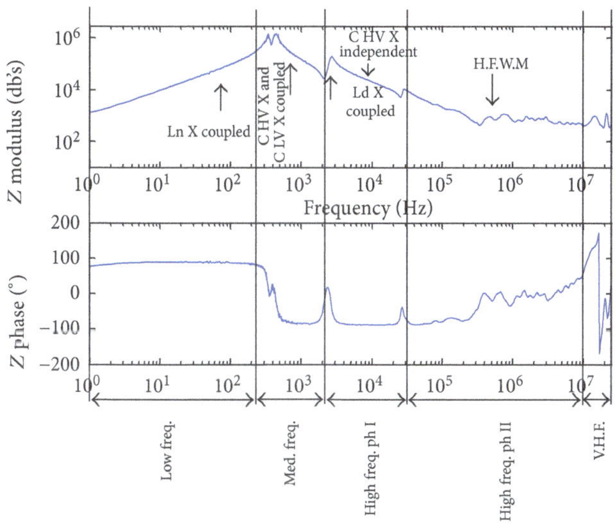

FIGURE 1: Identification of the most significant bandwidths in an end-to-end open circuit typical response.

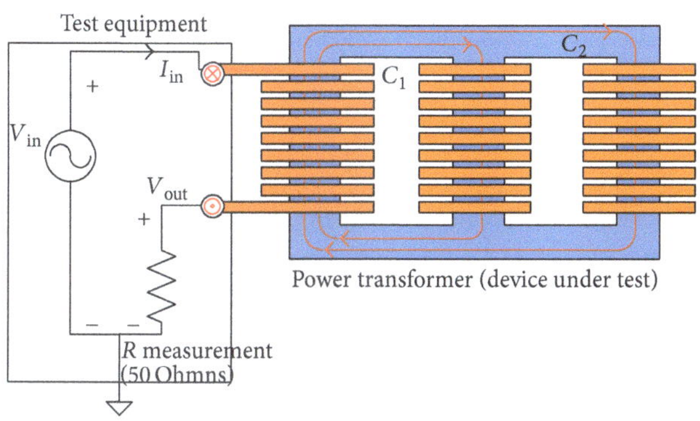

FIGURE 2: Electromagnetic field distribution at low frequency bandwidth. Mainly magnetic on the core.

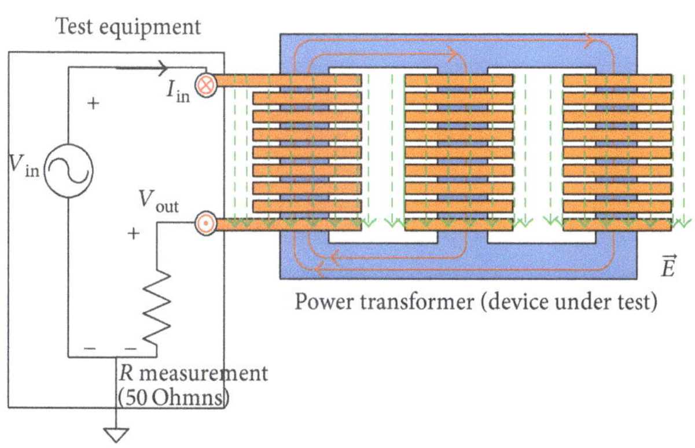

FIGURE 3: Electromagnetic field distribution at medium frequency bandwidth. Mainly electric field along the windings E.

reached [6]. However, new problems regarding losses or overheating and subsequent insulation damages and loss of transformer life expectancy could come across. Therefore, an accurate knowledge of the transformer in a wider frequency bandwidth is needed to tackle with new analysis.

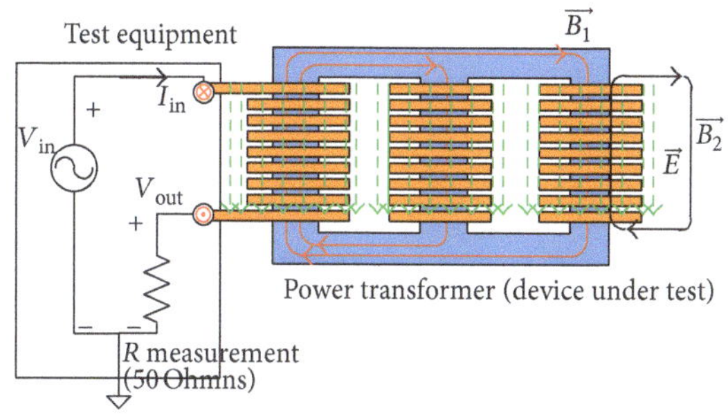

FIGURE 4: Electromagnetic field distribution at high frequency phase I bandwidth. Coexistence of electric field along the windings E, and magnetic field on core B_1 and core-dielectric B_2.

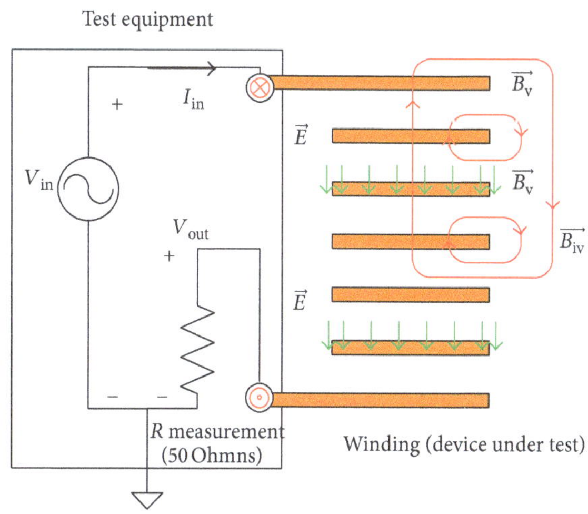

FIGURE 5: Electromagnetic field distribution at high frequency phase II bandwidth. Coexistence of electric E and magnetic B_v, B_{iv} field exclusively along the measured winding.

Transformer models have been used to interpret and analyze the transformer behavior, but they have been commonly limited to a narrow bandwidth centered in the power frequency.

In this paper, a novel procedure to obtain transformer models, that are able to represent the transformer behavior in a wide frequency bandwidth, is presented and applied to a 25 kVA 16 kV/420 V YNYn distribution transformer. The accuracy of the model is evaluated comparing the simulation and actual responses, obtained by an FRA test.

Finally, the transformer model is used as part of an AC-DC-AC power electronic converter. The exact knowledge of the frequency behavior allows taking advantage of the filtering characteristics associated with the leakage inductance and shunt capacitance of the windings, saving the use of the external LC filter in the final stage of the 3-phase setup.

2. Modeling Procedure

The proposed solution to obtain a wide frequency model is based on a novel modeling procedure developed in [7] whose main features are as follows.

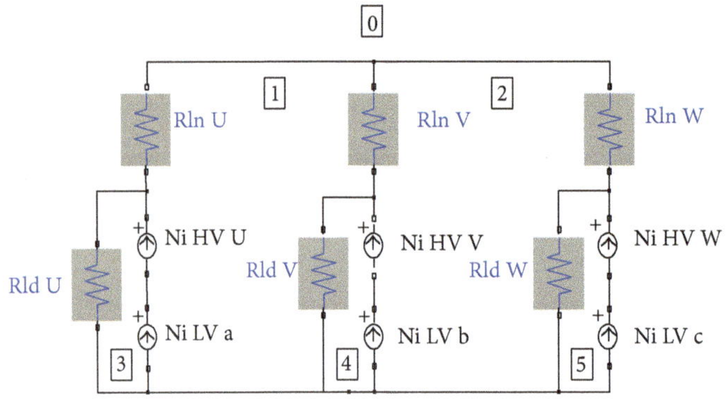

FIGURE 6: Magnetic circuit based on reluctances and magnetomotive forces for representing the magnetic field behaviour in the core and core-dielectric at low and high frequency phase I bandwidths.

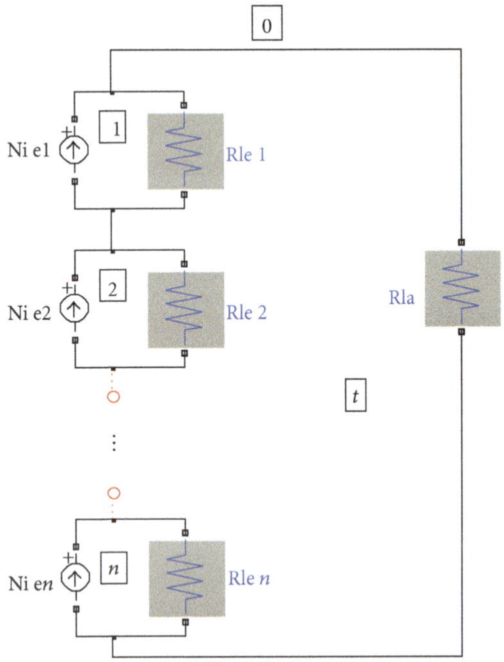

FIGURE 7: Magnetic circuit based on reluctances and magnetomotive forces for representing the magnetic field behaviour in the winding at high frequency phase II bandwidth.

(i) The input data are obtained by the Frequency Response measured in a wide frequency bandwidth on the FRA test.

(ii) The resulting model is based on a parametric structure developed using the Magnetic-Electric Duality Principle and Energy Storage Concept. These synthesizing tools allow the interpretation of the electromagnetic phenomena occurring not only on the magnetic core as other solutions proposed but also on the leakage path and dielectric media along the winding. Subsequently, the physical behavior is concentrated on an electrical circuit.

(iii) The structure is built in a modular way and constituted by plugging in different and independent blocks representing only the physical constitution of any internal element. The vector group connection of the windings (Star or Delta) is externally implemented on the model and is independent of the physical submodels or blocks.

(iv) The value of the parameters is calculated using an ad hoc optimization algorithm. It searches the solution that minimizes the difference between simulated and actual measured data. The algorithm is separately applied to the different blocks building the complete model and responsible for the response in a specific frequency bandwidth of the entire measurement sweep.

As a result, the modeling procedure follows three stages: the structure design (stage 1, Section 2.1), the algorithm design for calculation of the value of the parameters (stage 2, Section 2.2), and experimental validation (stage 3, Section 2.3).

2.1. Structure Design. The design of the structure is the first stage on the modeling procedure and it follows, for its part, other three steps.

The first one is the qualitative analysis of the electromagnetic field at different frequency bandwidths in the FRA test, divided in the partial bandwidths of Figure 1 and named as low frequency bandwidth, medium frequency bandwidth, and high frequency phase I and II bandwidths.

The constitution of the field at the low frequency bandwidth is depicted in Figure 2, where the measurement setup is included and the magnetic flux paths are represented by the C_1 and C_2 solid lines. The medium frequency bandwidth is defined as the range where the constitution of the analyzed field is mainly electrical (represented with the E dashed lines) and distributed along a structure defined along the windings (not only in the excited one but also in the rest due to the magnetic coupling) in the fashion depicted in Figure 3 with dashed lines.

The electromagnetic effects in the high frequency phase I and phase II bandwidths defined in the modeling procedure are represented in Figures 4 and 5, respectively.

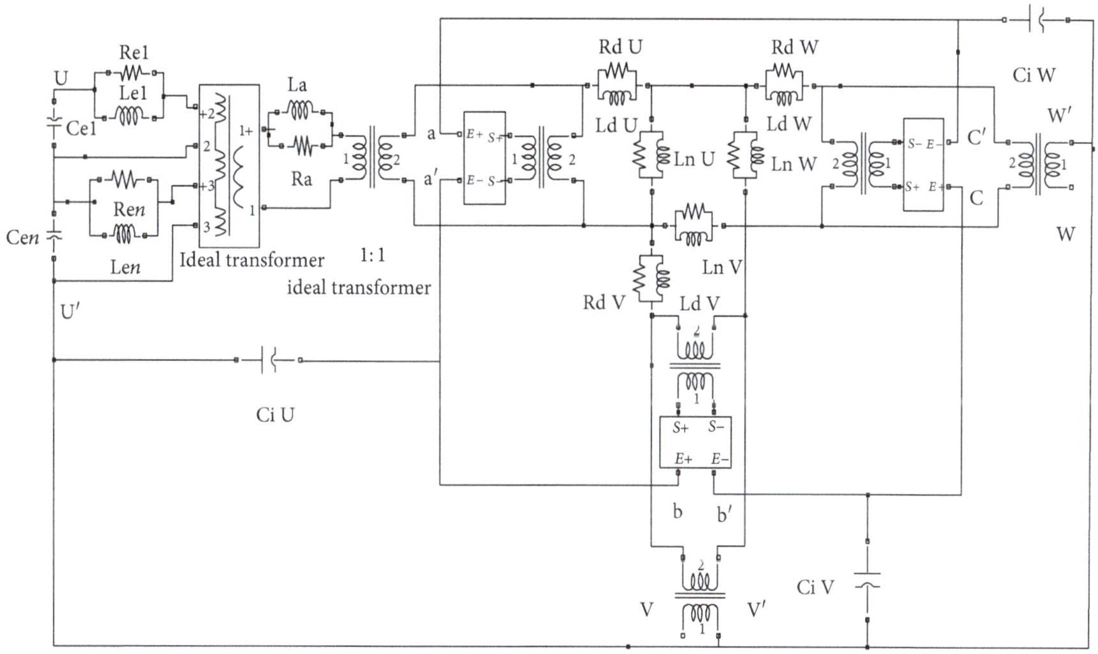

FIGURE 8: Complete electric circuit. Interwinding capacitances Ci U, Ci V, and Ci W added.

FIGURE 9: 25 kVA 16.125 kB/420 V YNyn distribution transformer measured and modeled.

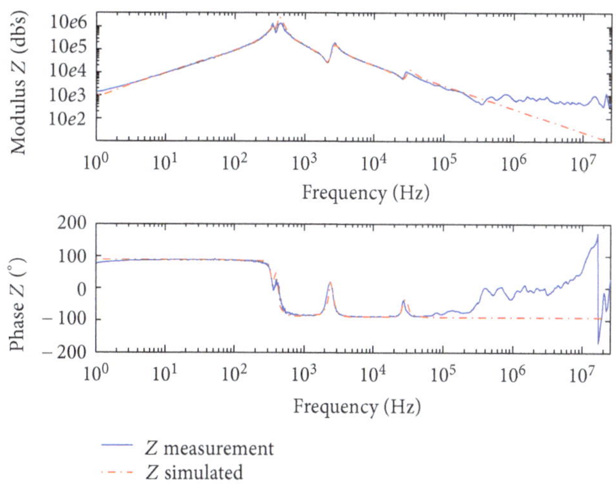

FIGURE 10: Actual versus simulated measurement. YNyn configuration end-to-end open circuit, phase U HV.

The second step designing the structure consists on combining the obtained results in a magnetic circuit that simulates the same magnetic field distribution.

The referenced magnetic field in the core (B_1 in Figure 4) and in the core-dielectric (B_2 in Figure 4) for low and high frequency phase I bandwidths can be represented by the well-known magnetic circuit of Figure 6. The parameters Rln X (where X is one of the U, V, or W phase of the 3 phase transformers indistinctly) represent the reluctances of the magnetic paths in the core and therefore depending on the core geometry and its permeability μ_c. The parameters Rld X represent the reluctances of the magnetic paths in the core-dielectric and depending on the winding geometry and mainly the dielectric permeability μ_d. The sources Ni HV X and Ni LV X represent the magnetomotive forces of the HV and LV windings, respectively. Using the same representation, the magnetic circuit depicted on Figure 7 represents the magnetic field distribution for the defined high frequency phase II bandwidth, represented in Figure 5, where Rla represents the reluctance of the magnetic path of the dielectric surrounding the N turns of the winding. Rle_j represents the same reluctance but for a group of j turns in the

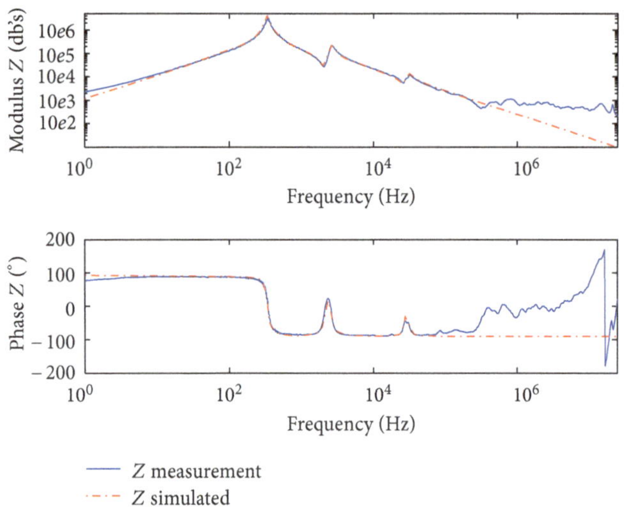

FIGURE 11: Actual versus simulated measurement. YNyn configuration end-to-end open circuit, phase V HV.

winding. Finally, Nie_j represents the magnetomotive force of a group of j turns in the winding.

The third step designing the structure consists in obtaining the final equivalent electric circuit using the well-established Magnetic-Electric Duality Principle to convert the magnetic circuit into electric one and the Energy Storage Concept, adding R and C elements to represent the losses and electric field effect, respectively. The resulting three submodels are

(i) LFCM submodel representing the magnetic core, at the low frequency bandwidth of Figure 1;

(ii) HFCM submodel representing the core and dielectric media at the high frequency phase I bandwidth of Figure 1;

(iii) HFWM submodel representing the winding at the medium frequency and high frequency phase II bandwidths of Figure 1.

The complete transformer model in Figure 8 is composed of the models representing the three previous submodels (for simplicity only U phase for HV winding is completely represented with 2 groups of turns) and the capacitors Ci X that represent the electric energy storage on the dielectric media between HV and LV windings on the generic X phase. The YNd configuration is represented in Figure 8 to illustrate how easily the vector group is implemented only connecting the output terminals of the windings. Any of these submodels is in charge of a specific bandwidth defined in Figure 1.

2.2. Algorithm Design. The second stage of the proposed modeling procedure consists in designing the mathematical algorithm in order to fit the actual measured response by the simulation of the model. In this case, the optimization algorithm is completely developed in [7] and its working principle is summed up in the objective function given by (1).

$$\frac{dE_t}{\partial P_i} = \sum_{j=1}^{j=m} \left\{ \left(\frac{1}{\left|Z_{\text{measured}\,j}\right|^2} \right) \cdot \frac{d\left(\left| Z_{\text{submodel}}\left(\omega_j\right) - Z_{\text{measured}\,j} \right|^2 \right)}{\partial P_i} \right\} = 0, \tag{1}$$

where E_t represents the total error when comparing the simulated frequency response of the model and the actual response of the measured transformer, P_i represents a generic R, L, or C parameter of a submodel, j is the index representing each frequency point in a $[1, m]$ bandwidth, $Z_{\text{measured}\,j}$ represents the complex impedance measured in the j frequency, $Z_{\text{submodel}}(\omega_j)$ is the complex impedance, depending of the R, L, and C parameters of the submodels and the angular pulsation ω_j for a frequency j.

2.3. Experimental Validation. The third and last stage in the proposed modeling procedure consists in the experimental validation of the resulting model. The model is approved if it is able to simulate the FRA measurements. For this contribution, a 25 kVA, 16.125 kV/420 V YNyn two-winding three-phase distribution transformer (Figure 9) has been measured with an FRA commercial device and modeled by implementing the mathematical algorithm in mathematical/simulation software. The calculated parameters are shown in Table 1 where they are split using the submodel division.

Figures 10 and 11 show the comparison between the actual measurement and the simulated response of the model implemented in MATLAB Simulink.

The modeling procedure is able to produce general models with the following features.

(i) The transformer behavior in a wide frequency bandwidth is related with the parameters of the model.

(ii) The submodels are obtained based on general and common physical principles, and particularizations for specific transformers are avoided. However, if particularization needed to be considered, it could be solved with a concrete analysis of the electromagnetic field, as shown in Section 2.1.

(iii) The vector group connection is independent of the electrical structure that represents the physical behavior of the transformer. As a result, any possible vector group can be modeled by changing the terminal connections of the model. This same conclusion can be applied for any FRA measurement setup.

(iv) Modeling of n-windings transformer is straightforward. Only the addition of n number of winding submodels is needed.

(v) Modeling of the autotransformer is straightforward. Only an external series connection of series and mutual windings is needed.

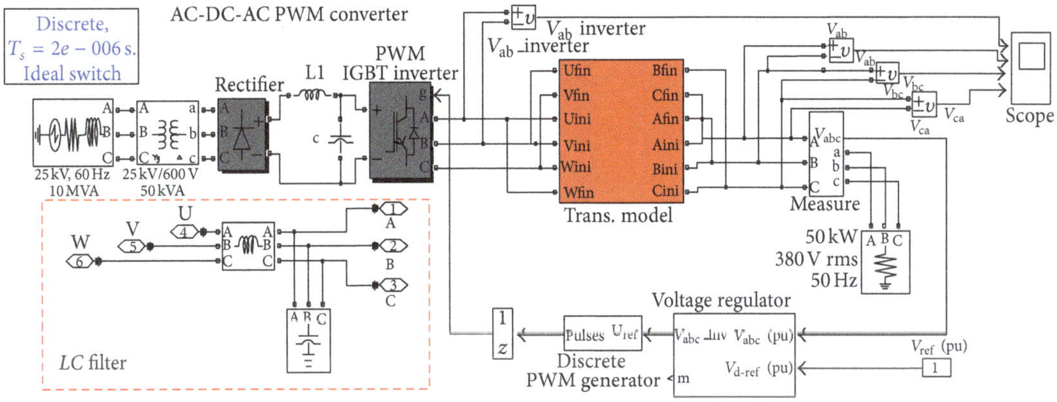

FIGURE 12: AC-DC-AC converter.

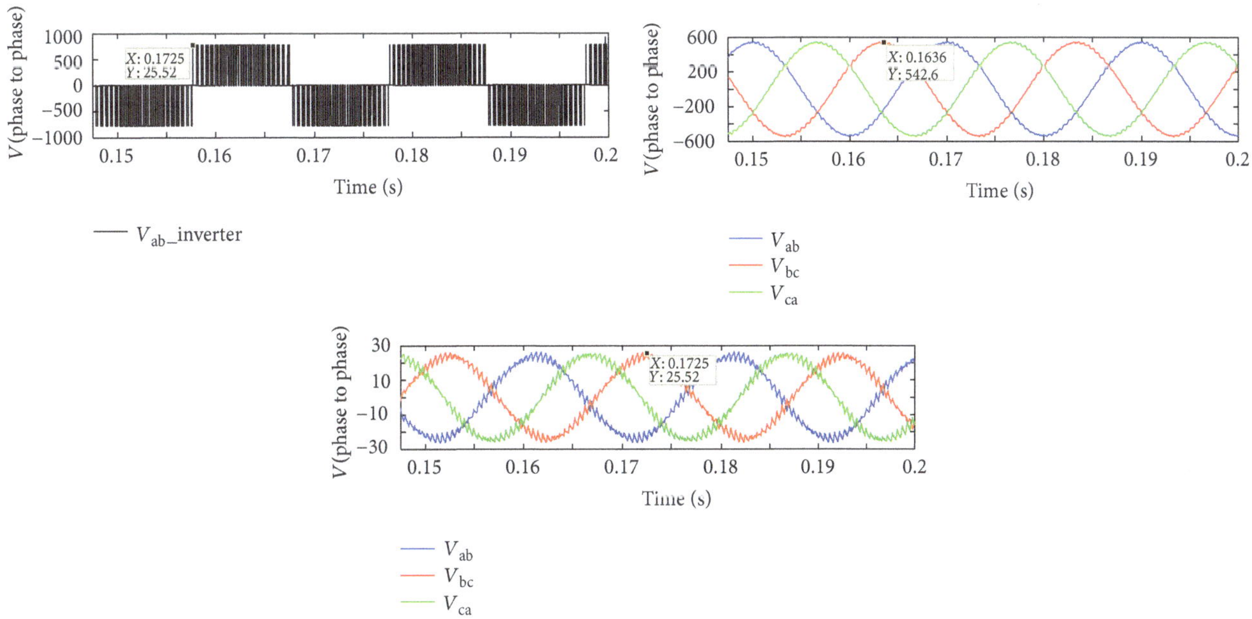

FIGURE 13: Upper (left) trace: PWM voltage; upper (right) trace: 3 phase to phase symmetrical voltages when *LC* filter is used; lower trace: 3 phase to phase symmetrical voltages when transformer model is used.

The capability of the resulting models can be exploited for sensitivity analysis, FRA traces interpretation, and implementation of simulation systems. That allows knowing exactly the transformer behavior in a wide frequency bandwidth in order to be used in the design process of power electronics converters as it is shown in the following Section 3.

3. AC-DC-AC Converter Application

The DC-AC conversion needs a filter in order to obtain a sinusoidal waveform from a PWM voltage waveform, avoiding the high frequency components. This task usually can be solved with the addition of an *LC* filter on the last stage of the converter. However, that solution implies an economic cost and total volume increase (critical for transport applications) due to the addition of *L* and *C* components. Moreover, an additional adaptation stage could be needed in order to increase/reduce the voltage amplitude of the inverter.

The use of a PET could solve at the same time the filtering and adaptation necessities. However, an accurate transformer model, as the one presented in previous paragraphs, is needed to assure a complete comprehension of its frequency response and therefore the behavior (losses, voltage regulator, and so on) when working as part of a DC-AC converter.

In the following example, the transformer model is implemented and simulated as part of an AC-DC-AC converter (Figure 12) based on the power electronic model named "power_bridges.mdl" in MATLAB SimPowerSystems.

The converter consists in a first stage where a DC voltage of 1000 volts is obtained. The second stage consists in a PWM inverter to obtain 3 phase to phase PWM voltages. One of the three voltages in the 3-phase converter, named as V_{ab}_inverter, has been plotted in the upper side of Figure 13.

TABLE 1: Value of the parameters of the experimental model.

	Phase U	Phase V	Phase W
		L.F.C.M sub-model	
LFB (Hz)	[1–2000]	[1–2000]	[1–2000]
Number° of points	201	201	201
Ln X (H)	183.1504	399.8045	185.3597
C HV (nF)	0.7452	0.6648	0.7356
Rn X (MΩ)	2.400	12.124	2.3
		H.F.C.M sub-model	
HFPIB (kHz)	[2–2.89]	[2–2.89]	[2–2.89]
Number° of points	17	17	17
Ld X (H)	4.0227	5.5128	5.0427
Rd X (kΩ)	11.690	10.154	11.690
		H.F.W.M sub-model (all phases HV side)	
HFPIIB (kHz)	[10–335]	[10–335]	[10–335]
Number° of points	239	239	239
Le_j, $j = 1$ (mH)	3.9218	3.8539	3.4211
Ce_j, $j = 1$ (nF)	1.0838	1.0602	1.0721
Re_j, $j = 1$ (kΩ)	7.412	7.895	7.296
Le_j, $j = 2$ (mH)	3.9414	4.0031	3.9612
Ce_j, $j = 2$ (nF)	2.5919	2.5219	2.5528
Re_j, $j = 2$ (kΩ)	5.339	5.762	5.391
C HV (nF)	0.7452	0.6648	0.7356
		Ci X parameters	
Bandwidth (Hz)	[4083.48–10023.8]	[4083.48–10023.8]	[4083.48–10023.8]
Number° of points	39	39	39
Ci X (nF)	0.1186	0.1170	0.1181

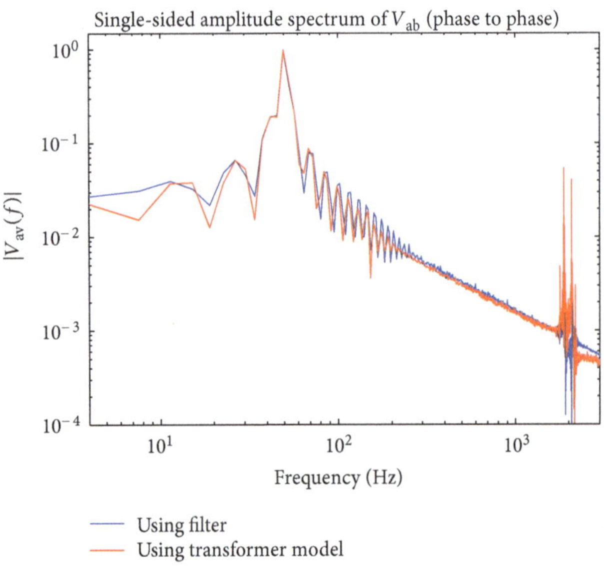

FIGURE 14: Spectrum of the V_{ab} voltages for LC and transformer simulation.

The filtering and adaptation stages have been solved using two alternatives. The first one is an LC filter (pointed by red dashed lines in Figure 12) used to get symmetrical 3 phase to phase sinusoidal voltages V_{ab}, V_{bc}, and V_{ca}. The second one is a power electronic transformer that has been simulated using the presented model.

The plots of Figure 13 show the PWM voltage signal in the upper side (the input of the LC or PET) and the comparison of the filtered voltage signals using the LC filter (results in the middle side of Figure 13) and the model of the PET (results in the lower side of Figure 13).

It has been possible to simulate the PET because its behavior is completely represented till 500 kHz using the model shown in Section 2. The analysis of the model (completely represented in Figure 8) and its simulated response (showed in Figures 10 and 11) demonstrate that the leakage inductance (Ld X in Table 1) and the shunt capacitance of the windings (C_{HV} in Table 1) conform an LC filter whose values give a resonant frequency of 2600 Hz. That means that the transformer filter can be used as the element needed in the inverter, saving the use of the external LC component. This concept is developed in [8] where the bases for designing LC filters integrated in the transformer is explained. The accurate agreement between actual and simulated responses until 500 kHz demonstrates that the simulation bandwidth is enough to cover the 2600 Hz behavior when the cut-off frequency of the filter appears.

The use of the transformer in the complete system in spite of the LC filter offers not only the filtering capability but also the voltage adaptation. For this reason, the 3 phase to phase voltages V_{ab}, V_{bc}, and V_{ca} (represented in the lower side of Figure 13) present a sinusoidal shape with 18 Volts rms in spite of the 383 Volts rms obtained when the external *LC* filter is used.

The filtering capability of the transformer is demonstrated in the comparison of Figure 14. The frequency spectrum of the V_{ab} phase to phase sinusoidal voltage is shown for both cases, when the *LC* and the transformer model are used. Both graphics are fairly equal, which demonstrates that the transformer filtering capability can be considered as accurate as the *LC* filter.

Finally, the efficiency and voltage regulation of the PET can be both calculated and simulated. The calculation can be done because the values of the resistive parameters representing the losses for the complete frequency bandwidth in the model are known, as represented in the example of Table 1 in the Rn X, Rd X, Re_j parameters. As a result, an efficiency of 93.74% and a voltage regulation of 94.7% were calculated after simulation.

4. Conclusions

The new development of smart grids involves the use of power electronic equipment and new challenges in the operation of the power transformers. Therefore, the knowledge of its behavior in a wide frequency bandwidth, not limited to the power frequency (50/60 Hz), is a necessity.

A novel modeling procedure to obtain the transformer model in a wide frequency bandwidth is presented. The main feature of the model is its capability of interpreting the physical behavior by means of electrical parameters. That allows developing simulations, design revision, or sensitivity analysis in order to take advantages of the capabilities of the transformer in power electronics design or avoid risky situations as additional losses, overheating, or overvoltages due to high frequency components.

Acronyms

FRA: Frequency response analysis
HFCM: High frequency core model
HFWM: High frequency winding model
LFCM: Low frequency core model
PET: Power electronic transformers.

Conflict of Interests

The authors would like to clarify that there is not financial gain and/or commercial interest related with MEGGER and MATLAB registered trademarks. The only reason these names have been included in the paper is for the sake of precision and to determine, in the more exact way, the procedure to simulate the model and measure the prototype transformer.

References

[1] International Electrotechnical Commission, *Power Transformers-Part 18: Measurement of Frequency Response*, vol. 60076, International Electrotechnical Commission, Geneva, Switzerland, 2012.

[2] Z. Wang and K. Yu, "The research of Power Electronic Transformer (PET) in smart distribution network," in *Proceedings of the International Conference on Power System Technology: Technological Innovations Making Power Grid Smarter (POWERCON '10)*, pp. 1–7, October 2010.

[3] J. M. Carrasco, L. G. Franquelo, J. T. Bialasiewicz et al., "Power-electronic systems for the grid integration of renewable energy sources: a survey," *IEEE Transactions on Industrial Electronics*, vol. 53, no. 4, pp. 1002–1016, 2006.

[4] D. Dujic et al., "Laboratory scale prototype of a power electronic transformer for traction applications keywords power electronic transformer topology," in *Proceedings of the 14th European Conference on Power Electronics and Applications (EPE '11)*, vol. 1, pp. 1–10, September 1 2011.

[5] P. S. Moses, M. A. S. Masoum, and K. M. Smedley, "Harmonic losses and stresses of nonlinear three-phase distribution transformers serving plug-in electric vehicle charging stations," in *Proceedings of the IEEE PES Innovative Smart Grid Technologies (ISGT '11)*, pp. 1–6, January 2011.

[6] M. Kang, P. N. Enjeti, and I. J. Pitel, "Analysis and design of electronic transformers for electric power distribution system," *IEEE Transactions on Power Electronics*, vol. 14, no. 6, pp. 1133–1141, 1999.

[7] C. G. García, *Procedimiento de modelado basado en el análisis de la respuesta en frecuencia y aplicación en transformadores trifásicos de potencia para su caracterización y diagnóstico. volume 1 [Ph.D. thesis]*, 2012.

[8] V. Valdivia, J. Pleite, P. Zumel, and C. Gonzalez, "Improving design of integrated magnetics for power electronics converters," *Electronics Letters*, vol. 44, no. 11, pp. 693–694, 2008.

6

Transformer Magnetization Losses Using a Nonfiltered Voltage-Source Inverter

Rickard Ekström, Senad Apelfröjd, and Mats Leijon

Division for Electricity, Swedish Centre for Renewable Electric Energy Conversion, Uppsala University, P.O. Box 534, 751 21 Uppsala, Sweden

Correspondence should be addressed to Rickard Ekström; rickard.ekstrom@angstrom.uu.se

Academic Editor: Jose Pomilio

Results from the magnetization of an 80 kVA power transformer, using a directly coupled nonfiltered three-phase voltage-source inverter (VSI), are presented. The major benefits of this topology are reduction in switching filter size as well as filter losses. Drawbacks include higher stress on the transformer windings and higher transformer magnetization losses. In this paper, the total magnetization losses are presented for different voltage levels. The transformer has been magnetized with the rated frequency of 50 Hz at various voltage levels. The saturation characteristics as well as the magnetizing resistance are derived as functions of the voltage. These are used as inputs for the simulations. The magnetization losses have been experimentally measured and simulated for three different DC levels. Results from the simulations show good agreement with the experimental results. As expected, the pulsed voltage waveforms generate larger magnetization losses than the corresponding 50 Hz case. The losses are strongly dependent on the DC level.

1. Introduction

Many grid connected voltage source inverters (VSIs) have an associated LCL filter to eliminate the switching frequencies from the output [1–3]. If a step-up transformer is placed after the filter, its leakage inductance can reduce the size of the second inductor leg or completely replace it [4]. Alternatively, the transformer can be put directly after the inverter, with the LCL filter on the transformer secondary side. This will reduce the ampacity of the filter inductors, reducing both size and losses in the filter. The drawback of this is an increased voltage transient stress on the transformer windings that has to be taken into account in the transformer design. However, the total efficiency of the system may be improved, and this topology could be considered for grid connection of renewable energy sources. Stray capacitance between windings, which normally is neglected at steady state, may play a more central role. Semiconductor switching into magnetic components is not unusual in power electronics, for example, the flyback converter. However, these circuits are based on ferrite cores, optimized for the switching frequencies in the system. Little research is found in the area of continuous pulse modulation into laminated steel-core transformers and its associated losses.

This paper presents a VSI with bipolar sinusoidal pulse-width modulation (SPWM) directly connected to a three-phase power transformer and the measured total magnetization losses. To get a reference case, the transformer is magnetized with 50 Hz at different voltage levels. The transformer saturation curve and magnetizing resistance are derived as functions of the induced voltage. A simulation model of the system is developed and uses the measured data at 50 Hz to set the internal parameters. Experiments with variations in DC voltage level and modulation index are conducted. The simulation model is verified, and the magnetization losses are evaluated for different amplitude modulations.

2. Theory

Starting from Faraday's law, the Magnetic field intensity, $\overline{B}$, in the transformer core is derived from the induced voltage $\overline{V}_{\mathrm{ind}}$

according to

$$B(t) = \frac{1}{NA}\int_0^t V_{\text{ind}}(t)\,dt + B_r(0), \quad (1)$$

where N is the number of winding turns, A is the average cross-sectional area of the magnetic core, and $B_r(0)$ is the remanescent flux of the magnetic core. B is coupled to the magnetic field strength $\overline{H}$ according to

$$\overline{H} = \mu_r \mu_0 \overline{B}, \quad (2)$$

where $\mu_0 = 4\pi \cdot 10^{-7}\,V\,s/Am$ is the magnetic permeability of free space, and μ_r is the relative permeability constant, a nonlinear material-dependent parameter. If the magnetizing current $\overline{I}_m$ is known, $\overline{H}$ is calculated by

$$H(t) = \frac{\overline{I}_m}{N\mathfrak{R}}, \quad (3)$$

where $\mathfrak{R}$ is the magnetic reluctance of the core.

The core losses of any magnetic component are distributed between eddy current losses, excess eddy current losses, and hysteresis losses. All of these have nonlinear behaviour and depend on the magnetic material properties and the operating temperature, as well as the waveform and frequency of the induced voltage. There are two established ways of calculating the core losses, either from the theoretical derivations of each loss separately [5, 6] or empirically by the Steinmetz equation [7] as follows:

$$P = kf^{\alpha}B^{\beta}, \quad (4)$$

where k is a material constant and f is the frequency of the magnetic flux. The exponents α and β are nonlinear functions of the other parameters.

Steinmetz equation is preferable, since the needed parameters usually are provided by the manufacturers data sheets. However, it is limited to sinusoidal waveforms and specific operating conditions. Also, having nonsinusoidal waveforms may cause large deviations in the loss calculations [8–10]. Splitting the waveform into its Fourier components and calculating the losses at the individual frequencies does not give very accurate results due to the complex interdependence of the losses. Also, it is difficult to accurately obtain all the Steinmetz parameters for a broad range of frequencies. Modifications of the Steinmetz equation are found in, for example, [11]. In [12, 13], attempts have been made to generalize the core loss calculations for an arbitrary current shape.

The classical eddy current losses may be described by [14]

$$P_{\text{eddy}} = \frac{A}{8\pi\rho}\left(\frac{dB(t)}{dt}\right)^2, \quad (5)$$

while hysteresis losses are calculated by integrating the B-H-curve over one cycle as

$$P_{\text{hys}} = f\int_t^{t+T} B(t)\,H(t)\,dt. \quad (6)$$

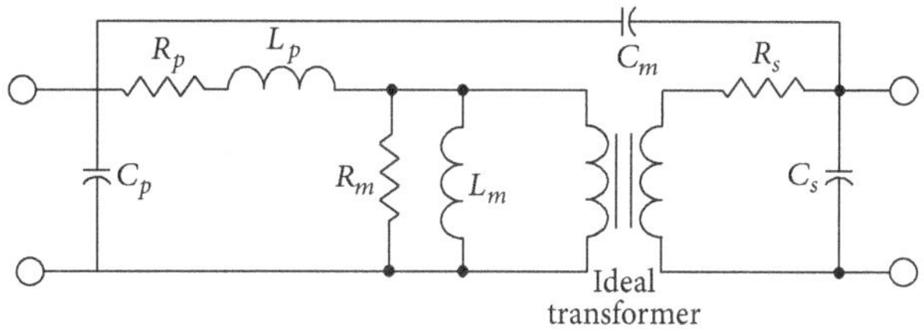

FIGURE 1: Transformer high-frequency equivalent circuit.

The high-frequency model of a single-phase transformer is shown for the no load case in Figure 1. R_p and R_s are the winding resistances on the primary and secondary sides, respectively. Only the leakage inductance of the primary side, L_p, is included, since the secondary leakage flux is so small at no load. Publications on high-frequency modelling of power transformers mainly discuss the pulse response, as in, for example, [15].

The internal stray capacitances of the windings, C_p and C_s, as well as the stray capacitance between the primary and secondary winding, C_m, become very important in transient analysis. Detailed calculations on transformer stray capacitance is found in [16, 17]. The stray capacitance will be in the order of pF and is neglected throughout this work since it will not significantly alter the power dissipation.

The magnetizing resistance, R_m, represents the total core losses, and the magnetizing inductance, L_m, represents the magnetic flux in the core. Assuming sinusoidal input voltage, both R_m and L_m, for rated voltage and frequency of the transformer, can be found in the manufacturers data sheets. However, these results are not valid at voltages and frequencies deviating from the transformer ratings.

One common way of inverter control signal generation is by the sinusoidal pulse-width modulation (SPWM), where a sinusoidal reference signal is compared with a carrier waveform. Assuming symmetry, the SPWM output voltage of the inverter can be expressed by its Fourier components according to

$$V_{\text{SPWM}}(t) = \sum_{h=1}^{N} V_h \sin(h\omega t), \quad (7)$$

where h is the order of the harmonic. Assuming ideal switching with no delay time between switching states, the three-phase rms voltage of the fundamental frequency is calculated as (remembering the neutral is floating)

$$V_1 = \frac{V_{\text{DC}}}{\sqrt{2}} m_a \quad (8)$$

and may be considered independent of the switching frequency, f_s. Equation (8) is only valid in the linear range (0-1) of the amplitude modulation index, m_a. In SPWM, the significant output harmonics are predominantly grouped around multiples of f_s [18].

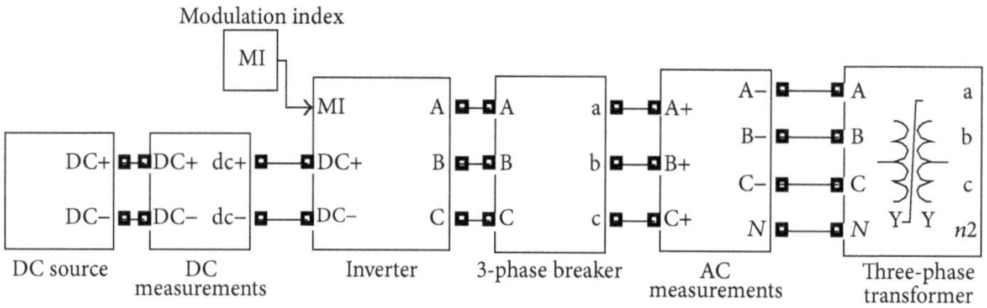

FIGURE 2: Simulink model block diagram.

The rms voltage of the unfiltered V_{SPWM} can be calculated as

$$V_{\mathrm{SPWM,rms}} = \sqrt{\sum_{h=1}^{N} V_{h,\mathrm{rms}}}, \quad (9)$$

where rms values of the harmonics can be found in [18].

3. Simulations

The Simulink model of the system is illustrated in Figure 2. Each block, from left to right, is described in the following. In the model, an ideal DC voltage source is used, and the DC power drawn by the circuit is measured. The three-phase inverter uses six IGBT blocks in parallel with six free-wheeling diodes. Device parameters come from the data sheets for the devices used in the experimental set-up. The inverter is operated with SPWM control, comparing a perfect sine wave to a sawtooth carrier wave at $f_s = 6$ kHz.

To limit the inrush currents at start-up and allow the system to settle, damping resistors are initially connected in series with the inverter and the transformer. When the system has settled, in approximately 100 ms, a breaker is closed to bypass the resistors in the circuit.

The three-phase transformer is modelled as three linear transformers in accordance with the equivalent model shown in Figure 1, but neglecting the stray capacitance of the system. The transformer parameters from Table 1 are used as input together with the saturation curve interpolated in Figure 4. A detailed description of the saturation model used is found in [19, 20]. By estimation of $V_{\mathrm{SPWM,rms}}$ from (9), a value of R_m is taken from the interpolated curve in Figure 3. R_m is kept constant for each full simulation together with m_a and the DC-link voltage.

The model does not include any temperature dependence, since the power loss is considered low for the no load tests. Also, the model assumes uniformly distributed flux in the core.

Initially, the transformer at no load is magnetized with the 50 Hz grid voltage at different voltage levels. The total losses are shown in Table 2. From this, a value of R_m is obtained, as shown in Figure 3. A curve fit is done by piece-wise shape preserving cubic interpolation (pchip). This curve is used as simulation input, using the derived rms value of the inverter voltage from (9). R_m varies widely over the measured range due to the magnetic properties and the transformer design. The interpolated curve is considered valid within the range of measured data points, that is, for $V_{\mathrm{rms}} \in [0 \quad 1.15]$. The value of the magnetizing resistance represents for both hysteresis losses and eddy current losses.

TABLE 1: Transformer characteristics.

Power rating	80 kVA
Voltage ratio	345 V Y/1 kV Y
R_p	0.0030535 pu
R_s	0.0096 pu
L_p	0.19858 pu
L_s	0.57303 pu
L_m (50 Hz, 1 kV)	500 pu
R_m (50 Hz, 1 kV)	89 pu

TABLE 2: Magnetization losses at normal 50 Hz magnetization.

V_s [pu]	Experimental P_{loss} [pu]	Simulated P_{loss} [pu]
0.130	0.00010	0.000088
0.425	0.00040	0.00042
0.700	0.0011	0.0011
1.00	0.0037	0.0037
1.15	0.0098	0.0098

To get the saturation curve of the transformer, no load phase currents and voltages were measured at various voltage levels. The induced voltage is proportional to the induced flux, and they coincide at 1 pu. The results are used to plot the upper part of the saturation curve, shown in Figure 4. Symmetry is assumed for the lower quadrant, and the saturation curve is extrapolated by the use of an arctan function.

4. Experimental Set-Up

A single-line diagram of the experimental set-up is depicted in Figure 5. The rectified grid voltage is used as DC source in the experiment. By changing tap on the tap transformer, different DC levels are obtained. The DC-capacitor bank is rated at 19 mF. Additional capacitors with small equivalent series resistance (ESR) are also used to reduce the high-frequency ripple in the DC voltage. The three-phase inverter consists of six 400GB126D IGBTs with 2SC0108T2Ax-17 Concept Driver Boards, as shown in Figure 6(a). Bipolar

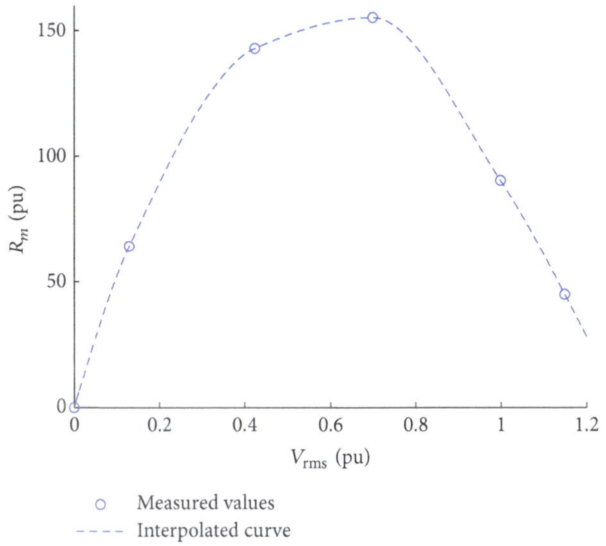

FIGURE 3: Magnetizing reluctance versus V_{rms}.

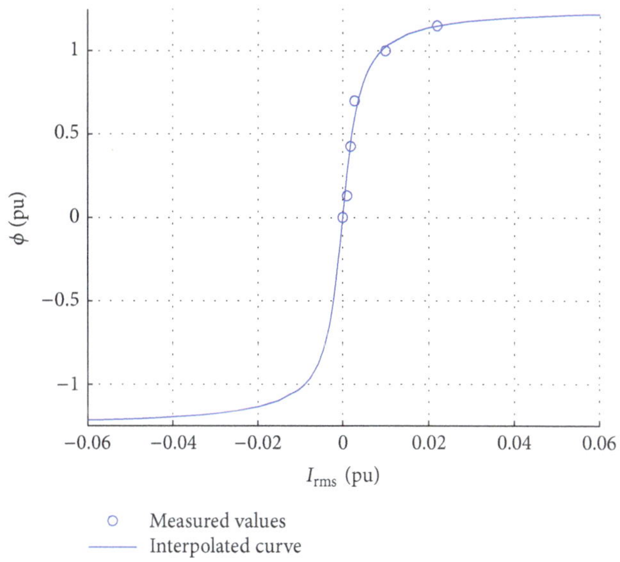

FIGURE 4: Saturation curve used in simulations.

SPWM control is used with a sawtooth wave as the carrier waveform. The switching frequency, f_s, is set to 6 kHz, and the delay time between switching transitions is set to 2.5 μs. The inverter is connected to the low side of the YY 345/1 kV 80 kVA three-phase transformer shown in Figure 6(b). The neutrals of the transformer are left floating to avoid the flow of zero sequences. The transformer ratings at rated voltage and frequency are displayed in Table 1. Base power and base voltages used for per-unit calculations are taken from this. The three-phase voltages into the magnetized transformer are measured by high-frequency differential voltage probes SI-9002, with an accuracy of ±1%. To calculate the total power losses dissipated in the transformer for no load. The three-phase currents are measured with current clamps Fluke i310s, with an accuracy of ±2%. Measurements are sampled at 400 kHz, and the power dissipation per phase is calculated. Transformer temperature was not measured during the experiment.

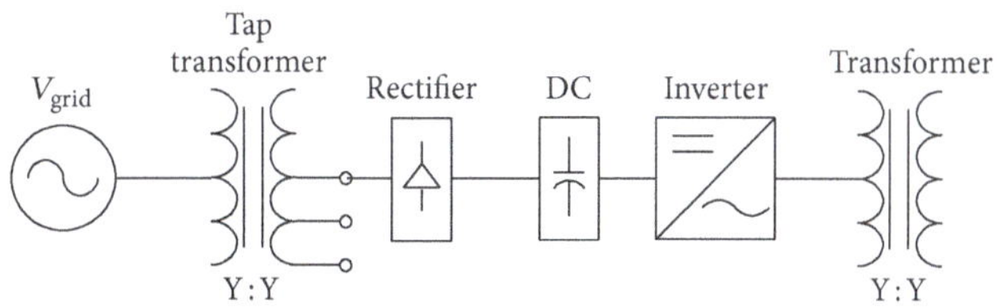

FIGURE 5: Single-line diagram of the experimental set-up. Magnetization losses are measured on the transformer to the right.

5. Results and Discussion

Transformer data at rated voltage and 50 Hz frequency are given by the manufacturer data sheet. The transformer is magnetized in no load at voltages lower than rated for 50 Hz. Measured and simulated power losses are displayed in Table 2. Simulations are performed for the same set-up, using the saturation curve in Figure 4 and the value of R_m in Figure 3. The winding parameters given for rated voltage are assumed constant and independent of frequency for the entire voltage range at no load. There is very good agreement between simulations and experiments, and the magnetization losses are slightly less than 1% of the transformer rating at rated voltage.

The saturation curve shows the classical trendline of an arctan function and will not affect magnetization losses significantly, as the winding resistance is small. For more accurate simulations, more measured values of R_m are required to give a better curve fit, especially at lower and higher voltage levels, where the curve is expected to stagnate. The approximated value of R_m puts a limit on the simulation range, which is kept within (0.2–0.8). The estimation of R_m is of course not exact, as it represents frequency-dependent eddy current losses and hysteresis losses. However, it will give a fair approximation of the system losses that is simple to implement.

Next, the model is used to simulate the magnetization losses with the nonfiltered inverter. $V_{SPWM,rms}$ is estimated for the different scenarios and used to get an average value of R_m. The true value of $V_{SPWM,rms}$ will deviate slightly from the estimated, as the voltage will be partially filtered by the transformer. The measured magnetizing current and induced phase voltage for one phase are displayed in Figure 4. In Figure 7(a), 50 Hz magnetization is shown. In Figure 7(b) the same fundamental voltage is induced using the SPWM-controlled inverter instead. Excluding switching frequency harmonics, the harmonic content of the two currents is very similar. However, the inverter current is almost twice as large as the 50 Hz current. The pulsed current does not induce magnetic flux in the magnetic core as efficiently as the 50 Hz current. Thus it has to be higher for the same flux, and the magnetic losses will be higher.

The DC voltage is set at three different levels, and the total power loss is measured, while the amplitude modulation

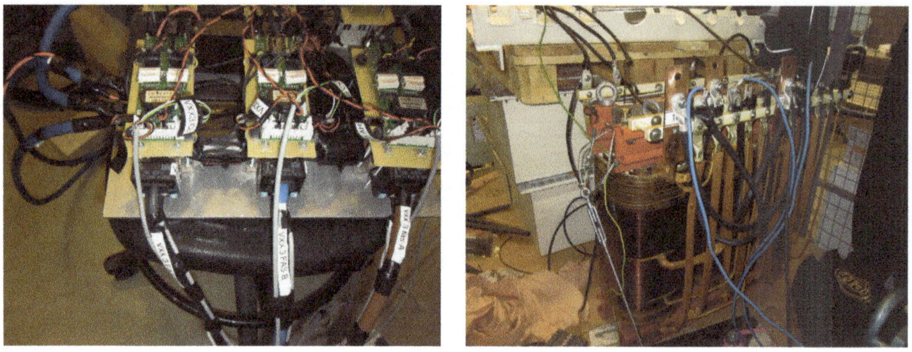

(a) (b)

FIGURE 6: (a) Inverter, (b) three-phase transformer.

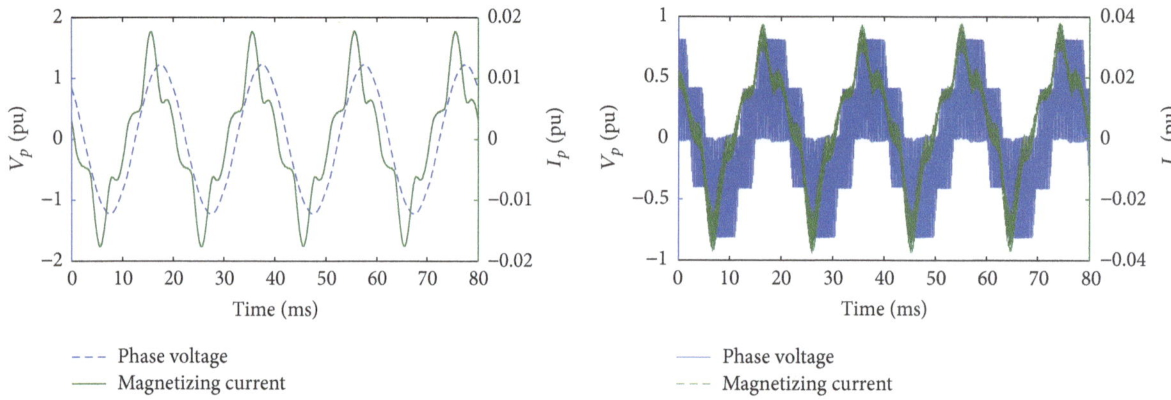

(a) (b)

FIGURE 7: (a) Magnetizing current for 50 Hz, (b) magnetizing current for SPWM.

is varied. In Figure 8(a), the DC level is set to 145 V, in Figure 8(b) is set to 335 V, and in Figure 8(c) is set to 550 V, respectively. There is rather good agreement between simulations and experiments. The limitation of the model is the estimation of R_m by the use of $V_{\mathrm{SPWM,rms}}$. For V_{DC} = 145 V, the simulated power is slightly higher than the experimental. This is also true for V_{DC} = 550 V. Here, the "knee" seen in the power curve is sharper in the experiments than the simulations. The possible explanation to this is the approximation of the R_m curve and using $V_{\mathrm{SPWM,rms}}$ to represent the voltage input. A squared interpolation of the R_m curve could possibly give more accurate simulation results.

For the lower DC voltages, the power loss is approximately proportional to m_a, whereas it starts to increase with the square of m_a for the highest voltage. This is reasonable, since the losses increase with the square of the rms voltage. Also, the magnetic core will be in its linear region for lower voltages and magnetic fluxes, whereas it will go into partial saturation, as the induced voltage is increased. In Figure 8(c), it can be seen that the losses at an induced voltage of 0.8 pu are less than half the losses at 1 pu. This indicates that a significant improvement in the magnetization losses can be made by shifting the operating point for the transformer, which is equivalent to having a larger core at nominal voltage. When the transformer is SPWM-magnetized at its rated voltage, (V_{DC} = 550 V, m_a = 1), the magnetization losses are ten times higher than the 50 Hz magnetization at rated voltage given in Table 2.

In Figure 9, the overlap of the experimental results are plotted from the data in Figure 8. The same fundamental voltage output is obtained for different DC levels by varying the amplitude modulation. The corresponding losses at 50 Hz-magnetization are also plotted as a reference. Nonfiltered inverter magnetization always dissipate higher losses in the transformer compared to the sine wave. This makes perfect sense, since the transformer is optimized for this frequency. Also, both eddy current losses and hysteresis losses increase with frequency. In this experiment, the lower DC voltages are not sufficient to give the transformer full magnetization, as can be seen in Figure 9. However, there is a general trend that the losses increase for higher DC voltages with the same V_s. This is in agreement with that $V_{\mathrm{SPWM,rms}}$ increase for an increased V_{DC} while maintaining the product $m_a \cdot V_{\mathrm{DC}}$ constant. At rated voltage, least magnetization losses are obtained for the lowest possible V_{DC}, that is, when $m_a = 1$.

6. Conclusions

In this paper the losses of a three-phase power transformer, magnetized by a nonfiltered inverter, have been evaluated.

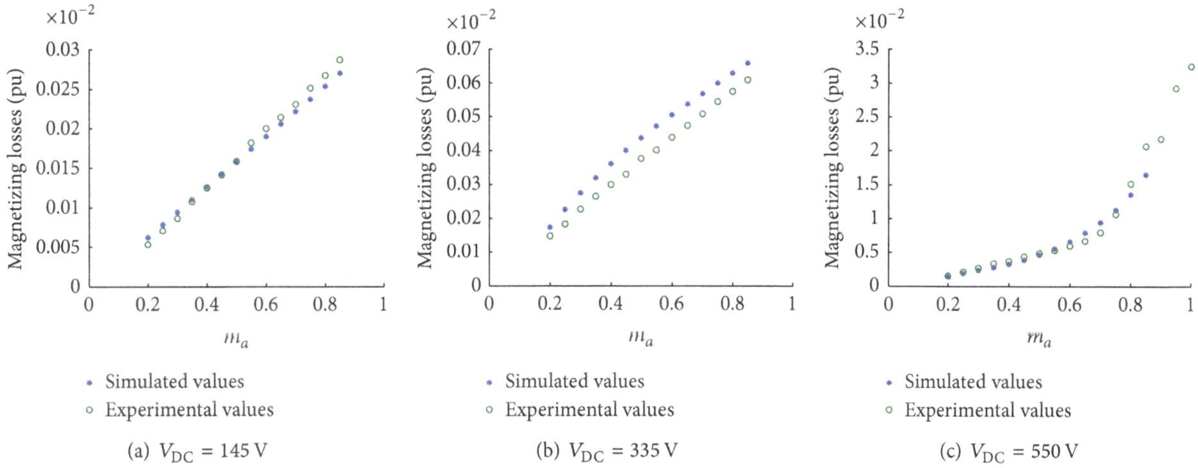

FIGURE 8: Comparison of magnetization losses between simulations and experiments, at different values of V_{DC}.

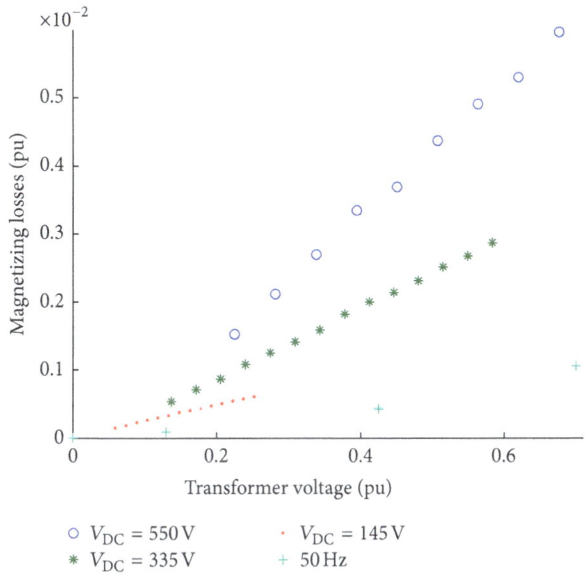

FIGURE 9: Magnetization losses for the same transformer voltage with different V_{DC}.

A simulation model has been developed based on the measured parameters of the system. The total core losses are lumped into an equivalent magnetizing resistance based on the rms value of the induced voltage, and this method shows good resemblance with the experimental measurements. For the studied cases, the magnetization losses are always higher using direct VSI magnetization compared to normal 50 Hz voltages. For VSI magnetization, the losses increase rapidly close to the transformer nominal voltage. This is because the switched voltage waveform pushes the transformer to operate more in its saturated region. At rated voltage, the magnetization losses are approximately ten times higher for SPWM than for normal 50 Hz. The results show that by shifting the transformer operating point, the magnetization losses can be significantly reduced.

To minimize the magnetization losses for a given secondary voltage, the amplitude modulation should be kept as high as possible. The increased transformer losses have to be evaluated in conjunction with decreased LCL filter losses to evaluate the total system efficiency in this set-up compared with more conventional topologies.

References

[1] H. R. Karshenas and H. Saghafi, "Performance investigation of LCL filters in grid connected converters," in *Proceedings of the IEEE PES Transmission and Distribution Conference*, 2006.

[2] T. C. Y. Wang, Z. Ye, G. Sinha, and X. Yuan, "Output filter design for a grid-interconnected three-phase inverter," in *Proceedings of the IEEE 34th Annual Power Electronics Specialists Conference*, pp. 779–784, June 2003.

[3] E. Twining and D. G. Holmes, "Grid current regulation of a three-phase voltage source inverter with an LCL input filter," *IEEE Transactions on Power Electronics*, vol. 18, no. 3, pp. 888–895, 2003.

[4] M. Prodanović and T. C. Green, "Control and filter design of three-phase inverters for high power quality grid connection," *IEEE Transactions on Power Electronics*, vol. 18, no. 1, pp. 373–380, 2003.

[5] W. Roshen, "Ferrite core loss for power magnetic components design," *IEEE Transaction on Magnetics*, vol. 27, no. 6, pp. 4407–4415, 1991.

[6] M. L. Hodgdon, "Applications of a theory of ferromagnetic hysteresis," *IEEE Transaction on Magnetics*, vol. 24, no. 1, pp. 218–221, 1988.

[7] C. P. Steinmetz, "On the law of hysteresis," *American Institute of Electrical Engineers Transactions*, vol. 9, pp. 3–64, 1892.

[8] A. Boglietti, P. Ferraris, M. Lazzari, and F. Profumo, "Iron losses in magnetic materials with six-step and PWM inverter supply," *IEEE Transactions on Magnetics*, vol. 27, no. 6, pp. 5334–5336, 1991.

[9] M. Amar and F. Protat, "Simple method for the estimation of power losses in silicon iron sheets under alternating pulse voltage excitation," *IEEE Transactions on Magnetics*, vol. 30, no. 2, pp. 942–944, 1994.

[10] A. V. Bossche, V. C. Valchev, and G. B. Geoorgiev, "Measurements and loss model of ferrites with nonsinusoidal waveforms," in *Proceedings of the 35th IEEE Power Electronics Specialists Conference*, pp. 4814–4818, 2004.

[11] J. Reinert, A. Brockmeyer, and R. W. A. A. De Doncker, "Calculation of losses in ferro- and ferrimagnetic materials based on the modified Steinmetz equation," *IEEE Transactions on Industry Applications*, vol. 37, no. 4, pp. 1055–1061, 2001.

[12] M. Albach, T. Duerbaum, and A. Brockmeyer, "Calculating core losses in transformers for arbitrary magnetizing currents a comparison of different approaches," in *Proceedings of the 27th Annual IEEE Power Electronics Specialists Conference (PESC '96)*, pp. 1463–1468, January 1996.

[13] W. A. Roshen, "A practical, accurate and very general core loss model for nonsinusoidal waveforms," *IEEE Transactions on Power Electronics*, vol. 22, no. 1, pp. 30–40, 2007.

[14] E. C. Snelling, *Soft Ferrites: Application and Properties*, Butterworths, London, UK, 2nd edition, 1988.

[15] A. Morched, L. Marti, and J. Ottevangers, "High frequency transformer model for the EMTP," *IEEE Transactions on Power Delivery*, vol. 8, no. 3, pp. 1615–1626, 1993.

[16] H. Y. Lu, J. G. Zhu, and Y. Hui, "Experimental determination of stray capacitances in high frequency transformers," *IEEE Transactions on Power Electronics*, vol. 18, no. 5, pp. 1105–1112, 2003.

[17] J. Biela and J. W. Kolar, "Using transformer parasitics for resonant converters—a review of the calculation of the stray capacitance of transformers," *IEEE Transactions on Industry Applications*, vol. 44, no. 1, pp. 223–233, 2008.

[18] N. Mohan, T. Undeland, and W. Robbins, *Power Electronics—Converters, Applications and Design*, John Wiley & Sons, New Dehli, India, 5th edition, 2007.

[19] S. Casoria, G. Sybille, and P. Brunelle, "Hysteresis modeling in the MATLAB/Power System Blockset," *Mathematics and Computers in Simulation*, vol. 63, no. 3-5, pp. 237–248, 2003.

[20] J. G. Frame, N. Mohan, and T. Liu, "Hysteresis modeling in an electromagnetic transients program," *IEEE Transactions on Power Apparatus and Systems*, vol. 101, pp. 3403–3412, 1982.

7

Power-Integrated Circuit Active Leakage Current Detector

M. F. Bulacio, T. A. González, G. Marinelli, R. Alonso, and H. E. Tacca

LABCATyP, Department of Electronics, Faculty of Engineering, University of Buenos Aires, Av. Paseo Colón 850, C1063ACV, Buenos Aires, Argentina

Correspondence should be addressed to M. F. Bulacio, mbulacio@fi.uba.ar

Academic Editor: Francesco Profumo

Most of the failures of induction motors become insulation faults, causing a permanent damage. Using differential current transformers, a system capable of insulation fault detection was developed, based on the differential relay protection scheme. Both signal injection and fault detection circuitry were integrated in a single chip. The proposed scheme is faster than other existing protection and not restricted to protect induction motors, but several other devices (such as IGBTs) and systems. This paper explains the principle of operation of fault protection scheme and analyzes an integrated implementation through simulations and experimental results. A power-integrated circuit (PIC) implementation is presented.

1. Introduction

Nowadays, induction motors are the most frequently used drives for motion control systems in industrial applications [1]. The industry is particularly concerned about the failure of these devices since it could cause significant economical losses due to the stop of production for long time spans and the risk to the personal safety of the operators. Statistics show that annual down times of 0.5% to 4% may be expected [2].

The main causes of motor failure are the variation of the feeding phase voltages, mechanical overloads, and problems produced by inadequate or faulted electrical installations [3–5]. Other causes may include the aging of the motor or eventual adverse environmental conditions. Most of these events lead to an increase of the current and to a subsequent rise of motor temperature. An unacceptable overheating of the motor damages the insulation of the windings and puts the motor out of service [4–7]. High dv/dt, characteristic of PWM variable speed drives, is a source of further stress and deterioration of the insulation.

The objective of motor protection is to keep the mechanical and thermal conditions within the acceptable limits while ensuring the highest fault detection sensitivity.

In order to fulfill this objective, several protection and monitoring systems are introduced in the bibliography with a clear tendency to integrate them in the variable speed drives. The purpose of monitoring systems is to detect a failure in an early stage of its development, by checking some parameters such as parasitic capacitances and resistance or the dissipation factor, without letting the fault to cause irreversible damage to the machine or its driving devices. On the other hand, most protection methods detect a fault when it becomes severe. Monitoring can be classified as online or offline depending on whether the motor can be tested during normal operation or it must be disconnected first.

Offline maintenance analyses are performed every 3 to 6 years [8] and some of them require the motors to be fully unmounted. Other strategies use the inverter itself to inject signals so as to monitor sensitive components of the system [8]. Although this allows the user to test devices without the need of disassembly, the proposed strategies are suitable for monitoring components that deteriorate gradually to failure. Other components such as power semiconductor devices or insulation, which fail abruptly, cannot be monitored using these methods due to their offline nature.

On the contrary online methods are those used to monitor adjustable speed drive systems in a continuous fashion. They provide more accurate and reliable information about the state of the insulation and the remaining service life of the machine than the offline tests. Several online tests exist and can be classified as chemical or thermal monitoring, turns fault detection, and fault-to-ground circuit breakers [9]. While the first two seek the early detection of faults, the last is generally used to protect against severe phase-to-ground

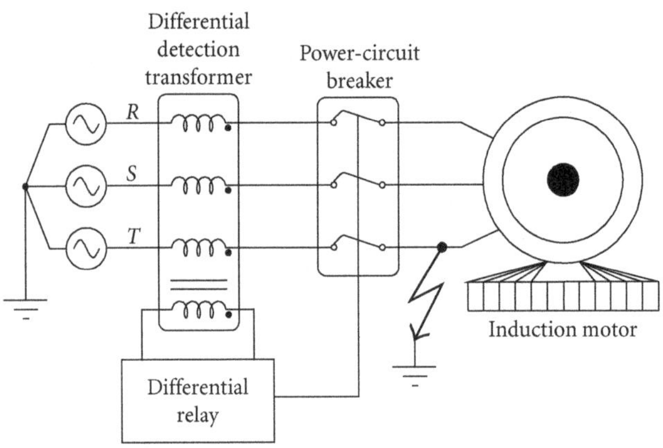

Figure 1: Differential protection scheme.

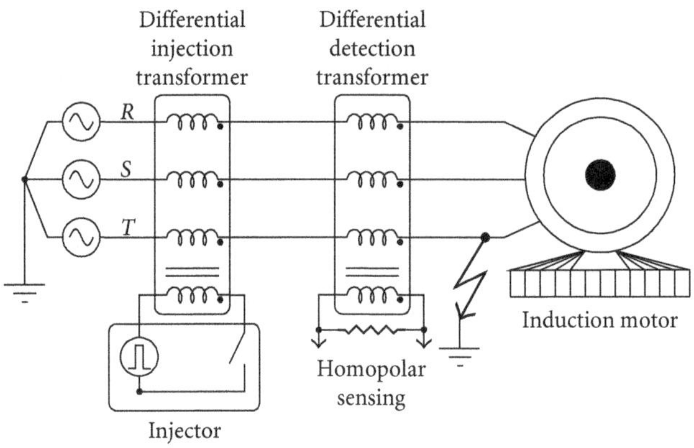

Figure 2: Injection and detection signal scheme.

faults. Protection schemes such as the one presented in [9, 10] are very effective methods of detecting small incipient faults (turns fault).

The weakness of online monitoring methods like the one presented in [8–10], which measures voltages and currents to calculate state defining parameters, are its slow response and its imprecision under severe failure conditions.

Consequently, this paper presents the development of an integrated protection system, which follows the protection principle of a differential relay but detects the fault and trip in a few microseconds. Thus, it could be used to protect sensible power modules (e.g., IGBTs from a three-phase bridge), which usually need to be disabled before approximately 10 μs.

All the subsystems required in a modern adjustable speed drive are being developed at LABCATyP using Flash-based FPGA devices [11–13]. Thus, a phase-to-ground current protection circuit was implemented with the objective to protect both the motor and the power module. In contrast to other methods presented in the bibliography, it was mandatory to provide a fast detection response.

The possibility to test the performance of the leakage current protection subsystem was given by an integrated circuit design award. Moreover, this allows the future use of this IP together with the HDL IP developed [12, 13] in the past at LABCATyP to integrate a complete controller.

Although there are various chips with this functionality in the market [14], the novelty of this device lies in the possibility of detecting failures in very short time, as it treats the detected signal at hardware level.

It is shown that this system could be used as a preventive offline protection system and online monitoring of the homopolar current component to detect sudden insulation faults from phase to ground. Simulation and experimental results are presented in order to validate the functionality of the design.

2. Line-to-Ground Insulation Fault Protection

2.1. Basic Protection Scheme. A common protection scheme used in the industry is the one found in differential relays [15, 16]. These devices switch the motor off whenever a zero sequence current in the feed lines appears. In home appliances they are used to avoid the circulation of current levels that could be dangerous for the human body. The effects on humans depend on the value and duration of the fault current [17].

Differential protection systems are conceptually based on the property of single-phase and three-phase systems, which states that the sum of the line currents must be equal to zero. Thus, if a current flows into the earth, the device detects the residual current and it causes a trip.

The leakage current level which makes the device trip defines the sensitivity of this type of protection (6 mA to several Amperes sensitivities are normally found in the market).

Three distinct parts constitute a differential protection: a differential sensor, a differential relay and a power circuit breaker. Figure 1 shows these parts and their connection in a simple real-life application.

The sensor is a transformer which adds the motor currents. If the currents do not add to zero, a voltage proportional to the residual current is induced in the secondary winding. The relay determines whether this voltage is enough to trip depending on the sensitivity of the protection. This sensitivity can be adjusted through an internal threshold which fixes the minimum current considered as a fault. The power circuit breaker should have an adequate breaking capability for the specific load and could be an electromagnetic switch or a solid-state device. Figure 1 depicts a short-to-earth event in which the homopolar current component produces an induced secondary voltage.

IEC 61008-1 standard establishes that differential breakers must trip when the value of the leakage current reaches from 50 to 100% of the selected sensitivity. The implementation of this type of protection by means of electronic devices allows the detection at 100% of the sensitivity, preventing erroneous trips.

2.2. Protection Scheme with Signal Injection. If the scheme shown in Figure 1 is built using small-core transformers, a failure could only be detected if high-frequency currents were to circulate the lines (e.g., between a PWM generator and the motor). In order to solve this problem without increasing the core size, an additional current transformer is used so as to provide the needed high-frequency signal (Figure 2).

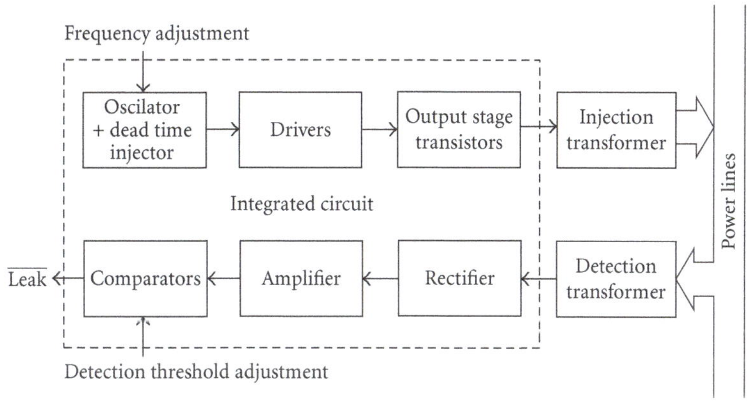

FIGURE 3: Block diagram of the integrated circuit.

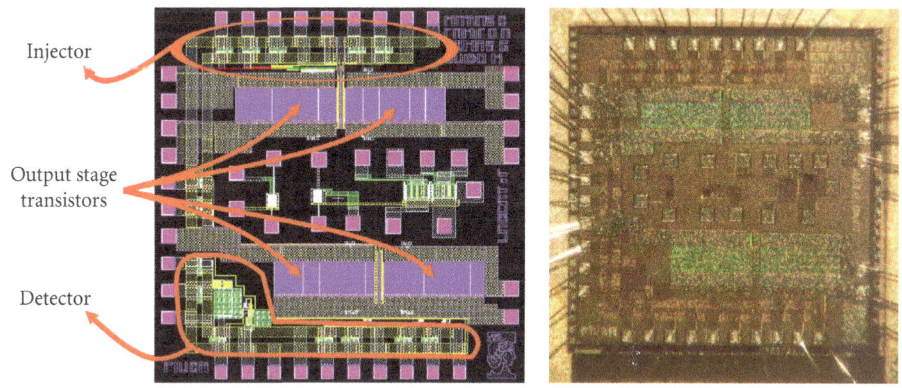

FIGURE 4: Layout and microscopic image of the integrated circuit.

The signal injection circuit will not produce a high-frequency circulating current as long as the motor does not have a neutral connection, which is the usual case. In this situation, the secondary of the differential current detection transformer presents no induced voltage. The signal will be effectively injected and detected when a ground or neutral fault occurs because only the homopolar current could produce a net flux in the transformer core to induce a detectable secondary voltage.

If the neutral terminal is accessible to the operator, both line-to-ground and line-to-neutral faults can be discriminated between each other by using two detection circuits and passing the neutral wire through one of the detection transformer cores.

2.3. Integrated Protection System. A circuit was developed in order to generate the high-frequency signal which is introduced on the power line through the injection differential transformer. On the other side, another circuit (designed to detect load insulation failures) processes the signal induced in the detection differential transformer.

Both circuits were integrated on a single die using the process XC06 of XFAB. The packaging was provided by APTASIC. Figure 3 shows a block diagram with the most important parts of the integrated circuit, and Figure 4 is a picture of the IC itself.

The oscillator generates the base frequency that is used to obtain two complementary command signals with dead time. They excite the push-pull-integrated output transistors, which are connected to the external differential injection transformer, in order to inject a high-frequency square signal to the power line.

A Schottky-based full diode bridge is connected at the differential detection transformer output to rectify the detected signal. This signal is amplified and then compared against an externally adjustable threshold, which determines the sensibility of the protection. Two separate comparators with configurable external thresholds were designed. The output of the amplifier must be connected externally to both comparators inputs. Then two external RC lowpass filters could be designed to offer low current/slow detection and a high current/fast leakage detection so as to provide protection to human life and power electronic devices, respectively. The connection of the fast filter detection is shown in Figure 8(a).

2.4. Simulations. In order to verify the circuit's correct functionality, simulations were made using XFAB design kit models. Figure 5 shows a simplified circuit including signal generation, signal detection, analog signal processing, and leakage model.

Under normal operation, without fault, the zero sequence current circulating through the power line is zero,

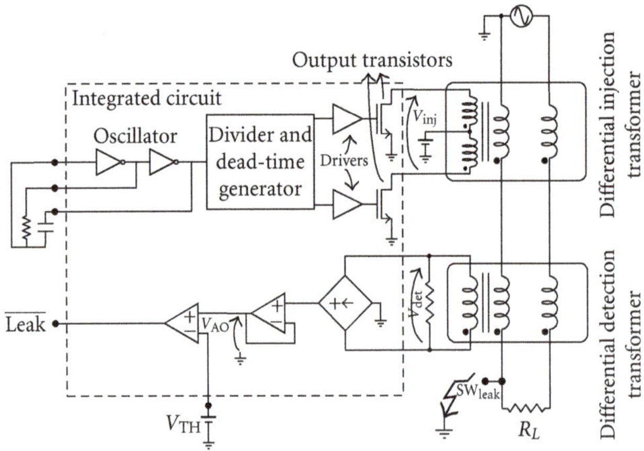

FIGURE 5: Simulated circuit.

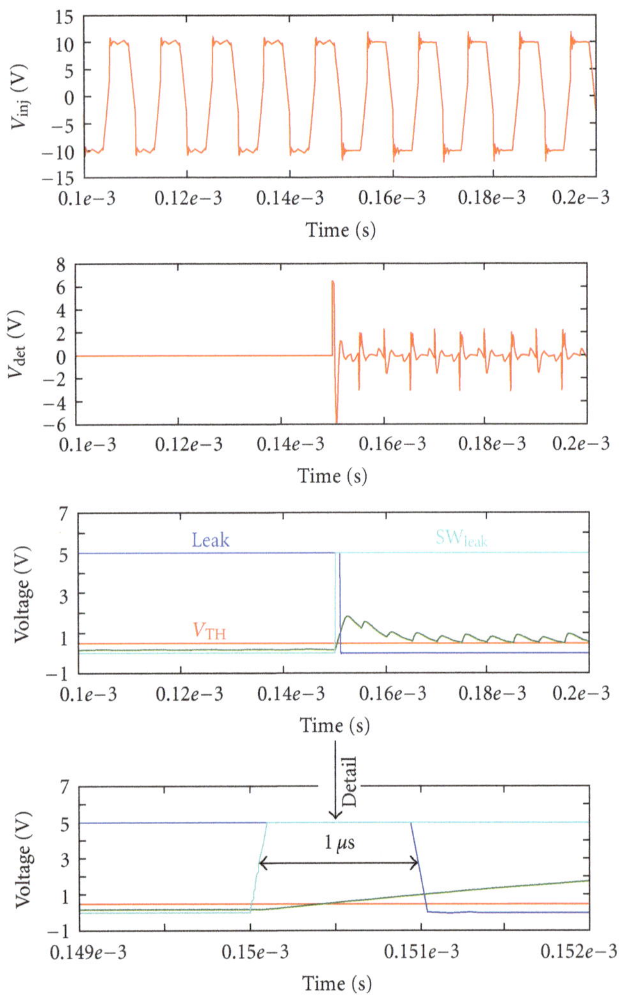

FIGURE 6: Simulation of an insulation fault.

and no voltage signal is induced at the detection transformer. When a failure occurs, the current is not zero anymore, and a signal appears at the detection transformer. If this signal is greater than the threshold after the amplification and filtering, the output of the comparator goes to a low level, indicating a failure condition.

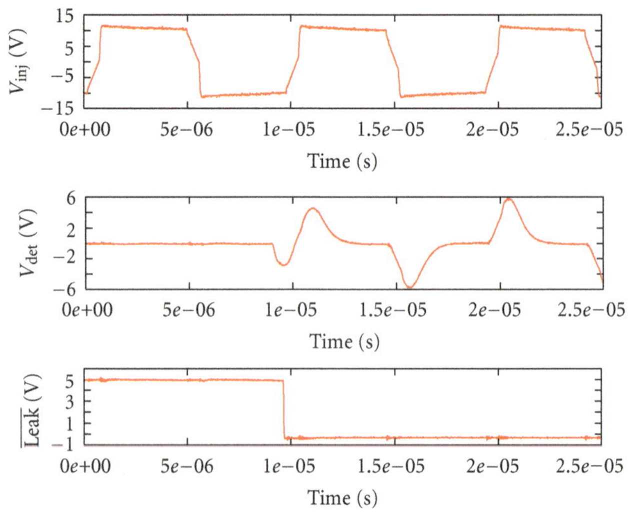

FIGURE 7: Oscilloscope capture of leakage fault.

The results of the simulation are displayed in Figure 6, where each voltage reference corresponds to those shown in Figure 5. The first two signals are the primary voltage of the injection transformer and the secondary voltage of the detection transformer. While the first signal is always present, the second one is only induced from the moment the simulated fault begins (at 0.15 ms). The other signals plotted on Figure 6 show the detector performance: A delay of 1 μs is measured from the moment SW_{leak} is tripped to the time in which the output signal leak presents the low-level indication. This delay is shorter than that with other detection methods and could help not only protect humans from injury but also avoid the destruction of sensitive power devices.

2.5. Experimental Results. Firstly, in order to test the injector circuitry and the current and voltage capability of the integrated output transistors, the circuit was excited with a function generator. Then, it was easy to change the base-working frequency used to generate the command signals. The power transistors command signals were complementary and had the expected dead-time, matching with the simulations (not shown).

The output transistors were excited with these signals and connected to pull-up resistors to corroborate correct commutation and current rating. Transistors can successfully withstand 10 V during cut-off, which is the peak voltage that is generated in this push-pull configuration with the intended 5 V supply on the transformer's middlepoint. A current of approximately 300 mA through the output transistors was achieved. Hence, the output-stage can work from about 20 KHz using high-frequency ferrite cores. Further testing proved the possibility of supplying the push-pull stage from up to 15 V with inductive loading (each transistor blocking 30 V).

Secondly, the injection and detection scheme was tested with a bench similar to the one presented in Figure 5. The oscilloscope captures are shown in Figure 7. They show both

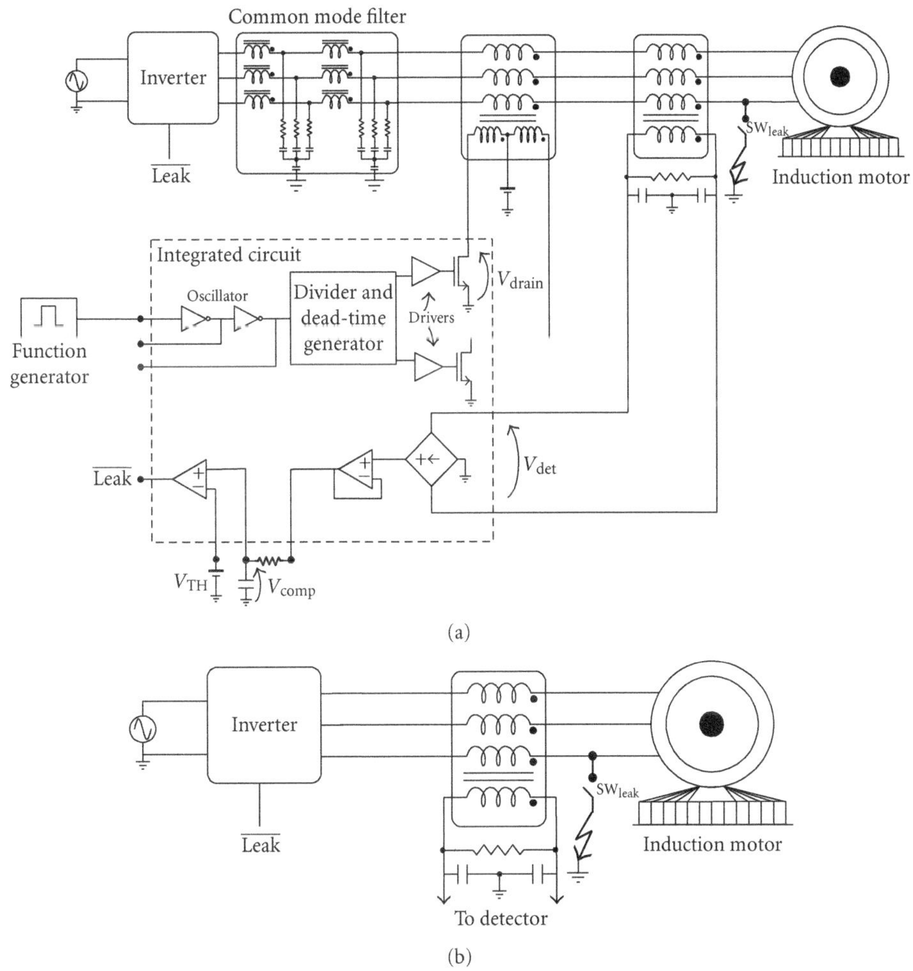

FIGURE 8: (a) Testbench for preventive offline protection and online protection. (b) Modified testbench for only online protection. The ground loop closes only when the inverter is on.

the injected and detected signals and prove the fast detection, with a delay of the order of 1 μs.

Even though this test was made with a single-phase resistive load, it is representative of the use of the designed PIC as fault to ground protection of generic polyphase circuitry with any kind of linear load and sinusoidal low frequency or DC supply voltage referenced to ground.

Finally, two additional tests were performed using the testbench shown in Figure 8 to show real life proofs. The simulated impedance leak consisted of a 22-ohm resistor in series with 4.7 nF capacitor. The injected signal was configured to approximately 135 kHz. The comparator output was connected to the HITACHI X200-002SFEF inverter through the logic input that disables the turn-off.

The first test presents the use of the chip in an offline preventive application. The monitoring system detects a fault and avoids the turn-on of the inverter or drive system. In this case, the filter connected at the inverter output closes the ground loop. Both the injection and detection subsystems must be used. Figure 9 shows that the detection is accomplished within few microseconds with an adjusted threshold of 0.32 V. In the final application a small three-phase RC network should be added at the inverter output to close the loop instead of the presented filter.

The second test shows the use of the detector as part of an inverter-fed system. In this case, the injector and filter are not needed because the motor is supplied by a PWM high-frequency signal as explained in Section 2. When the simulated fault condition appears and closes the ground loop, the PWM signal is detected as shown in Figure 10 and the inverter stops its operation. The threshold previously adjusted was kept for this oscilloscope capture but further tests without the filter showed that it should be increased to 3 V in order to avoid false positives.

3. Future Work: Communications over the Power Lines of the Motor

Another line of work which is currently under development is to use the components of the protection system to send and receive communication signals [13]. In this concept design, the power lines become a communication channel used for command and status signaling [18–20]. The advantage of

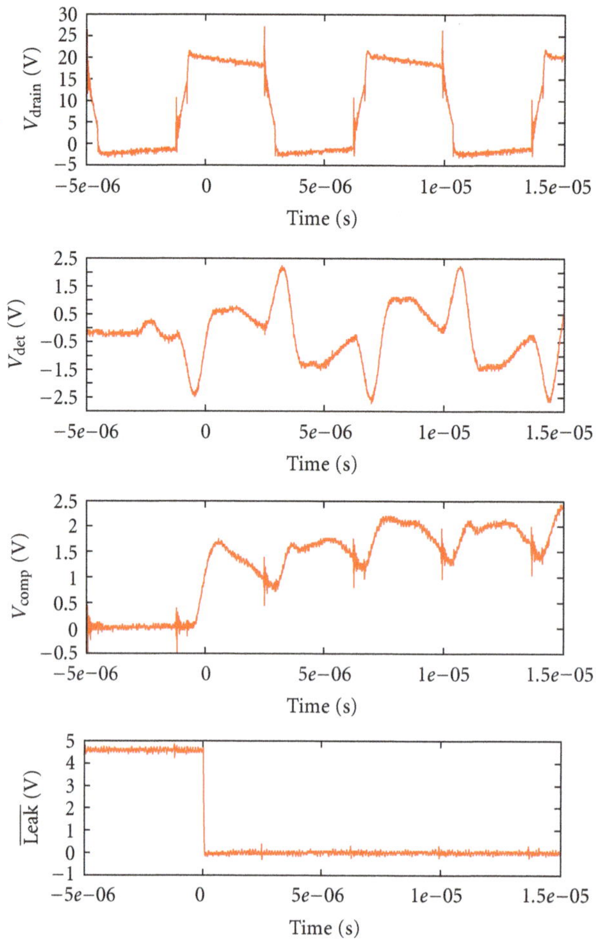

FIGURE 9: Scope captures of offline preventive detection.

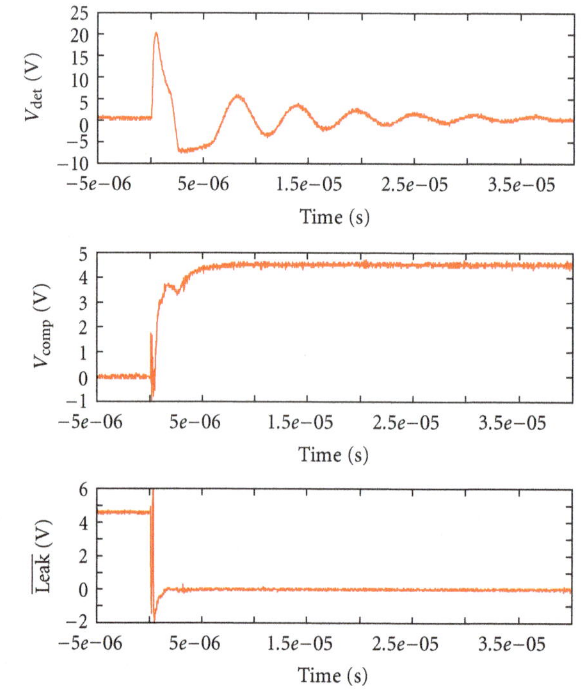

FIGURE 10: Scope captures of the online detection.

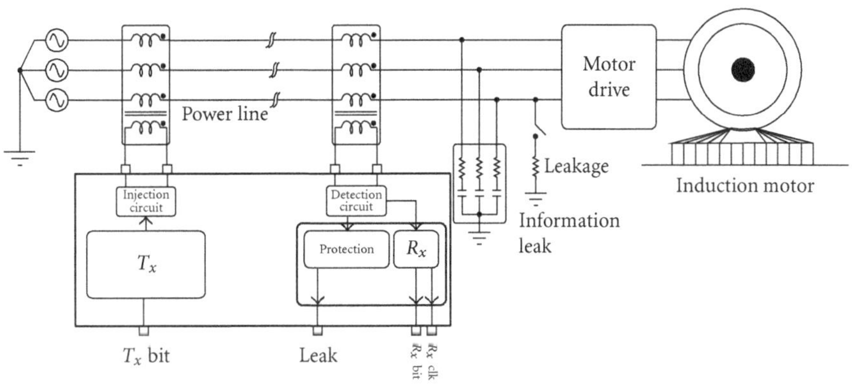

Figure 11: Communication and protection scheme.

this system is that no additional wiring is needed which usually makes it cheaper and easier to implement [18–21].

Figure 11 shows the general scheme used for the transmission (injection) and reception (detection) of commands in the power lines of the motor, where the signal injection is identical to the previous protection system. An FSK modulator produces the excitation signals with a frequency that depends on the information bit to send. On the other side, a conditioning circuit filters the detected signal and recovers the information injected in the motor power lines.

A proper design of the added RC networks (information leak) favors the existence of a communication channel without the presence of an insulation fault. The protection threshold should be carefully selected in order to continue providing a reliable protection.

A second version of the IC that implements the communication and protection functions is being developed.

4. Conclusions

The proposed differential scheme was validated with computed simulation and discrete experimental prototypes. Once the correct functioning of this protection scheme was verified, an integrated circuit was designed, sent to XFAB for fabrication, and then bonded and encapsulated by APTASIC.

The circuit proposed was extensively simulated and measurements were made to verify the output-stage ratings and the range of frequencies of the oscillator. All test benches proved a fast detection, with a delay of the order of a few microseconds. This delay is shorter than that with other detection methods but also avoids the destruction of electronic power devices.

While the application proposed in this paper is intended to protect electric motors and its control systems, this circuit is not limited to these devices. It can be used to protect regulators, matrix converters, power factor correctors, and even emergency power inverter systems.

Future test should be done and design procedures must be established to adjust the two comparator thresholds and transformers to avoid the destruction of sensitive power devices and at the same time protect humans from injury.

Finally, a new application of the integrated circuit was proposed not only capable of insulation fault detection but also able to send and receive command and diagnostic signals to the motors or devices connected to the power lines.

Acknowledgments

This work was made possible by the invaluable help of the IIIE at UNS, the DIE at UCU, and José Lipovetzky from the Microelectronics and Device Physics Laboratory at UBA. The authors would like to also thank Motortech S.A., Carlos Stahr of XFAB, Tanner EDA, and the people of APTASIC who very generously donated the packaging to the project.

References

[1] R. S. Navarro, *Máquinas Eléctricas*, McGraw-Hill, 1989.

[2] Rockwell Automation, "Basics for practical operation," Motor protection, 1997.

[3] R. Bayindir, I. Sefa, I. Colak, and A. Bektas, "Fault detection and protection of induction motors using sensors," *IEEE Transactions on Energy Conversion*, vol. 23, no. 3, pp. 734–741, 2008.

[4] A. H. Bonnett and G. C. Soukup, "Cause and analysis of stator and rotor failures in three-phase squirrel-cage induction motors," *IEEE Transactions on Industry Applications*, vol. 28, no. 4, pp. 921–937, 1992.

[5] M. Chafai, L. Refoufi, and H. Bentarzi, "Reliability assessment and improvement of large power induction motor winding insulation protection system using predictive analysis," *WSEAS Transactions on Circuits and Systems*, vol. 7, no. 4, pp. 184–193, 2008.

[6] Anon, "Report of large motor reliability survey of industrial and commercial installations, Part I," *IEEE Transactions on Industry Applications*, vol. 21, no. 4, pp. 853–864, 1985.

[7] Anon, "Report of large motor reliability survey of industrial and commercial installations, Part II," *IEEE Transactions on Industry Applications*, vol. 21, no. 4, pp. 865–872, 1985.

[8] S. B. Lee, J. Yang, J. Hong et al., "A new strategy for condition monitoring of adjustable speed induction machine drive systems," *IEEE Transactions on Power Electronics*, vol. 26, no. 2, pp. 389–398, 2011.

[9] F. Briz, M. W. Degner, P. García, and A. B. Diez, "High-frequency carrier-signal voltage selection for stator winding fault diagnosis in inverter-fed AC machines," *IEEE Transactions on Industrial Electronics*, vol. 55, no. 12, pp. 4181–4190, 2008.

[10] K. Younsi, P. Neti, M. Shah et al., "On-line capacitance and dissipation factor monitoring of AC stator insulation," *IEEE Transactions on Dielectrics and Electrical Insulation*, vol. 17, no. 5, pp. 1441–1452, 2010.

[11] R. Arias, H. Mediote, and H. Tacca, "Flash FPGA-based numerical pulse-width modulator," *Advances in Power Electronics*, vol. 2011, Article ID 215376, 6 pages, 2011.

[12] H. Mediote, *Aplicaciones de circuitos integrados programables al control de motores de inducción*, thesis, Facultad de Ingeniería de la Universidad de Buenos Aires, 2008.

[13] M. F. Bulacio, *Aplicaciones de circuitos FPGA al control de motores de inducción mediante comunicación a través de las líneas de alimentación*, thesis, Facultad de Ingeniería de la Universidad de Buenos Aires, 2011.

[14] "Fairchild Semiconductor, RV4145A, Low Power Ground Fault Interrupter," Rev. 1.0.5, 2004.

[15] T. K. Chatterjee, D. K. Mittra, S. Mahata, and S. Kareddy, "A novel solid-state integrated protection system for three phase induction motors," in *International Conference on Power Systems (ICPS '09)*, December 2009.

[16] IEEE Std C37.96-2000, "IEEE Guide for AC Motor Protection," Marzo 2000.

[17] I. Chatterjee, D. Wu, and O. P. Gandhi, "Human body impedance and threshold currents for perception and pain for contact hazard analysis in the VLF-MF band," *IEEE Transactions on Biomedical Engineering*, vol. 33, no. 5, pp. 486–494, 1986.

[18] A. Kosonen, M. Jokinen, J. Ahola, M. Niemelä, and J. Toukonen, "Ethernet-based broadband power line communication between motor and inverter," *IET Electric Power Applications*, vol. 2, no. 5, pp. 316–324, 2008.

[19] N. Ginot, M. A. Mannah, C. Batard, and M. MacHmoum, "Application of power line communication for data transmission over PWM network," *IEEE Transactions on Smart Grid*, vol. 1, no. 2, pp. 178–185, 2010.

[20] A. Kosonen and J. Ahola, "Communication concept for sensors at an inverter-fed electric motor utilizing power-line communication and energy harvesting," *IEEE Transactions on Power Delivery*, vol. 25, no. 4, pp. 2406–2413, 2010.

[21] T. Brooks, "Wireless technology for industrial sensor and control networks," in *Proceedings of the ISA/EEE Sensors for Industry Conference*, pp. 73–77, Rosemont, Ill, USA, November 2001.

8

Controller Design Considerations for ACM APFC Systems

Alexander Abramovitz

Department of Electrical Engineering and Computer Science, University of California, Irvine, CA 92697, USA

Correspondence should be addressed to Alexander Abramovitz, alabr@hotmail.com

Academic Editor: C. M. Liaw

This paper is concerned with performance of the current shaping network in Average Current Mode (ACM) Active Power Factor Correction (APFC) systems. Theoretical expressions for the ripple components are derived. Then, ripple interaction and impact on the current loop reference signal are investigated. A modification of the controller network is suggested that results in an improved Total Harmonic Distortion (THD). Design guidelines are suggested. The theoretical predictions were validated by simulation.

1. Introduction

Over the past few years, a variety of current shaping methodologies were developed for Active Power Factor Correction (APFC) [1–3]. Each approach undertaken is a compromise in between the performance indexes and the circuit complexity and cost. The Critical Conduction Mode (CrCM) APFC operating on the CCM-DCM boundary [4–7] shapes the average input current by comparator/zero detector and is unconditionally stable. However, the natural simplicity, robustness, and stability of CrCM APFC are offset by high ripple current which cause increased conduction and core losses. Difficulties in filtering the variable frequency current ripple and poor efficiency restrict this technique to low-power and low-cost applications. Other Discontinuous Conduction Mode (DCM) based designs, which upside is simplicity suffer from similar problems exhibiting also higher harmonic distortion of the line current [8, 9].

APFC without line voltage sensing, [10–13] stands out as a robust, technologically simple, and cost-effective solution. A simple and clear physical insight into the principle of operation of the current loop of this class of APFCs was suggested in [14]. All of the mentioned above APFCs with no input voltage sensing make use of a hidden current loop inside a DC-DC converter. The designs mainly differ in their method of realization of the transresistive feedback, PWM, and supplementary current loop control circuitry. Regardless of the implementation, however, the duty cycle programming is implemented according to the converters input port ideal average relationships and ideal modulator ramp signal. As a result, accurate current loop operation can only be attained in the Continuous Conduction Mode (CCM) under negligible current ripple conditions. In practice, however, the CCM-DCM mode changes, current ripple, and ramp carrier imperfections cause the duty cycle to deviate from the ideal relationship resulting in distortion in the average input current.

The very proper average current mode three-loop APFC [15–18] achieves its control objectives by a control scheme shown at Figure 1. The three-loop APFC uses a slow outer voltage loop to control the output voltage and a fast current loop for active shaping of the average input current. The reference signal for the inner current loop is derived from the rectified power line voltage by the multiplier-squarer-divider circuit. Additional outer feed-forward loop compensates for the line voltage variations. Such an APFC could be designed to operate in CCM most of the cycle but could tolerate also DCM intervals sustaining good current tracking.

The two feedback loops of the APFC of Figure 1 have conflicting objectives. In particular, a strong outer loop that manages to stabilize the output voltage will deteriorate the power factor by dictating an input current that ensures a fixed output voltage rather than the desired sinusoidal-shaped current. The outer loop is usually designed with

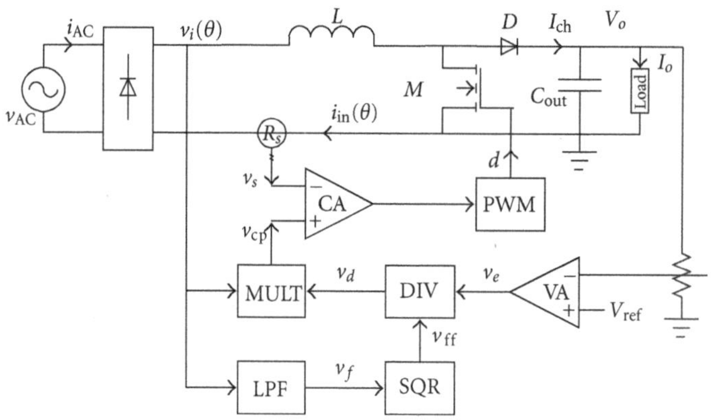

Figure 1: Three-loop ACM APFC system block diagram.

a limited bandwidth as well as with a low gain so to heavily attenuate the output ripple. Otherwise, the ripple will distort the input current by penetrating into the reference of the current loop. As a result the response of the outer loop to load variations is rather slow. The outer loop error amplifier output voltage contains a second harmonic ripple component which modulates the current amplitude and thus distorts the current reference signal. Obviously, this also results in distorted line current.

This paper suggests that one of the causes of the major current distortion in most, if not all, APFC systems is the second harmonic ripple components that are brought into the current reference by the outer loops. The proposed solution to the problem can be found by compensating the output ripple rather than filtering. This alternative was investigated in this study analytically and by simulation. It was found that such a strategy is possible and furthermore it can easily be implemented by simple additional circuitry to commercially available APFC controllers. This paper presents the theory of the proposed approach and demonstrates the improvement in Total Harmonic Distortion (THD) that can be gained and provides design guidelines.

2. Basic Assumptions and Strategy

The analyses to follow are carried under the following assumptions:

(a) the line voltage is a pure sinusoidal wave;

(b) the rectifiers are ideal;

(c) the power stage is linear and 100% efficient;

(d) the current loop is ideal, that is, it forces the input current to follow the reference signal;

(e) the holdup capacitor is large and hence the output ripple is small as compared to the dc output voltage.

The following approach is adopted: initially it is assumed that the input current is a pure sine wave. Next, the outer loop ripple components propagation mechanism into the current reference circuit is investigated. The assumption of the ideal current loop helps identify the current reference distortion components, which originate in the outer loops. Based on the results, the requirements for maintaining a distortion-free current reference are formulated. Finally, under the assumption of the ideal current loop, the residual distortion caused by the proposed ripple cancellation scheme is examined.

3. General Considerations: Review of APFC Output Current and Voltage

Under the assumptions above, the line voltage and current of the APFC of Figure 1 are in phase and of a sinusoidal shape. To simplify the notation the analyses are confined to one-half of the line period for which the rectified inputs can be expressed as

$$v_{in}(\theta) = V_m \sin\theta, \quad i_{in}(\theta) = I_m \sin\theta, \tag{1}$$

where $\theta = \omega t$ is the power line angle ($0 \leq \theta \leq \pi$). The instantaneous input power of the APFC is given by

$$p_{in}(\theta) = v_{in}(\theta) i_{in}(\theta) = V_m I_m \sin^2\theta = P_{av}(1 - \cos 2\theta), \tag{2}$$

where $P_{av} = (1/2)V_m I_m$ is the average power over the power line cycle. The instantaneous charging power at the output of the APFC is a function of the output voltage, $v_{out}(\theta)$, and the output charging current, $i_{ch}(\theta)$:

$$p_{out}(\theta) = v_{out}(\theta) i_{ch}(\theta) \approx V_{DC} i_{ch}(\theta). \tag{3}$$

Here, the output ripple is neglected, that is, the output voltage, $v_{out}(\theta)$, of the APFC is assumed to be equal to its average value, V_{DC}. For an ideal power stage with 100% efficiency, instantaneous input and output power are equal:

$$p_{in}(\theta) = p_{out}(\theta). \tag{4}$$

Applying (2)–(4), the charging current, $i_{ch}(\theta)$, is approximated to

$$i_{ch}(\theta) = \frac{P_{av}}{V_{DC}}(1 - \cos 2\theta). \tag{5}$$

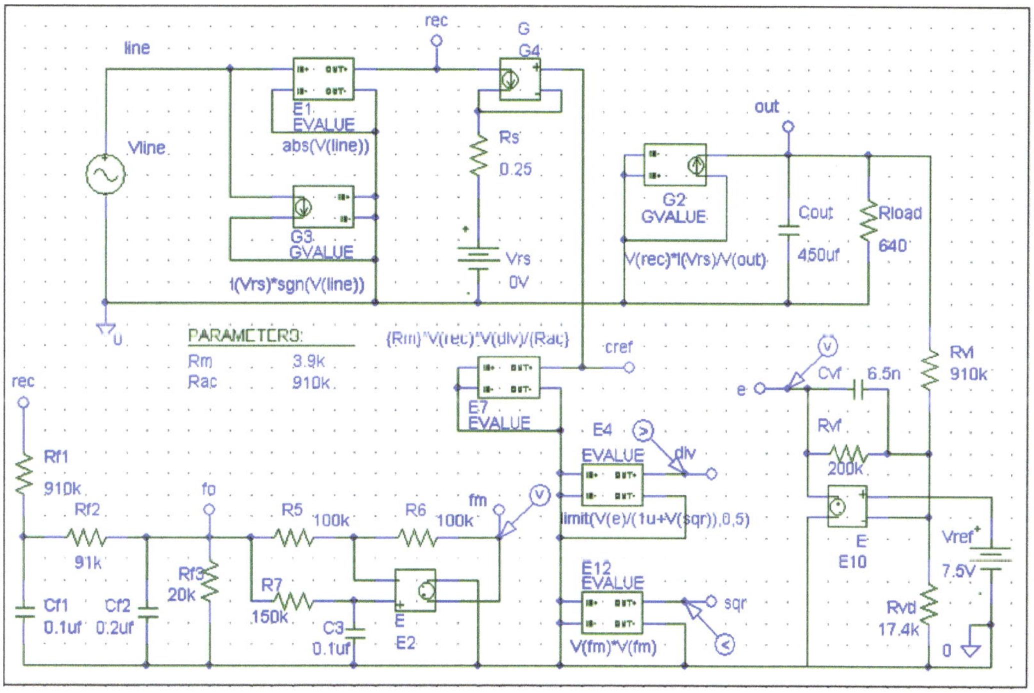

Figure 2: Simulation diagram of the modified three-loop APFC based on the behavioral model.

The holding capacitor current, $i_c(\theta)$, is the difference between the charging current and the DC load currents:

$$i_c(\theta) = i_{ch}(\theta) - I_{DC} = -\frac{P_{av}}{V_{DC}}\cos 2\theta. \quad (6)$$

The AC current of the holdup capacitor, $i_c(\theta)$, generates a second harmonics ripple component, $v_{o2}(\theta)$, which appears at the output:

$$v_{o2}(\theta) = \frac{1}{\omega C}\int i_c(\theta)d\theta = \frac{-P_{av}}{2\omega C V_{DC}}\sin 2\theta. \quad (7)$$

Clearly, the output ripple is a linear function of power. The instantaneous output voltage including the second harmonic ripple can now be written as

$$v_{out}(\theta) = V_{DC} + v_{o2}(\theta). \quad (8)$$

Using (7) yields

$$v_{out}(\theta) = V_{DC}\left(1 - \frac{P_{av}}{2\omega C V_{DC}^2}\sin 2\theta\right). \quad (9)$$

The second harmonic frequency peak-to-peak ripple, γ, is derived from the above to be

$$\gamma = \frac{\Delta V_{out_{pp}}}{V_{DC}}100\% = \frac{P_{av}}{\omega C V_{DC}^2}100\%. \quad (10)$$

4. The Feed-Forward Path Signal

The rectified input voltage fed to the feed-forward path filter (LPF) of the PFC topology under study, see Figure 1, can be represented by the Fourier series:

$$v_{in}(\theta) = V_{i0} + V_{i2}\cos 2\theta + V_{i4}\cos 4\theta + \cdots, \quad (11)$$

where $V_{i0} = (2/\pi)V_m$, $V_{i2} = -(4/3\pi)V_m$, $V_{i4} = -(4/15\pi)V_m$, and so forth are the Fourier coefficients. The relative phases of the harmonics are zero. When this signal is fed to the LPF feed-forward network of Figure 2, the LPF output voltage, $v_f(\theta)$, appears as

$$v_f(\theta) = V_{i0}H_{f0} + V_{i2}H_{f2}\cos(2\theta + \phi_{f2}) + V_{i4}H_{f4}\cos(4\theta + \phi_{f4}) + \cdots, \quad (12)$$

where H_{fn} and ϕ_{fn} are the gain and phase of the LPF transfer function at the nth harmonic of the line frequency. Since the amplitude of the harmonics of the rectified voltage decreases with their frequency and since the LPF provides sufficient attenuation at high frequencies, harmonic components higher than the second one may be neglected at the output of the LPF. Therefore, the response of the LPF is approximated to

$$v_f(\theta) = V_{i0}H_{f0}\left(1 + \frac{V_{i2}}{V_{i0}}\frac{H_{f2}}{H_{f0}}\cos(2\theta + \phi_{f2})\right). \quad (13)$$

The $v_f(\theta)$ signal is squared with the gain of k_s by the squarer circuit, see Figure 1. Applying the same reasoning as above the resulting quadratic term could be neglected. Therefore, the approximated feed-forward signal is

$$v_{ff}(\theta) = k_s v_f^2(\theta) \approx k_s V_{i0}^2 H_{f0}^2\left(1 + 2\frac{V_{i2}}{V_{i0}}\frac{H_{f2}}{H_{f0}}\cos(2\theta + \phi_{f2})\right). \quad (14)$$

Substituting the values of the Fourier coefficients, the expression for the feed-forward signal fed to the divider is obtained as

$$v_{ff}(\theta) = V_{ffo} v_{ffn}(\theta), \tag{15}$$

where the term, $V_{ffo} = k_s(4/\pi^2)V_m^2 H_{fo}^2$ and $v_{ffn}(\theta)$ is the normalized feed-forward voltage:

$$v_{ffn}(\theta) = \frac{v_{ff}(\theta)}{V_{ffo}} = 1 - \frac{4}{3}\frac{H_{f2}}{H_{fo}}\cos\left(2\theta + \phi_{f2}\right). \tag{16}$$

Equation (15) describes a waveform having a DC value proportional to the square of the amplitude of the input voltage, V_m^2, scaled by the DC gain of the feed-forward path: $k_s(4/\pi^2)H_{fo}^2$. The normalized feed-forward signal (16) contains an undesirable second harmonic component propagating into the current reference. Interesting to note is that the normalized ripple (16) at the output of the LPF is constant and depends on the attenuation at the second harmonic frequency, $(4/3)(H_{f2}/H_{fo})$. The amount of attenuation is determined by LPF configuration and its comer frequency. Lowering the corner frequency increases the attenuation, lowering also the harmonic contents of the feed-forward signal. This results in better line current quality. However, lowering the corner frequency slows the APFC response to line variations.

5. The Voltage Feedback Path Signal

To establish the operating point of the voltage error amplifier the issue of the input current generation should be addressed first. The input current of the PFC circuit of Figure 1 is generated by the following algorithm:

$$i_{\text{in}}(\theta) = I_m \sin\theta = K(v_e(\theta))\left(\frac{1}{V_m^2}\right)(V_m \sin\theta). \tag{17}$$

where, $v_e(\theta)$ is the voltage feedback error amplifier signal and K is the system's gain constant. Thus the amplitude of the input current and the average input power are

$$I_m = K\left(\frac{v_e(\theta)}{V_m}\right), \tag{18}$$

$$P_{\text{av}} = \frac{I_m V_m}{2} = \frac{K}{2} v_e(\theta). \tag{19}$$

The advantage of the feed-forward path is apparent; the APFC power level (19) is independent of the line voltage.

The steady-state output voltage of the error amplifier, $v_e(\theta)$, contains the DC term, V_{e0}, as well as the second harmonics ripple component, $v_{e2}(\theta)$:

$$v_e(\theta) = V_{e0} + v_{e2}(\theta). \tag{20}$$

The steady state DC error amplifier voltage, V_{e0}, required to maintain the average power level, P_{av}, could be derived from (19) neglecting the ripple component as

$$V_{e0} = \frac{2}{K} P_{\text{av}}. \tag{21}$$

The second harmonic component, $v_{e2}(\theta)$, appears as the response of the error amplifier to the output voltage ripple, $v_{o2}(\theta)$, given by (7)

$$v_{e2}(\theta) = -\frac{H_{v2}}{2\omega C}\frac{P_{\text{av}}}{V_{\text{DC}}}\sin(2\theta + \phi_{v2}). \tag{22}$$

Here, the error amplifier's gain and phase, at the frequency of the second harmonic, are denoted H_{v2} and ϕ_{v2} respectively. Combining (20), (21), and (22), yields the expression for the error amplifier output voltage delivered to the input of the divider as

$$v_e(\theta) = V_{e0} v_{en}(\theta), \tag{23}$$

where $v_{en}(\theta)$ is the normalized feedback voltage:

$$v_{en}(\theta) = \frac{v_{e(\theta)}}{V_{e0}} = 1 - \frac{K}{4V_{\text{DC}}}\frac{II_{v2}}{\omega C}\sin(2\theta + \phi_{v2}). \tag{24}$$

Note that the normalized second harmonic ripple component at the output of the error amplifier is constant and independent of the power level output voltage of the APFC. This observation forms the basis for the following development of the proposed ripple cancellation method.

6. The Current Programming Signal

To produce the correct current programming signal, the APFC circuit of Figure 1 uses the divider output, $v_d(\theta)$, to modulate the line voltage waveform. The divider output signal is calculated using (15) and (23) yielding

$$v_d(\theta) = k_d\left(\frac{v_e(\theta)}{v_{ff}(\theta)}\right) = \left(\frac{\pi^2}{2}\frac{k_d}{k_s K H_{fo}^2}\frac{P_{\text{av}}}{V_m^2}\right) v_{dn}(\theta). \tag{25}$$

Here, k_d is the divider gain constant and $v_{dn}(\theta)$ is the normalized divider output:

$$v_{dn}(\theta) = \frac{v_{en}(\theta)}{v_{ffn}(\theta)} = \frac{1 - (K/4V_{\text{DC}})(H_{v2}/\omega C)\sin(2\theta + \phi_{v2})}{1 - (4/3)\left(H_{f2}/H_{fo}\right)\cos\left(2\theta + \phi_{f2}\right)}. \tag{26}$$

The current programming signal, $v_{\text{cp}}(\theta)$, produced by modulation of the line voltage (1) with the divider output (25) is given by

$$\begin{aligned} v_{\text{cp}}(\theta) &= k_m v_d(\theta) v_{\text{in}}(\theta) \\ &= \left(\frac{\pi^2}{2}\frac{k_m k_d}{k_s}\frac{1}{K H_{fo}^2}\frac{P_{\text{av}}}{V_m}\right) v_{dn}(\theta)\sin(\theta), \end{aligned} \tag{27}$$

where k_m is the multiplier gain constant. The current programming signal amplitude is therefore

$$v_{\text{cp}m} = \left(\frac{\pi^2}{2}\frac{k_m k_d}{k_s}\frac{1}{K H_{fo}^2}\frac{P_{\text{av}}}{V_m}\right) v_{dn}(\theta). \tag{28}$$

Equations (27) and (28) reveal the mechanism by which distortion components of the outer loops are introduced into the current reference. The term $v_{dn}(\theta)$ contains harmonic components which modulates the current programming signal amplitude (28). Consequently, the input current of the APFC could not possibly be any less distorted than (27).

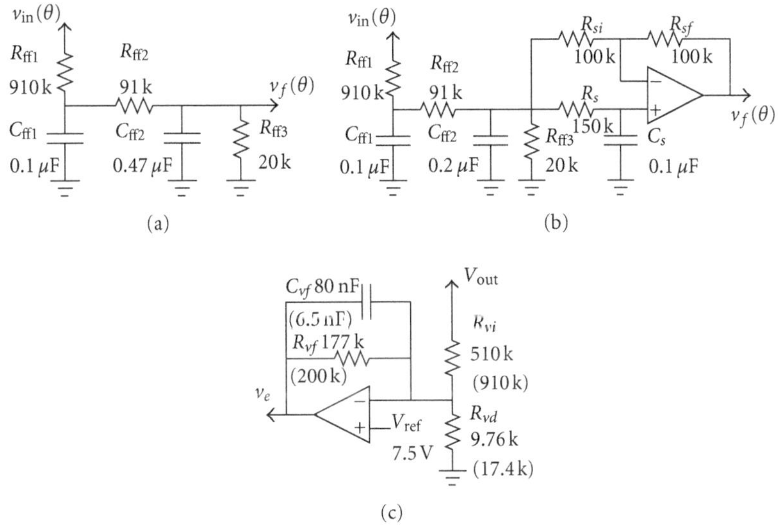

FIGURE 3: APFC feed-forward filters: original design (a); modified (b); voltage error amplifier configuration with original and modified (in brackets) values (c).

7. The Power Gain Constant

When the current loop tightly regulates the input current, the output voltage of the current sensing network, $v_{\text{sense}}(\theta)$, see Figure 1, is forced to follow the current programming signal, $v_{\text{cp}}(\theta)$:

$$v_{\text{sense}}(\theta) = v_{\text{cp}}(\theta). \tag{29}$$

In the most common case, the current sensing network is just a series resistance, R_s, so the amplitude of the sensed voltage is

$$V_{\text{sense}_m} = I_m R_s = 2\frac{P_{\text{av}}}{V_m} R_s. \tag{30}$$

The power gain constant K may be found substituting (28), (30) into (29) and assuming an ideal $v_{dn}(\theta)$, that is, (26) equals unity. This gives

$$K = \frac{\pi^2}{4} \frac{k_m k_d}{k_s} \frac{1}{R_s H_{f0}^2}. \tag{31}$$

This relationship remains valid also for the general case, when the current sensing network transfer function low frequency gain is denoted by R_s.

8. Filter Design for Minimum Line Current Distortion

Examination of (27) and (28) reveals that any harmonic disturbances at the divider output will produce cross products with the sine term. This will appear as harmonics in the current programming signal and cause distortion in the input current. Traditionally, the designers [15–18] minimize the distortion by providing large attenuation of the second harmonic, that is, minimizing the H_{v2} and H_{f2} terms in (16), (24), and (26). However, this has only a limited success for the following reasons. To ensure stability and adequate phase margin, the error amplifier usually has a single-pole transfer function of −20 db/decade gain roll off beyond its corner frequency and its filtering action is poor. Since the ripple frequency is quite low the bandwidth of the error amplifier is severely restricted resulting in poor transient response. A high-order low-pass filter may be used to achieve efficient filtering of the average component of the rectified input voltage. However, once again due to the low second harmonic frequency the filter has a narrow bandwidth and its transient response is rather poor. Here an alternative approach is proposed. Equations (27) and (28) suggest that it is possible to achieve distortion-free current programming signal by making the normalized divider output (26) equal unity:

$$v_{dn}(\theta) = \frac{1 - (K/4V_{\text{DC}})(H_{v2}/\omega C)\sin(2\theta + \phi_{v2})}{1 - (4/3)\left(H_{f2}/H_{f0}\right)\cos\left(2\theta + \phi_{f2}\right)} = 1. \tag{32}$$

To satisfy (32) requires that the normalized ripple components of the error and the feed-forward signals be equal both in amplitude and phase:

$$\frac{K}{4V_{\text{DC}}} \frac{H_{v2}}{\omega C} \sin(2\theta + \phi_{v2}) = \frac{4}{3} \frac{H_{f2}}{H_{f0}} \cos\left(2\theta + \phi_{f2}\right). \tag{33}$$

Since the normalized ripple components are of the same frequency and of a constant amplitude, it is possible to fulfill (34) by proper controller circuit design. For this purpose both amplitude and phase conditions must be satisfied. The amplitude condition defines the LPF attenuation needed to equate the amplitudes of the ripple components:

$$\frac{H_{f2}}{H_{f0}} = \frac{3K}{8V_{\text{DC}}} \frac{H_{v2}}{(2\omega)C}. \tag{34}$$

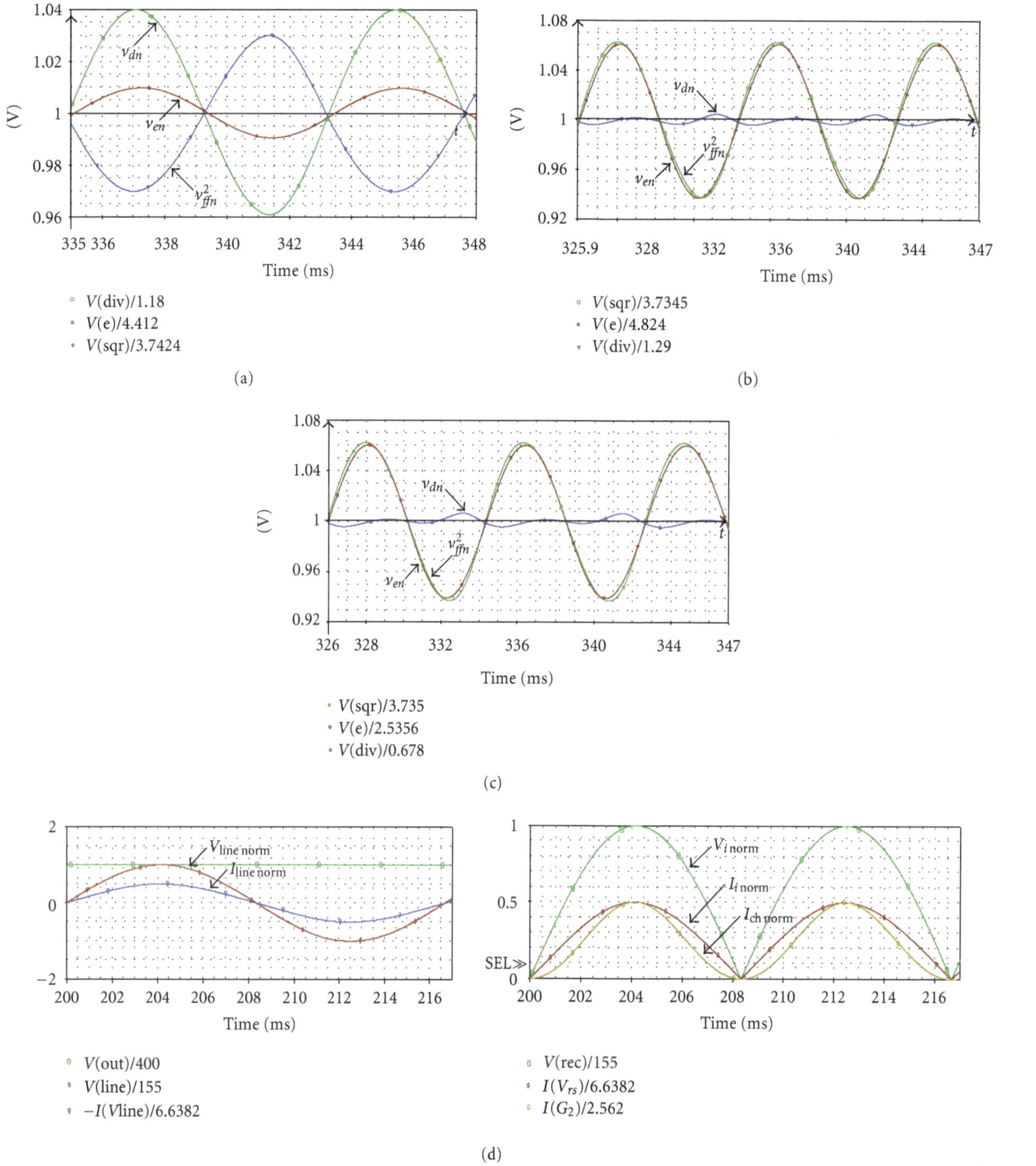

Figure 4: Normalized voltages of the error amplifier v_{en}, the feed-forward network v_{ffn}, and the divider output v_{dn}, in the original design (250 W) (a); modified design (250 W) (b), Modified design (125 W) (c), Modified design waveforms of the normalized rectified voltage V_{in}, converter's input current I_{in}, and the APFC's charging current I_{ch} (d).

This could be rearranged using (31) into the following form:

$$H_{f2} = \frac{3\pi}{128 f C R_s H_{f0} V_{\text{DC}}}\left(\frac{k_m k_d}{k_s}\right) H_{v2}, \quad (35)$$

where, f is the line frequency.

The phase condition could be derived by equating the sine and cosine terms in (33) and using elementary trigonometry relationship:

$$\cos\left(2\theta + \phi_{f2}\right) = \sin\left(2\theta + \phi_{v2}\right) = \cos\left(2\theta + \phi_{v2} - \frac{\pi}{2}\right). \quad (36)$$

TABLE 1: Analysis printout of the original circuit, full power (250 W).

Fourier components of transient response
DC component = 2.758333E − 04

Harm no.	Freq (Hz)	Fourier comp	Normalized comp	Phase (Deg)	Norm. phase (Deg)
1	6.000E + 01	3.198E + 00	1.000E + 00	−1.797E + 02	0.000E + 00
2	1.200E + 02	4.295E − 04	1.343E − 04	1.456E + 02	5.050E + 02
3	1.800E + 02	5.949E − 02	1.860E − 02	1.742E + 01	5.565E + 02
4	2.400E + 02	1.831E − 04	5.724E − 05	−1.779E + 02	5.408E + 02
5	3.000E + 02	1.372E − 03	4.290E − 04	2.705E − 01	8.987E + 02
6	3.600E + 02	1.248E − 04	3.903E − 05	1.740E + 02	1.252E + 03
7	4.200E + 02	1.649E − 04	5.157E − 05	6.782E + 00	1.265E + 03
8	4.800E + 02	8.522E − 05	2.665E − 05	1.764E + 02	1.614E + 03
9	5.400E + 02	1.090E − 05	3.409E − 06	1.197E + 02	1.737E + 03
10	6.000E + 02	7.085E − 05	2.215E − 05	−1.798E + 02	1.617E + 03

Total harmonic distortion = 1.860710E + 00 percent

TABLE 2: Analysis printout of the modified circuit, full power (250 W).

Fourier components of transient response
DC component = 3.755277E − 04

Harm no.	Freq (Hz)	Fourier comp	Normalized comp	Phase (Deg)	Norm. phase (Deg)
1	6.000E + 01	3.424E + 00	1.000E + 00	1.800E + 02	0.000E + 00
2	1.200E + 02	1.028E − 03	3.002E − 04	1.585E + 02	−2.015E + 02
3	1.800E + 02	4.195E − 03	1.225E − 03	1.890E + 01	−5.210E + 02
4	2.400E + 02	3.668E − 04	1.071E − 04	1.752E + 02	−5.447E + 02
5	3.000E + 02	4.809E − 03	1.405E − 03	−1.639E + 02	−1.064E + 03
6	3.600E + 02	2.352E − 04	6.870E − 05	1.797E + 02	−9.001E + 02
7	4.200E + 02	1.009E − 03	2.947E − 04	−1.640E + 02	−1.424E + 03
8	4.800E + 02	1.685E − 04	4.922E − 05	−1.782E + 02	−1.618E + 03
9	5.400E + 02	3.725E − 04	1.088E − 04	−1.705E + 02	−1.790E + 03
10	6.000E + 02	1.309E − 04	3.824E − 05	−1.777E + 02	−1.977E + 03

Total harmonic distortion = 1.919198E − 01 percent

Thus, the required phase shift of the LPF at the second harmonic frequency is

$$\phi_{f2} = \phi_{v2} - \frac{\pi}{2}. \quad (37)$$

The error amplifier transfer function is determined primarily by the stability and performance considerations of the outer loop. Consequently, to make the solutions of (34), (37) unique, the design of the feed-forward filter should be carried out after the complete knowledge of the error amplifier characteristics is gained. Note that ϕ_{v2}, as defined by (22), is the overall error amplifier phase shift. Since the phase shift of a single-pole error amplifier lags from the initial +180 degrees, due to the inverting configuration, to about +90, applying (37) requires about 0/360 degrees of LPF phase shift. Designing the feed-forward path filter according to the amplitude and phase conditions stated above will ensure minimum distortion of the PFC input current. It should be emphasized that, under the assumptions made, this approach will render a perfect current programming signal independent of the APFC power level.

9. Simulation Results

To study the performance of the APFC circuit of Figure 1, the system's model was simulated by PSPICE circuit simulator. The simulation was focused on the performance of the feedback and feed-forward loops under the assumption that the inner loop is ideal. The inner loop and power stage models were characterized by (4), (5), and (16) and implemented in PSPICE according to these behavioral relationships, see Figure 2. The design of the feed-forward LPF and voltage feedback amplifier followed the procedure discussed in [15, 16]. The program simulated a closed-loop PFC system fed by a rectified 110 V/60 Hz line and delivers $P_{av} = 250$ W at $V_{DC} = 400$ V to a resistive load of 640 Ω. The holdup capacitor of 450 uF was chosen. The overall squarer-multiplier-divider constant, defined by external resistors, was set to $k_m k_d/k_s = R_m/R_{AC} = 3.9k/910k = 4.286 \cdot 10^{-3}$. Current sense resistor used was $R_s = 0.25\,\Omega$. According to Figure 3(b), the feed-forward filter transfer function was: $H_{f0} = (R_{ff3}/(R_{ff1} + R_{ff2} + R_{ff3})) = 19.59 \cdot 10^{-3}$. For these conditions, the numerical value of expression (31) equals $K = 110.2$ [W/V].

Table 3: Analysis printout of the modified circuit, half power (125 W).

Fourier components of transient response					
DC component = $3.755277E-04$					
Harm no.	Freq (Hz)	Fourier comp	Normalized comp	Phase (Deg)	Norm. phase (Deg)
1	$6.000E+01$	$1.801E+00$	$1.000E+00$	$1.799E+02$	$0.000E+00$
2	$1.200E+02$	$5.410E-0$	$3.004E-04$	$1.584E+02$	$-2.015E+02$
3	$1.800E+02$	$1.431E-03$	$7.949E-04$	$-2.125E+01$	$-5.611E+02$
4	$2.400E+02$	$1.938E-04$	$1.076E-04$	$1.753E+02$	$-5.444E+02$
5	$3.000E+02$	$2.529E-03$	$1.405E-03$	$-1.637E+02$	$-1.063E+03$
6	$3.600E+02$	$1.239E-04$	$6.883E-05$	$1.798E+02$	$-8.998E+02$
7	$4.200E+02$	$5.320E-04$	$2.955E-04$	$-1.638E+02$	$-1.423E+03$
8	$4.800E+02$	$8.877E-05$	$4.930E-05$	$-1.781E+02$	$-1.618E+03$
9	$5.400E+02$	$1.962E-04$	$1.090E-04$	$-1.704E+02$	$-1.790E+03$
10	$6.000E+02$	$6.895E-05$	$3.829E-05$	$-1.776E+02$	$-1.977E+03$
Total harmonic distortion = $1.677760E-01$ percent					

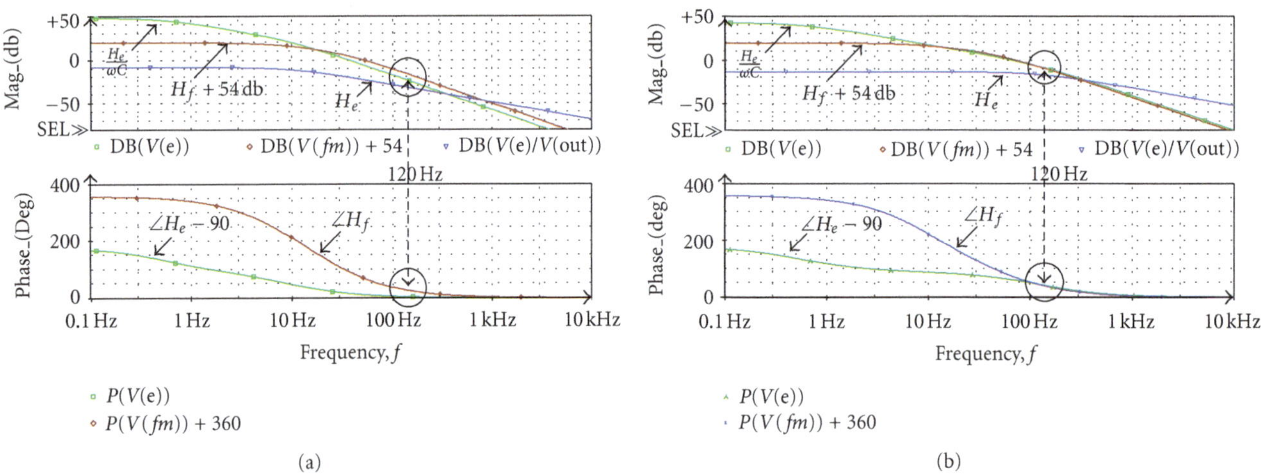

Figure 5: Frequency response of the error amplifier, H_e, and the feed-forward network, H_f, in the original design (a) and the modified design (b). Notice that the magnitude and phase conditions are satisfied at the vicinity of the second-line harmonic frequency.

To demonstrate the validity of the proposed ripple cancellation method, the LPF of the feed-forward network was modified to follow (34) and (37) as shown at Figure 3(b). New values were given to the error amplifier parameters as depicted in Figure 3(c). The simulation results of the original and proposed circuits are presented in Figures 4(a) and 4(b). Here, the normalized outputs of the voltage error amplifier, feed-forward, and divider circuits are shown. Comparison of the results reveals that the amplitude of the ripple voltage at the divider output is greatly reduced by the proposed method, reducing also the THD of the simulated APFC system. The normalized rectified line voltage and current as well as the APFC's charging currents are shown in Figure 4(c). Fourier analysis was carried out by SPICE on the simulated line current. Fourier analysis results are given in Tables 1, 2 and 3. The analysis reveals that the distorted current programming signal of the original design, see Table 1, contributes 1.86% to the total harmonic distortion of the APFC system, which is rather good performance index, whereas, using the proposed method the distortion is reduced to a negligible level of only 0.19%, see Table 2. That is about one order of magnitude improvement of THD.

10. Conclusions

One of the main sources of input current distortion in high-frequency APFC systems is the output voltage ripple which propagates into the feedback loop as well as the second harmonic of the line voltage injected through the feed-forward loop. These signals adversely affect the current reference signal quality.

This paper examined the propagation of the ripple components in the feedback and feed-forward loops of the APFC controller circuits and their interaction within the current shaping network. Based on this analysis, a ripple cancellation method is proposed. Analytical results are confirmed by simulation and show good agreement with predicted theoretical results. The simulation results suggest that the proposed method can reduce the THD of the input current of PFC system by more than an order of magnitude. The present analysis and simulation were carried out under some simplifying assumptions, yet, second-order effects proved to be negligible.

Another important advantage of the proposed method is a dramatic increase of the error amplifier bandwidth, see Figure 5. This is possible in view of the fact that effective ripple cancellation requires a lesser amount of error amplifier attenuation of the second-line harmonic. Therefore, the outer feedback loop bandwidth becomes dominated by the relatively large holding capacitor, which value is determined primarily by the energy storage requirements.

The discussion of this paper is confined to the topology of one family of APFC controllers (Figure 1); however, the proposed approach could be expanded to other APFC schemes by following the idea of ripple compensation developed here.

References

[1] K. N. Sakthivel, S. K. Das, and K. R. Kini, "Importance of quality AC power distribution and understanding of EMC standards IEC, 61000-3-2, IEC, 61000-3-3 & IEC, 61000-3-11," in *Proceedings of the 8th International Conference on Electromagnetic Interference and Compatibility*, pp. 423–430, 2003.

[2] B. Singh, B. N. Singh, A. Chandra, K. Al-Haddad, A. Pandey, and D. P. Kothari, "A review of single-phase improved power quality AC-DC converters," *IEEE Transactions on Industrial Electronics*, vol. 50, no. 5, pp. 962–981, 2003.

[3] M. M. Jovanović and Y. Jang, "State-of-the-art, single-phase, active power-factor-correction techniques for high-power applications—an overview," *IEEE Transactions on Industrial Electronics*, vol. 52, no. 3, pp. 701–708, 2005.

[4] B. Andreycak, "Power factor correction using UC3852 controlled on-time zero current switching technique," U-132 Application note, Unitrode Integrated Circuits.

[5] M. Marvi and A. Fotowat-Ahmady, "A fully ZVS critical conduction mode boost PFC," *IEEE Transactions on Power Electronics*, vol. 27, no. 4, pp. 1958–1965, 2012.

[6] T. F. Wu, J. R. Tsai, Y. M. Chen, and Z. H. Tsai, "Integrated circuits of a PFC controller for interleaved critical-mode boost converters," in *Proceedings of the 22nd Annual IEEE Applied Power Electronics Conference and Exposition (APEC '07)*, pp. 1347–1350, March 2007.

[7] J. R. Tsai, T. F. Wu, C. Y. Wu, Y. M. Chen, and M. C. Lee, "Interleaving phase shifters for critical-mode boost PFC," *IEEE Transactions on Power Electronics*, vol. 23, no. 3, pp. 1348–1357, 2008.

[8] K. De Gussemé, D. M. Van de Sype, A. P. M. Van den Bossche, and J. A. Melkebeek, "Input-current distortion of CCM boost PFC converters operated in DCM," *IEEE Transactions on Industrial Electronics*, vol. 54, no. 2, pp. 858–865, 2007.

[9] K. H. Liu and Y. L. Lin, "Current waveform distortion in power factor correction circuits employing discontinuous-mode boost converters," in *Proceedings of the 20th Annual IEEE Power Electronics Specialists Conference (PESC '89)*, pp. 825–829, June 1989.

[10] D. Maksimović, Y. Jang, and R. W. Erickson, "Nonlinear-carrier control for high-power-factor boost rectifiers," *IEEE Transactions on Power Electronics*, vol. 11, no. 4, pp. 578–584, 1996.

[11] J. P. Gegner and C. Q. Lee, "Linear peak current mode control: a simple active power factor correction control technique for continuous conduction mode," in *Proceedings of the 27th Annual IEEE Power Electronics Specialists Conference (PESC '96)*, pp. 196–202, January 1996.

[12] Z. Lai and K. M. Smedley, "Family of power-factor-correction controllers," in *Proceedings of the IEEE 12th Applied Power Electronics Conference (APEC '97)*, pp. 66–73, February 1997.

[13] J. Rajagopalan, F. C. Lee, and P. Nora, "Generalized technique for derivation of average current mode control laws for power factor correction without input voltage sensing," in *Proceedings of the IEEE 12th Applied Power Electronics Conference*, pp. 81–87, February 1997.

[14] S. Ben-Yaakov and I. Zeltser, "The dynamics of a PWM boost converter with resistive input," *IEEE Transactions on Industrial Electronics*, vol. 46, no. 3, pp. 613–619, 1999.

[15] P. C. Todd, "UC3854 controlled power factor correction circuit design," U-134 Application note, Unitrode Integrated Circuits.

[16] J. B. Williams, "Design of feedback loop in unity power factor AC to DC converter," in *Proceedings of the 20th Annual IEEE Power Electronics Specialists Conference (PESC '89)*, pp. 959–967, June 1989.

[17] L. H. Dixon, "High power factor preregulators for offline power supplies," in *Proceedings of the Unitrode Seminar*, 1993.

[18] L. H. Dixon, "High power factor preregulator design optimization," in *Proceedings of the Unitrode Seminar*, 1992.

9

Carrier-Based Common Mode Voltage Control Techniques in Three-Level Diode-Clamped Inverter

Pradyumn Chaturvedi,[1] Shailendra Jain,[2] and Pramod Agarwal[3]

[1] *Electrical Engineering Department, SATI, Vidisha 464001, India*
[2] *Electrical Engineering Department, MANIT, Bhopal 462051, India*
[3] *Electrical Engineering Department, IIT, Uttaranchal, Roorkee 247667, India*

Correspondence should be addressed to Pradyumn Chaturvedi, pradyumnc74@rediffmail.com

Academic Editor: Francesco Profumo

Switching converters are used in electric drive applications to produce variable voltage, variable frequency supply which generates harmful large *dv/dt* and high-frequency common mode voltages (CMV). Multilevel inverters generate lower CMV as compared to conventional two-level inverters. This paper presents simple carrier-based technique to control the common mode voltages in multilevel inverters using different structures of sine-triangle comparison method such as phase disposition (PD), phase opposition disposition (POD) by adding common mode voltage offset signal to actual reference voltage signal. This paper also presented the method to optimize the magnitude of this offset signal to reduce CMV and total harmonic distortion in inverter output voltage. The presented techniques give comparable performance as obtained in complex space vector-based control strategy, in terms of number of commutations, magnitude, and rate of change of CMV and harmonic profile of inverter output voltage. Simulation and experimental results presented confirm the effectiveness of the proposed techniques to control the common mode voltages.

1. Introduction

In the medium-voltage, high-power adjustable speed drive (ASD) system, AC supply is first converted into DC (known as DC buffer stage) and then converting back this DC into variable voltage variable frequency AC supply using inverter (voltage source or current source type). Figure 1 shows the general block diagram of AC-DC-AC induction motor drive system. The front-end converter may be uncontrolled or controlled voltage source or current source rectifier while the motor side converter can be conventional two-level or multilevel VSI, CSI. In VSI fed drive, the DC link capacitor C_d (C_1 and C_2) is sufficiently large and DC link inductor L_d is not required whereas in CSI fed drive, L_d is sufficiently large and C_d is not needed [1, 2].

The DC buffer stage consists of only large capacitor in the case of voltage source inverter and only large inductor in the case of current source inverter. This process involves switching power semiconductors to manipulate the magnitude and frequency of the output waveforms. The switching action of rectifiers and inverters results in common mode voltages which are essentially zero-sequence voltages superimposed with switching noise which will appear at rectifier, inverter, and motor terminals [1, 2]. If not mitigated, they appear on the neutral of the stator winding with respect to ground, which should be zero when the motor is powered by a three-phase balanced utility supply. Further, the motor line-to-ground voltage, which should be equal to the motor line-to-neutral (phase) voltage, can be substantially increased due to the common mode voltages, leading to premature failure of motor winding insulation system. As a consequence, the motor life expectancy is shortened. This phenomenon is different from the high *dv/dt* caused by the switching transients of the high-speed switches [3–6].

Therefore, it is necessary to control/eliminate the common mode voltage magnitude and its rate of change.

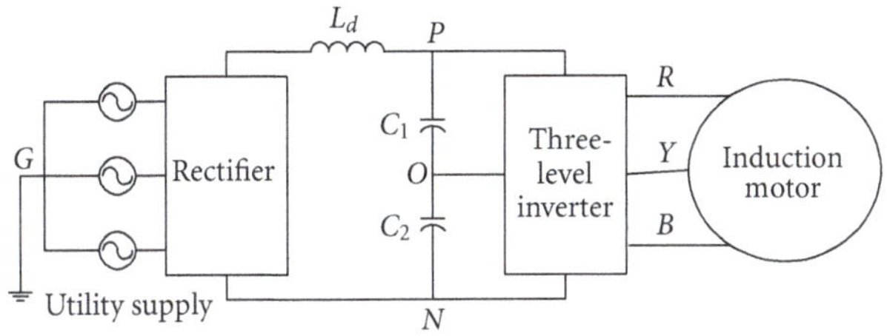

FIGURE 1: General block diagram of AC-DC-AC VSI or CSI fed ASD (only L_d in CSI and only C_1, C_2 in VSI).

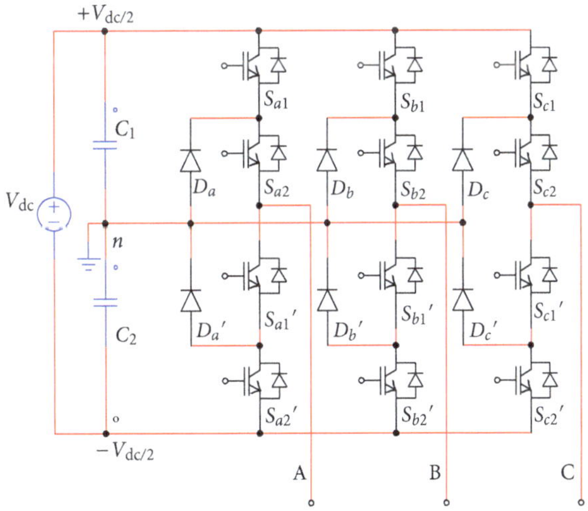

FIGURE 2: Schematic diagram of three-phase, three-level diode-clamped inverter.

Common mode voltage reduction/elimination techniques, available in literature, can be classified as given below [1–6]:

(A) using some extra hardware circuitry such as isolation transformer, zero-sequence impedance (common mode choke), active and passive filters, dual bridge inverter, using four-leg (four-phase) inverter, or

(B) using modification in control strategy employed based on space vector PWM technique (SVPWM) or sinusoidal PWM technique (SPWM).

Techniques mentioned in (A) above increases the system cost and complexity in control with reduced reliability. Techniques mentioned in (B) are commonly used. In SVPWM techniques, the switching state voltage vectors generating zero common mode voltage are only used [7, 8]. Various forms of SVPWM techniques are available to mitigate the common mode voltage problem, that is, nearest three vector selection (NTV) SVPWM, radial state SVPWM zero common mode voltage SVPWM, and so forth [8]. CMV reduction in adjustable speed drive system and its rate of change play an important role in designing electromagnetic interference (EMI) filters. SVPWM-based techniques for this purpose requires comparatively complex algorithm [7, 9–12].

In industrial applications requiring variable voltage, variable frequency supply to fed three-phase induction motors at medium- and high-voltage levels, multilevel inverters (MLI) have been found better counterpart to conventional two-level inverters. MLI produce lower common mode voltage stress across converter and motor terminals. Carrierbased modulation techniques are most widely used in MLI, especially in neutral point clamped (NPC) inverter. For cascaded H-bridge inverters, phase shifted carrier-based modulation techniques are most widely used with some derived strategies for improved performance [6, 8]. However, research is going on SVPWM-based technique to solve some inherent problems in NPC inverters, carrier-based SPWM techniques remain widely adopted because of their simplicity and reduced computational requirements as compared to SVPWM technique. Recent work reported in [6, 13–17] shows that carrier-based SPWM and SVPWM techniques for MLI are equivalent. Due to inherent problem of unequal properties of two DC link capacitors, inverter voltage gets unbalanced which can be made balanced easily by using the technique presented in [17, 18].

A current control strategy is presented in [19] to reduce the maximum common mode voltage levels of a three-phase inverter in adjustable speed drive applications. A hybrid selective harmonic elimination PWM scheme is given in [20] for common mode voltage reduction in three-level neutral-point-clamped inverter-based induction motor drive system. It automatically corrects the modulation index level to control the number of switching per cycle and also improves the load voltage and current waveform quality. Loh et al. presented the space vector PWM-based modulation

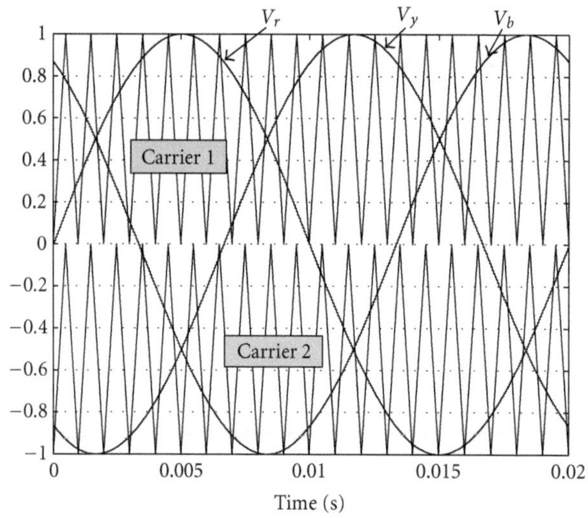

FIGURE 3: Generation of control pulses for 3-level inverter with PD SPWM technique.

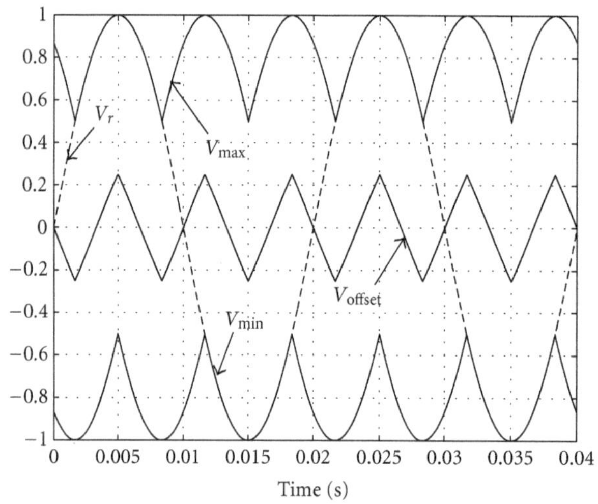

FIGURE 4: Signals "V_{max}," "V_{min}," "V_r," and offset voltage "V_{offset}" in offset PD SPWM technique.

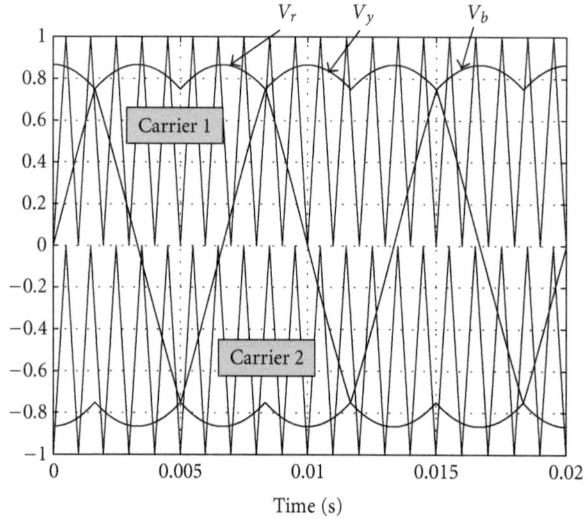

FIGURE 5: Generation of pulses for 3-level inverter with offset PD SPWM technique.

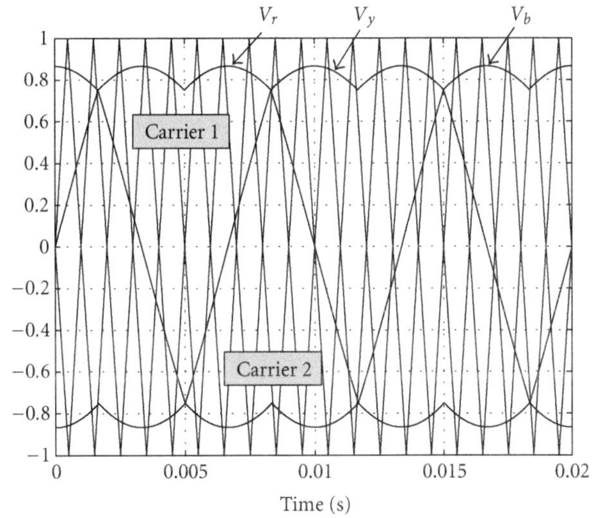

FIGURE 6: Generation of pulses for 3-level inverter with offset POD SPWM technique.

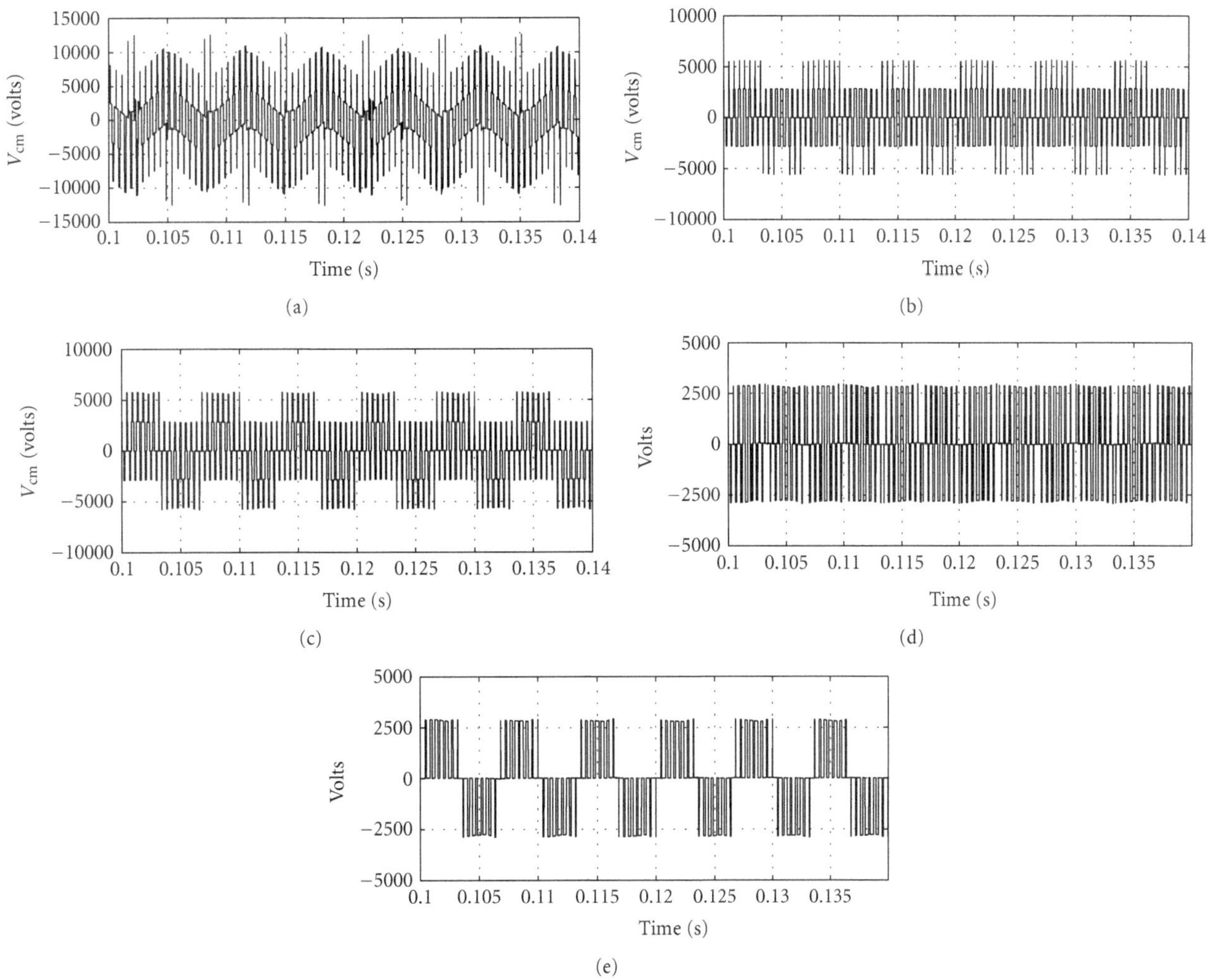

FIGURE 7: Common mode voltages in (a) 2-level inverter, (b) PD SPWM, (c) PD SPWM with offset voltage signal, (d) POD SPWM, and (e) POD with offset voltage signal.

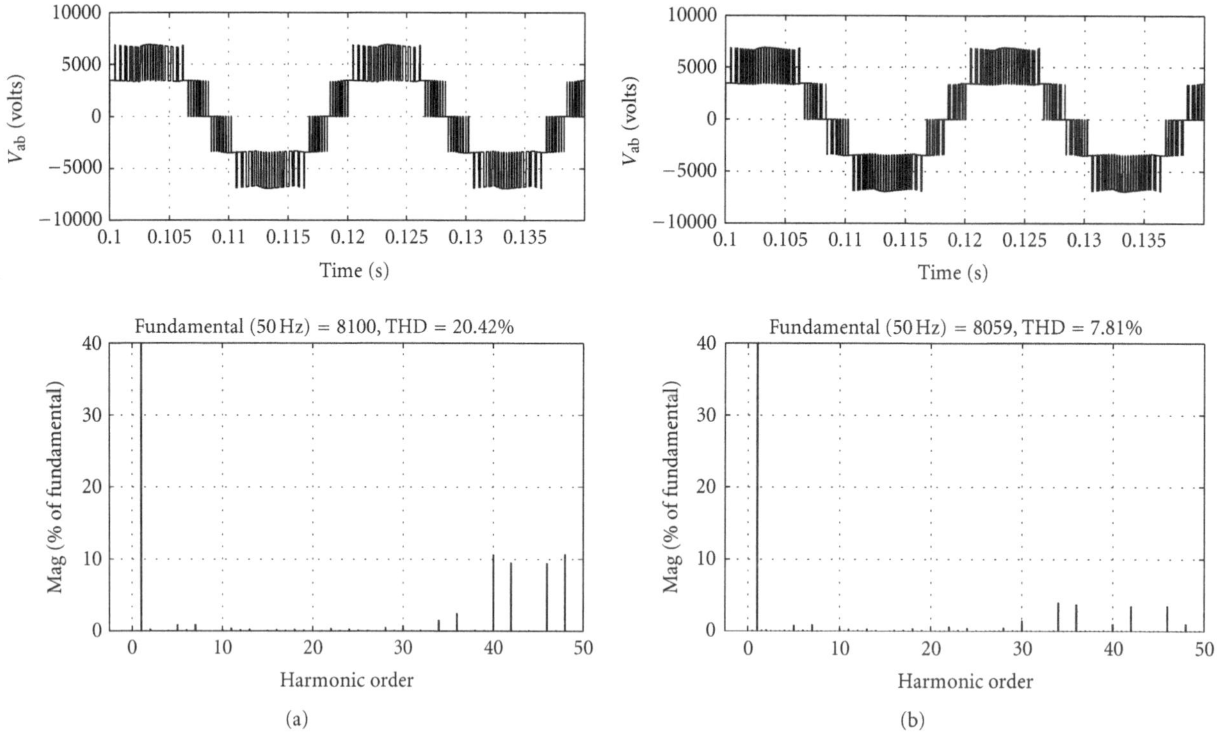

FIGURE 8: Inverter output voltage and its harmonic spectrum with (a) PD SPWM and (b) offset PD SPWM.

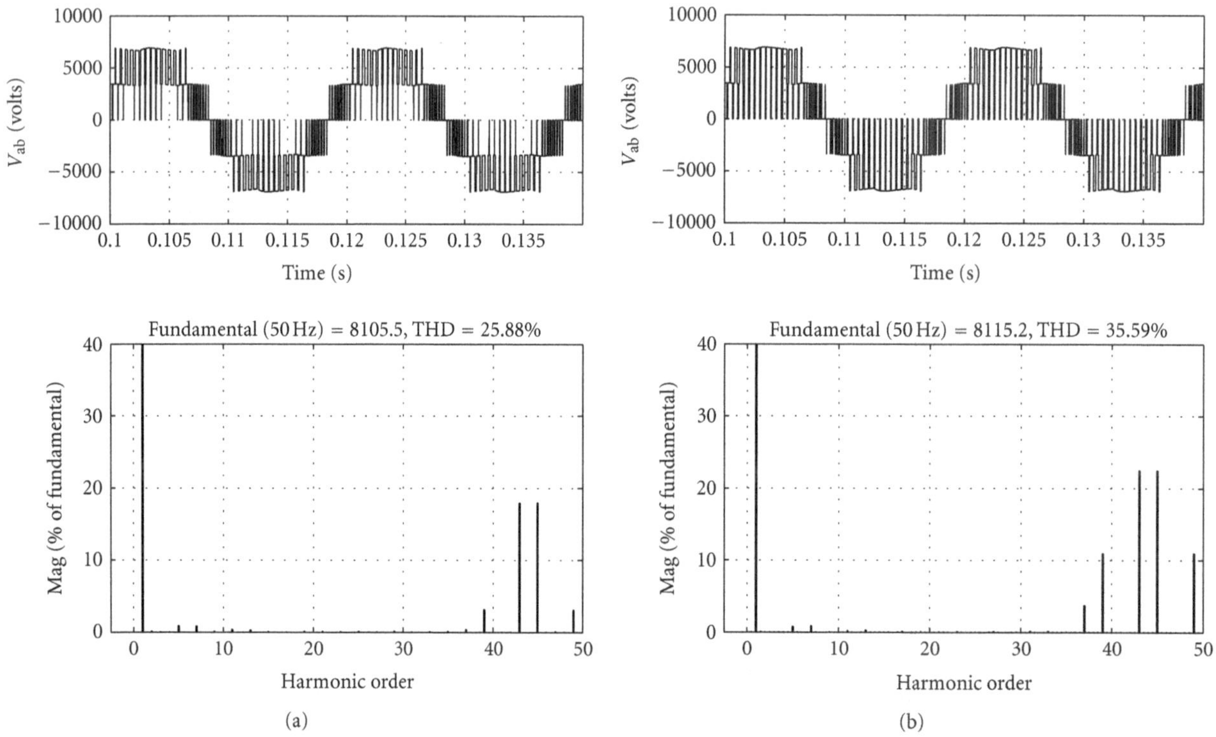

FIGURE 9: Inverter output voltage and its harmonic spectrum with (a) POD SPWM and (b) offset POD SPWM.

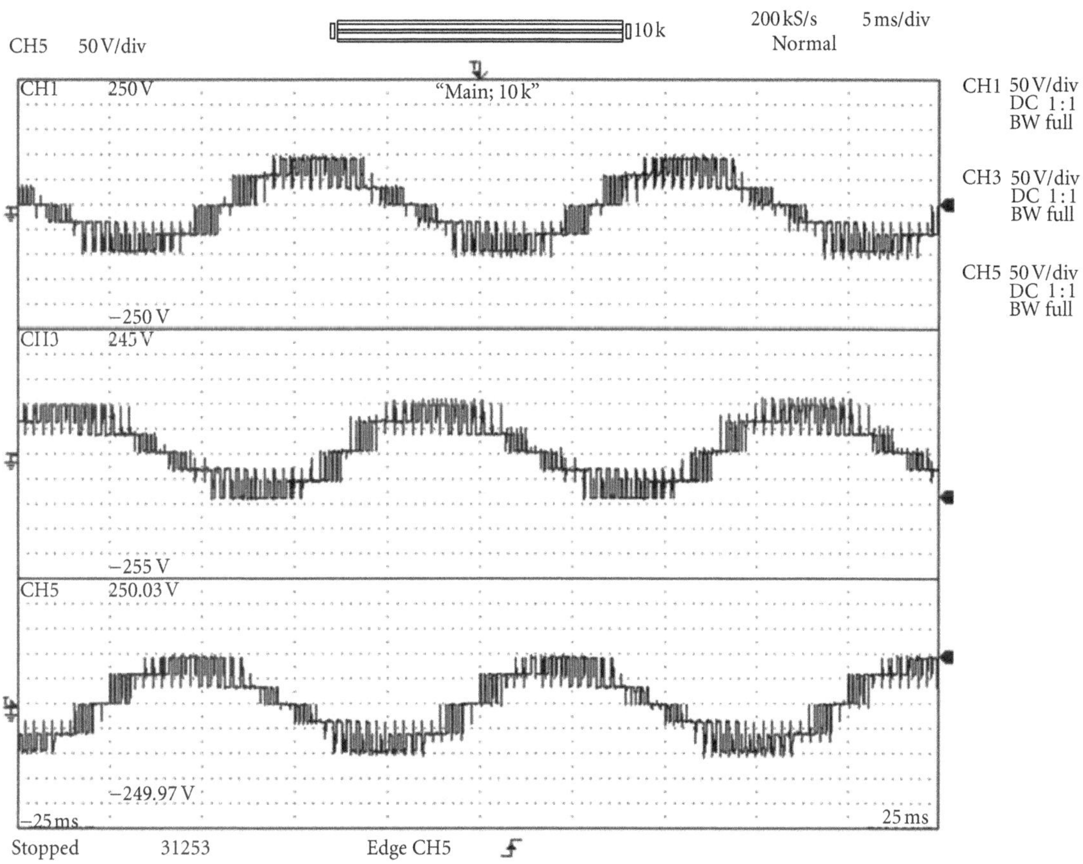

FIGURE 10: Inverter output voltages V_{ab}, V_{bc}, and V_{ca} with PD SPWM technique (x-axis: 5 ms/div, y-axis: 50 V/div).

techniques to reduce the common mode voltage in cascaded multilevel inverter [21]. The techniques used by Loh were phase disposition, phase opposition disposition, and alternate phase opposition disposition with phase shifting algorithm using space vector PWM.

Holmes and others [14–16] presented analytical expressions for different voltages using FFT analysis for cascaded and hybrid multilevel inverters. The modulation technique used was alternate phase opposition disposition and phase shifted carrier PWM for 5-level cascaded and hybrid multilevel inverters. Simple SPWM-based techniques have been proposed in the presented work to control the CMV in AC-DC-AC ASD system. The proposed SPWM-based control techniques are based on phase disposition (PD) and phase opposition disposition (POD) concepts with addition of offset voltage signal to actual reference signal.

The second section reviews the various common mode voltage control techniques and explains the proposed concept to control CMV. Simulation performance of the proposed control technique is given in section three. Experimental verification of proposed techniques is presented in section four. Section five gives conclusion.

2. Common Mode Voltage Control

Various common mode voltage reduction/elimination techniques can be classified as given below.

(A) Using some extra hardware circuitry such as

(i) isolation transformer;

(ii) zero-sequence impedance (common mode choke);

(iii) active and passive filters;

(iv) dual bridge inverter;

(v) using four-leg (four-phase) inverter.

(B) Using modification in control strategy employed such as, based on

(i) space vector PWM technique (SVPWM);

(ii) sinusoidal PWM technique (SPWM).

The proper choice of common mode choke provides high impedance to common mode current and can eliminate the CMV. Further, any practice to eliminate CMV may increase the common mode current [9, 11]. Thus, investigation is required on this issue. The main issue is that the THD and switching losses increase with controlling the CMV. During the control pulse generation in multilevel inverters, some switching states produce reduced/zero common mode voltage across motor windings and inverter output terminals.

In SVPWM techniques, the switching state voltage vectors generating zero common mode voltage are only used

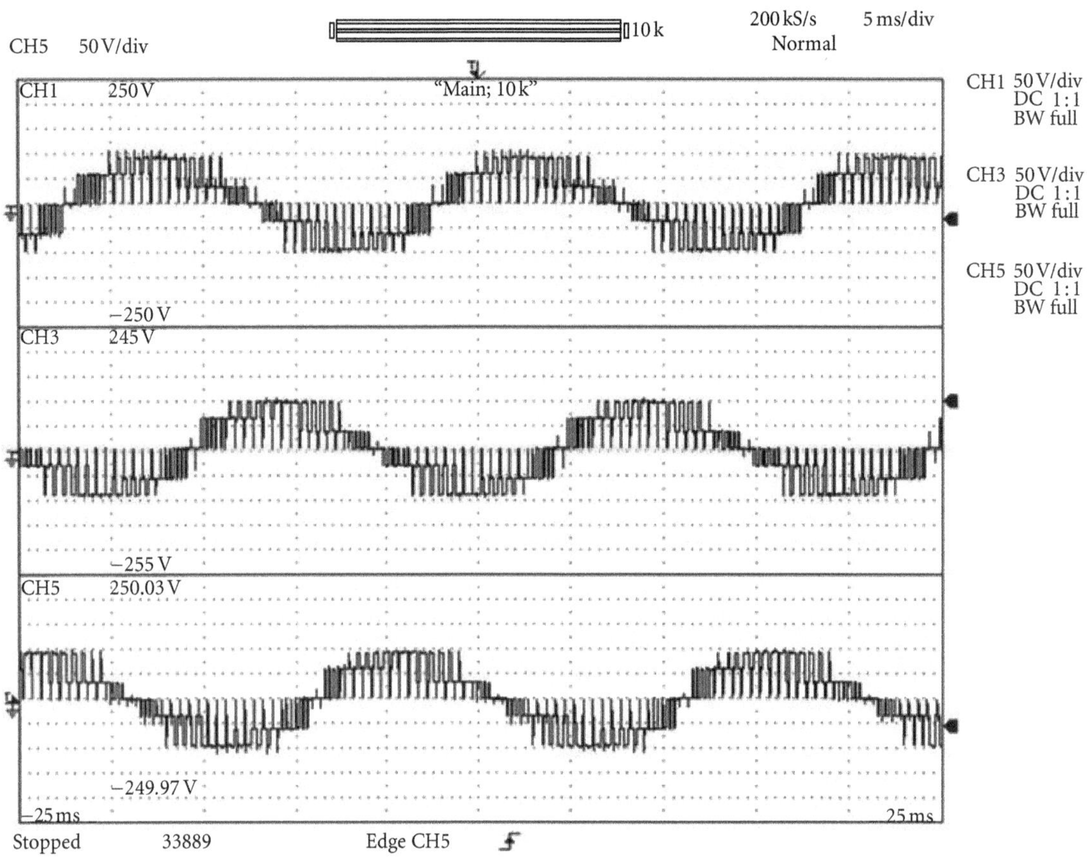

FIGURE 11: Inverter output voltages V_{ab}, V_{bc}, and V_{ca} with POD SPWM technique (x-axis: 5 ms/div, y-axis: 50 V/div).

TABLE 1: Simulation parameters.

DC link voltage	10.16 kV
Load power	2.45 MW
Load voltage	6.6 kV
Load current	256 amps
Load power factor	0.84
Carrier frequency	2.2 kHz

[7, 8]. Various forms of SVPWM techniques are available to mitigate the common mode voltage problem, that is, nearest three vector selection (NTV) SVPWM, radial state SVPWM, zero common mode voltage SVPWM, and so forth [8].

In ASD system, magnitude and rate of change of CMV play an important role in designing EMI filters. SVPWM-based techniques for this purpose require comparatively complex algorithm [7, 9–12]. On the other hand, SPWM-based techniques require less complex algorithm with ease to implement.

This section presents various simple SPWM-based techniques controlling the magnitude and rate of change of CMV. Different SPWM techniques can be classified as [14–16]

(i) phase disposition (PD) method,

(ii) phase opposition Disposition (POD) method,

(iii) phase-shifted (PS) method,

(iv) hybrid (H) method,

(v) Third harmonic injection (THIPWM) method.

Figure 2 shows the schematic diagram of three-phase, three-level diode-clamped inverter. Figure 3 shows the principle of generation of control pulses for 3-level inverter with conventional PD SPWM technique. Fundamental frequency three-phase sinusoidal reference waves V_r, V_y, and V_b are compared with two high-frequency triangular carrier waves "carrier 1" and "carrier 2". Each intersection gives rise to the control pulses for switching devices of inverter. These reference sinusoidal waves, $V(r, y, b)$ can be represented by

$$\begin{aligned} V_r &= V_m \sin(\omega t), \\ V_y &= V_m \sin(\omega t - 120^\circ), \\ V_b &= V_m \sin(\omega t - 240^\circ). \end{aligned} \tag{1}$$

The control pulses generated by comparison of reference voltage wave with "carrier 1" are given to upper switches (S_{a1}, S_{b1}, S_{c1}) of inverter shown in Figure 2. The corresponding complementary signals of these pulses are given to (S_{a1}', S_{b1}', S_{c1}'). The control pulses generated by comparison of reference voltage wave with "carrier 2" are given to switches (S_{a2}, S_{b2}, S_{c2}). The corresponding complementary signals of these pulses are given to (S_{a2}', S_{b2}', S_{c2}').

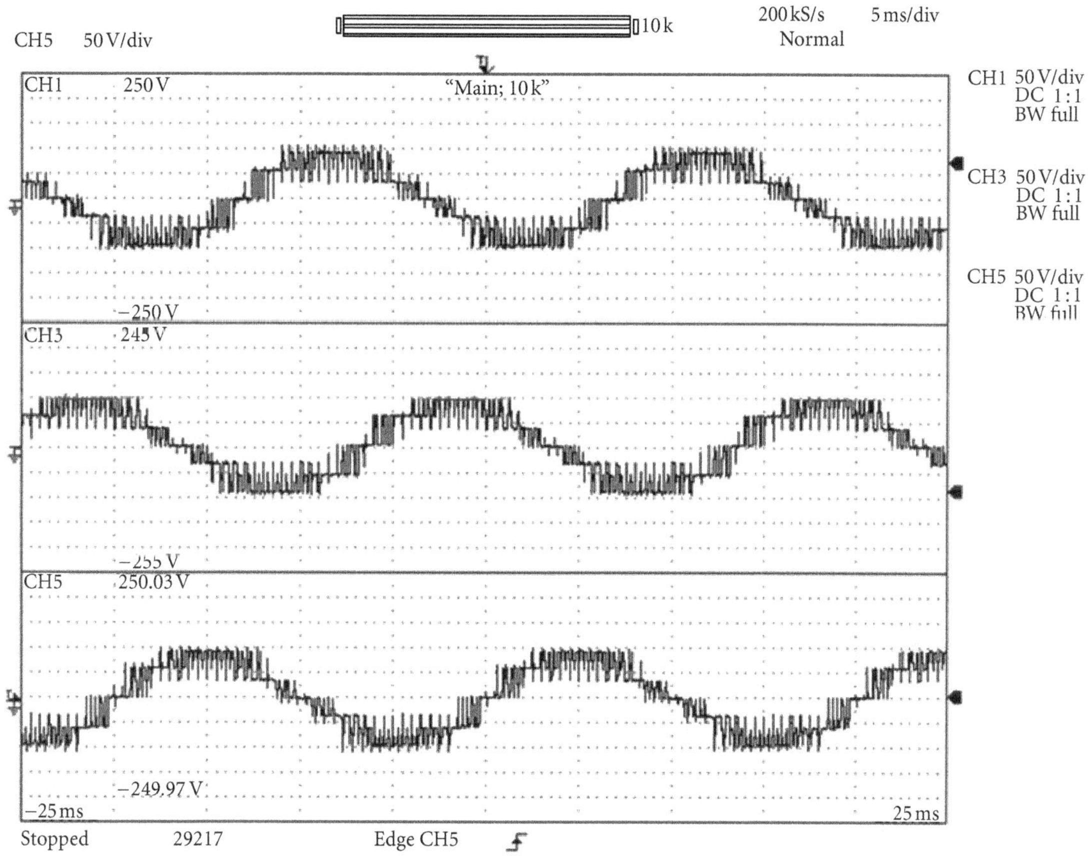

FIGURE 12: Inverter output voltages V_{ab}, V_{bc}, and V_{ca} with offset PD SPWM technique (x-axis: 5 ms/div, y-axis: 50 V/div).

2.1. Proposed Method of CMV Control. Common mode voltage magnitude and its rate of change depend on the control pulse width, number of commutation, and its frequency which can be modified by adding some offset signal to reference voltage wave. Offset voltage plays an important role in modifying the modulating signal. This section gives the basic idea of adding the offset signal to PD and POD SPWM techniques. Common mode voltage or zero sequence voltage in output voltage of inverter can be represented by

$$V_{cm} = \frac{(V_r + V_y + V_b)}{3}, \quad (2)$$

where, V_r, V_y, and V_b are the phase voltages of inverter. This voltage is around DC link voltage in conventional 2-level inverters and half of the DC link voltage in 3-level inverter. To reduce it, following common mode offset voltage is to be added

$$V_{off} = -\frac{[\min(V_r, V_y, V_b) + \max(V_r, V_y, V_b)]}{2}, \quad (3)$$

where $\min(V_r, V_y, V_b)$ and $\max(V_r, V_y, V_b)$ are the minimum and maximum functions of reference voltage wave. New reference or modulating wave can be calculated by adding this common mode offset voltage signal to original reference voltage wave as

$$V^*(r, y, b) = V(r, y, b) + V_{off}, \quad (4)$$

where, $V(r, y, b)$ is given by (1). The "max", "min", one-phase voltage V_r and offset voltage signal, according to (1) to (4), are shown in Figure 4 for PD SPWM technique. This common mode voltage offset signal controls the CMV in inverter output. Figures 5 and 6 give the new 3-phase reference waves as obtained from (4) in PD and POD SPWM techniques. In POD SPWM technique, the two carrier signals are in phase opposition as shown in Figure 6.

3. Simulation Results

The proposed CMV control techniques are now verified in simulation. The simulation parameters for common mode control are given in Table 1 for medium voltage, 3-phase, squirrel cage induction motor.

The CMV in 2-level inverter is compared with CMV in 3-level SPWM techniques using the proposed offset voltage SPWM techniques. Figure 7 shows the common mode voltages in 2-level and different 3-level SPWM techniques indicating the effect of adding the CMV offset signal. Figures 8 and 9 show the inverter output voltage and its harmonic spectrum with PD SPWM, Offset PD SPWM, POD SPWM, and offset POD SPWM.

The magnitude of CMV is reduced as we move from 2-level to 3-level SPWM technique or from PD to POD SPWM technique as shown. The frequency of CMV is three times the fundamental frequency of inverter output voltage.

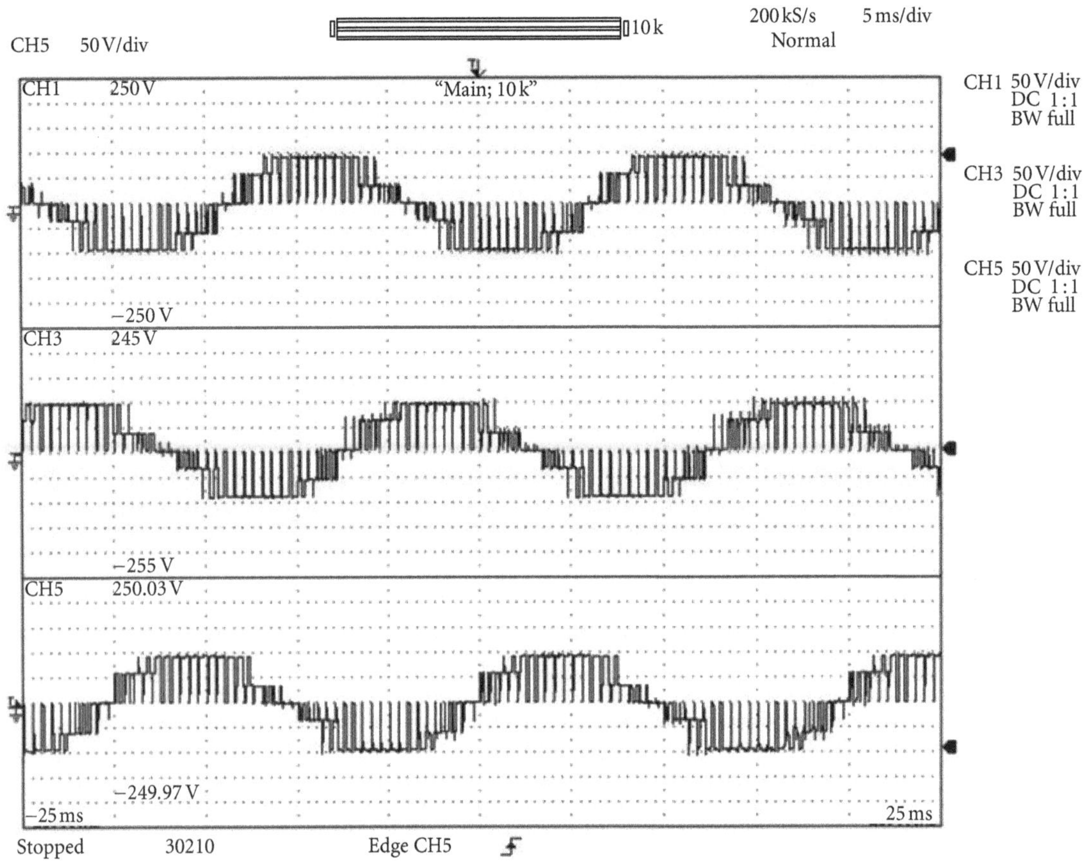

Figure 13: Inverter output voltages V_{ab}, V_{bc}, and V_{ca} with offset POD SPWM technique (x-axis: 5 ms/div, y-axis: 50 V/div).

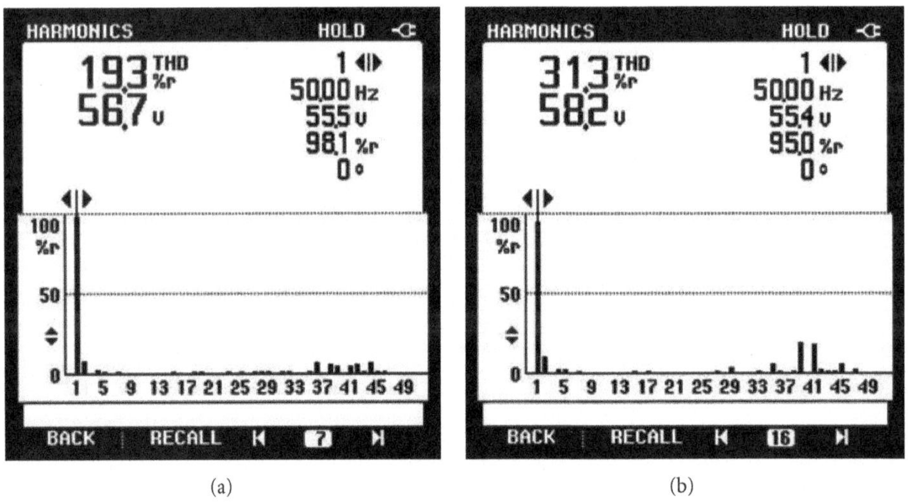

Figure 14: Frequency spectrums of inverter output line voltage (V_{ab}) with (a) PD SPWM and (b) POD SPWM techniques.

POD with offset signal gives better CMV behavior but at the cost of increased voltage THD. While PD SPWM with offset signal gives much better voltage THD performance (7.81% without any filter) with reduced switching transients in CMV as compared to PD SPWM and other techniques.

The magnitude, ripples, and number of commutations in CMV at inverter and load terminals in POD with offset signal are greatly reduced. Comparison of inverter voltage THD for these techniques is given in Figures 8 and 9, respectively.

The switching frequency is increased in PD with offset signal technique which can be controlled by the method

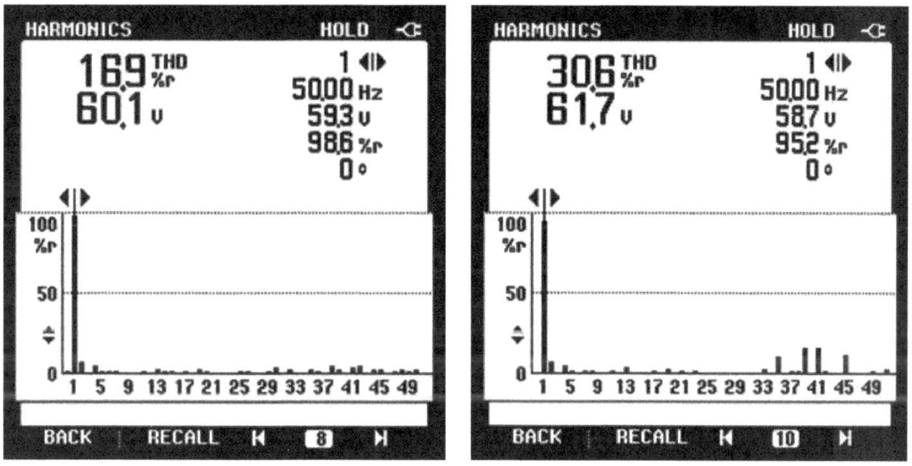

(a) (b)

FIGURE 15: Frequency spectrums of inverter output line voltage (V_{ab}) with (a) offset PD SPWM and (b) offset POD SPWM techniques.

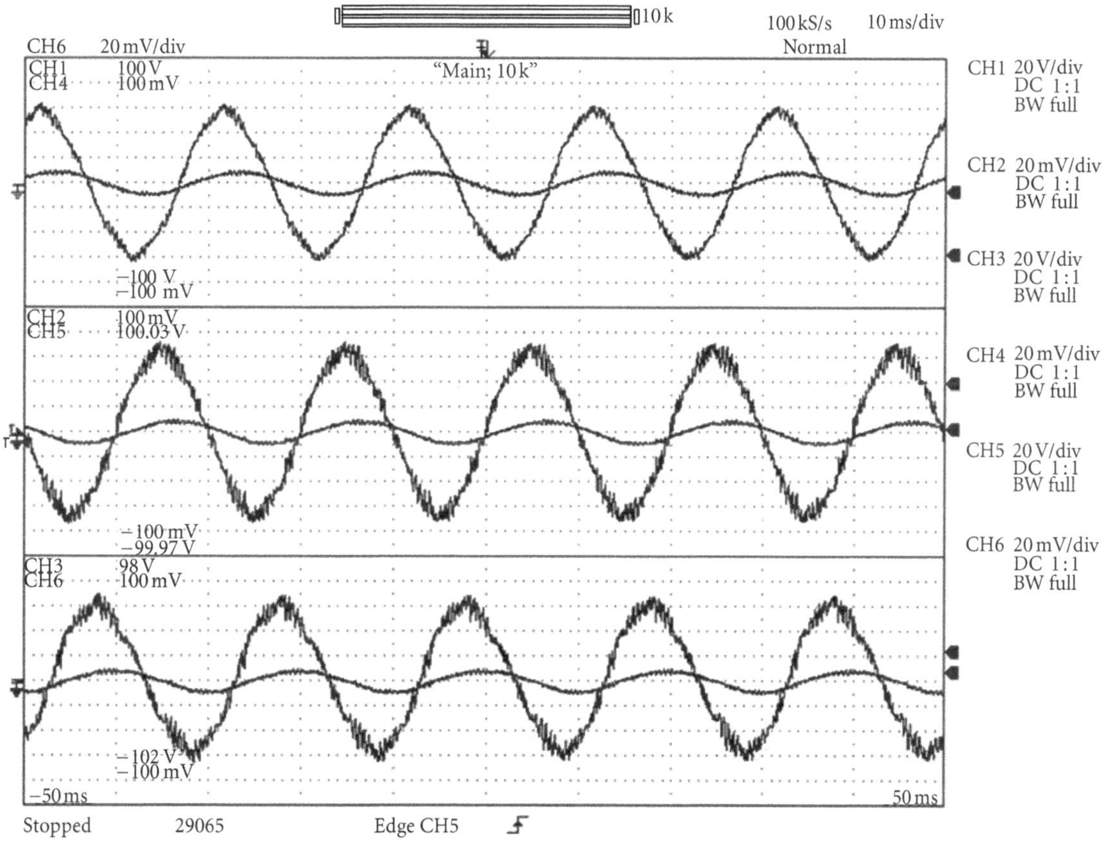

FIGURE 16: Load voltages and load currents PD SPWM technique (x-axis: 10 ms/div, y-axis: voltages-20 V/div, currents-5 A/div).

proposed in [17]. However, the change in line voltage across switching devices is two times in POD and offset POD method as compared to PD and offset PD method as given in Figure 9. It compensates any increase in switching frequency because switching losses depend on instantaneous voltage and current across and through the switch, respectively. Table 2 provides comprehensive comparison of these techniques based on inverter voltage/current THD (without/with passive filter), average CMV, and average neutral point potential (NPP).

4. Experimental Results

A laboratory prototype of three-phase, diode-clamped or neutral-point-clamped inverter has been developed for experimentation. The experimental parameters are listed in

Table 2: Comparison of CMV control techniques.

SPWM technique	Without filter				With filter ($L = 1.2$ mH, $C = 100$ uF)				Average NPP (Vnp) volts	Average CMV (Vcm) volts
	Line voltage (kV)		Line current (amps)		Line voltage (kV)		Line current (amps)			
	V_{1ab}	% THD	i_{1a}	% THD	V_{1ab}	% THD	i_{1a}	% THD		
PD	8.1	20.51	254	28.39	15.6	3.13	256	0.86	8.3	59.2
POD	8.1	25.97	250	32.57	15.6	2.92	254	0.78	16.1	69.1
PD + off-set voltage	8.0	**7.81**	254	**18.16**	15.6	**3.24**	258	**0.85**	**1.4**	**25.3**
POD + off-set voltage	8.1	35.63	257	43.43	15.6	3.39	259	0.77	8.3	59.6

Table 3: Experimental parameters.

DC link voltage	100 volts
DC link capacitor	2200 μF (each)
Load resistance	45 Ω
Load inductance	5.2 mH
Carrier frequency	2 kHz

Table 3. The dSPACE DS1104 has been used for real-time control of the inverter.

The inverter line voltages are shown in Figures 10, 11, 12 and 13 for PD, POD, offset PD, and offset POD techniques, respectively. It is clearly shown that the frequency of switching per half cycle and the rate of change of switch voltage is different in each case. Slight unbalance in line voltage is due to no load operation. The DC link voltage balancing has been given in [17, 18] which not only balances the DC link voltages but also reduces the switching losses.

Figure 14 shows the frequency spectrums of inverter output line voltage (V_{ab}) with (a) PD SPWM, and (b) POD SPWM techniques, respectively. Figure 15 shows the frequency spectrums of inverter output line voltage (V_{ab}) with (a) offset PD SPWM, and (b) offset POD SPWM techniques, respectively. It is clearly shown that minimum THD occurs in offset PD SPWM technique and maximum fundamental load voltage is available in offset POD SPWM technique. Second-order harmonics is present in each case because of slight unbalance in DC link voltages which can be easily balanced by using the technique given in [18]. Figure 16 shows the waveforms of load voltages and load currents with PD SPWM techniques. Figure 17 shows the frequency spectrums of load voltage (V_{ab}) and load current (i_a) with PD SPWM technique. These experimental results are comparable with the simulation results presented in Figures 8 and 9.

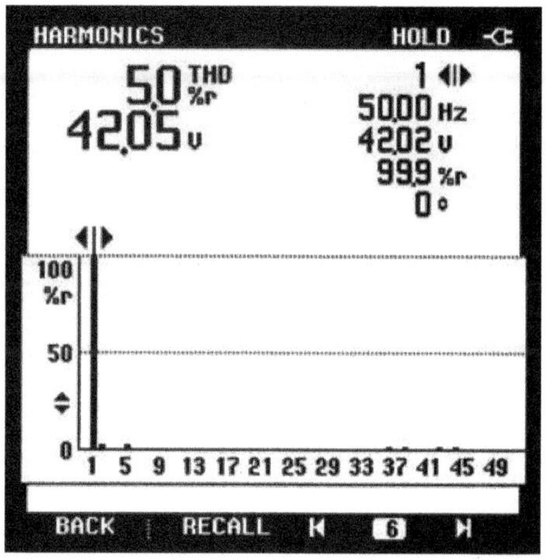

(a)

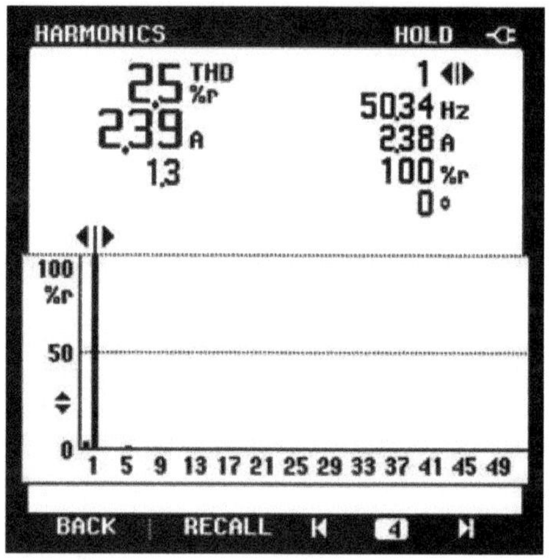

(b)

Figure 17: Frequency spectrums of load voltage (V_{ab}) and load current (i_a) with PD SPWM technique.

5. Conclusion

This paper explored the opportunities of common mode voltage (CMV) control using simple sine-triangle comparison techniques for 3-level diode-clamped inverter. CMV at inverter terminals has been obtained and compared for 2-level, 3-level PD SPWM, and 3-level POD SPWM techniques. Further, a new method for reduction of CMV based on SPWM technique has been derived based on the addition of variable CMV offset signal to the original modulating wave.

Simulation and experimental results show the effectiveness of the proposed techniques giving improved performance with the reduction of magnitude and number of commutations per cycle in CMV. It shows that offset PD SPWM gives drastically reduced voltage THD (7.81%) without using any filter. These values are much lower as compared to their counterpart 2-level (52.48%), 3-level PD SPWM (20.51%), and POD SPWM (25.97%). Average value of neutral point potential (NPP) and CMV are minimum in offset PD SPWM technique. The sharp edges in CMV with conventional SPWM technique are also reduced with the proposed method. Experimental results are comparable with the simulated results. The problem of slight unbalancing in DC link voltages as evident from experimental results can be solved by using the technique presented in [17, 18].

References

[1] J. Pontt, J. Rodríguez, and R. Huerta, "Mitigation of noneliminated harmonics of SHEPWM three-level multipulse three-phase active front end converters with low switching frequency for meeting standard IEEE-519-92," *IEEE Transactions on Power Electronics*, vol. 19, no. 6, pp. 1594–1600, 2004.

[2] D. A. Rendusara, E. Cengelci, P. N. Enjeti, V. R. Stefanovic, and J. W. Gray, "Analysis of common mode voltage-"Neutral shift" in medium voltage PWM adjustable speed drive (MV-ASD) systems," *IEEE Transactions on Power Electronics*, vol. 15, no. 6, pp. 1124–1133, 2000.

[3] J. Rodríguez, J. Pontt, P. Correa, P. Cortés, and C. Silva, "A new modulation method to reduce common-mode voltages in multilevel inverters," *IEEE Transactions on Industrial Electronics*, vol. 51, no. 4, pp. 834–839, 2004.

[4] J. Rodríguez, J. Pontt, R. Huerta, and P. Newman, "24-pulse active front end rectifier with low switching frequency," in *Proceedings of the 35th IEEE Annual Power Electronics Specialists Conference (PESC '04)*, pp. 3517–3523, June 2004.

[5] B. P. Schmitt and R. Sommer, "Retrofit of fixed speed induction motors with medium voltage drive converters using NPC three-level inverter high-voltage IGBT based topology," in *Proceedings of the IEEE International Symposium on Industrial Electronics Proceedings (ISIE '01)*, pp. 746–751, June 2001.

[6] F. Wang, "Motor shaft voltages and bearing currents and their reduction in multilevel medium-voltage PWM voltage-source-inverter drive applications," *IEEE Transactions on Industry Applications*, vol. 36, no. 5, pp. 1336–1341, 2000.

[7] A. Videt, P. Le Moigne, N. Idir, P. Baudesson, and X. Cimetière, "A new carrier-based PWM providing common-mode-current reduction and DC-bus balancing for three-level inverters," *IEEE Transactions on Industrial Electronics*, vol. 54, no. 6, pp. 3001–3011, 2007.

[8] F. Wang, "Sine-triangle versus space-vector modulation for three-level PWM voltage-source inverters," *IEEE Transactions on Industry Applications*, vol. 38, no. 2, pp. 500–506, 2002.

[9] A. K. Gupta and A. M. Khambadkone, "A space vector modulation scheme to reduce common mode voltage for cascaded multilevel inverters," *IEEE Transactions on Power Electronics*, vol. 22, no. 5, pp. 1672–1681, 2007.

[10] R. S. Kanchan, P. N. Tekwani, and K. Gopakumar, "Three-level inverter scheme with common mode voltage elimination and dc link capacitor voltage balancing for an open-end winding induction motor drive," *IEEE Transactions on Power Electronics*, vol. 21, no. 6, pp. 1676–1683, 2006.

[11] P. C. Loh, D. G. Holmes, Y. Fukuta, and T. A. Lipo, "A reduced common mode hysteresis current regulation strategy for multilevel inverters," *IEEE Transactions on Power Electronics*, vol. 19, no. 1, pp. 192–200, 2004.

[12] S. Wei, N. Zargari, B. Wu, and S. Rizzo, "Comparison and mitigation of common mode voltage in power converter topologies," in *Proceedings of the 39th IEEE Industry Applications Conference (IAS '04)*, pp. 1852–1857, October 2004.

[13] H. Akagi and T. Hatada, "Voltage balancing control for a three-level diode-clamped converter in a medium-voltage transformerless hybrid active filter," *IEEE Transactions on Power Electronics*, vol. 24, no. 3, pp. 571–579, 2009.

[14] B. P. Mcgrath and D. G. Holmes, "A comparison of multicarrier PWM strategies for cascaded and neutral point clamped multilevel inverters," in *Proceedings of the IEEE Power Electronics Specialists Conference (PESC '00)*, vol. 2, pp. 647–679, November 2000.

[15] B. P. McGrath and D. G. Holmes, "Multicarrier PWM strategies for multilevel inverters," *IEEE Transactions on Industrial Electronics*, vol. 49, no. 4, pp. 858–867, 2002.

[16] B. P. McGrath, D. G. Holmes, and T. Lipo, "Optimized space vector switching sequences for multilevel inverters," *IEEE Transactions on Power Electronics*, vol. 18, no. 6, pp. 1293–1301, 2003.

[17] P. K. Chaturvedi, S. Jain, and P. Agarwal, "A simple carrier based neutral point potential regulator for 3-level diode clamped inverter," *International Journal of Power Electronics*, vol. 3, no. 1, pp. 1–25, 2011.

[18] P. K. Chaturvedi, S. Jain, and P. Agarwal, "Reduced switching loss pulse width modulation technique for three-level diode clamped inverter," *IET Power Electronics*, vol. 4, no. 4, pp. 393–399, 2011.

[19] M. E. Adabi and A. Vahedi, "Common mode voltage reduction with a modified hysteresis current control strategy," in *Proceedings of the 2nd Power Electronics, Drive Systems and Technologies Conference (PEDSTC '11)*, pp. 415–420, February 2011.

[20] Z. Zhao, Y. Zhong, H. Gao, L. Yuan, and T. Lu, "Hybrid selective harmonic elimination PWM for common mode voltage reduction in three-level neutral point clamped inverters for variable speed induction drives," *IEEE Transactions on Power Electronics*, vol. 27, no. 3, pp. 1152–1158, 2012.

[21] P. C. Loh, D. G. Holmes, Y. Fukuta, and T. A. Lipo, "Reduced common-mode modulation strategies for cascaded multilevel inverters," *IEEE Transactions on Industry Applications*, vol. 39, no. 5, pp. 1386–1395, 2003.

10

Comparative Studies of Different Control Strategies of a Dynamic Voltage Restorer Based on Matrix Converter

Amin Shabanpour and Ali Reza Seifi

School of Electrical & Computer Engineering, University of Shiraz, Shiraz 7134851151, Iran

Correspondence should be addressed to Ali Reza Seifi, seifi@shirazu.ac.ir

Academic Editor: Pavol Bauer

A dynamic voltage restorer (DVR) with no energy storage is studied. By using a matrix converter instead of the conventional AC/DC/AC converters, elimination of the DC-link capacitor is possible. The switching algorithm of matrix converter is the space vector modulation. There are different compensation algorithms to control the conventional DVR. These methods have been analyzed in this paper for the proposed matrix-converter-based DVR. A deep analysis through different diagrams would show the advantages or disadvantages of each compensation method. Equations for all methods are derived, and the characteristics of algorithms are compared with each other.

1. Introduction

Advanced industrials usually use electronic devices such as programmable logic controllers or power electronic drives in their devices. Power quality problems may affect the function of these devices. They could malfunction or trip through voltage disturbances. One of the most important power quality problems is voltage sag which is considered to be the most occurring issue in the power systems [1]. This phenomenon is specified as a sudden decrease in the voltage from 90% to 10% of rated voltage which lasts for half cycle to one minute. The main source of voltage sags are system faults, switching of heavy loads, or starting of large motors [2].

One of the best solutions to improve power quality is the dynamic voltage restorer (DVR). DVR is a kind of custom power devices that can inject active/reactive power to the grids. This can protect critical loads from disturbances such as sags or swells.

Four DVR topologies are compared in [3]. In this paper, the DVR with no energy storage devices and load-side connected shunt converter is selected. By using a matrix converter instead of the AC/DC/AC converters in that topology, the DC-link and its bulky capacitor would be eliminated.

Generally a matrix converter is a direct AC/AC converter. A three-phase matrix converter has nine bidirectional switches that are able to connect any output phase of the system to any input phase at any time [4]. Some advantages of the matrix converter are bidirectional power flow, controllable sinusoidal output voltage, ability to control the input power factor, and a compact structure. However the complexity of its control is its main drawback [5].

Different DVR topologies based on matrix converters are presented before. A flywheel-supported DVR controlled by the matrix converter is proposed in [6]. Another topology is studied in [7] that the behavior of a DVR with matrix converter utilizing Alesina-Venturini modulation method is shown. In [8], a matrix converter-based DVR is analyzed which utilizes the space-vector modulation.

There are three compensation strategies for conventional DVR in literatures [9–11]. Comparisons between them are done in [12] for a general type of DVR but there is less study on compensation strategies for a matrix-converter-based DVR that is the aim of this paper. All compensation methods includes prefault, in-phase, and energy-optimized compensation algorithm are analyzed and compared in this paper for the matrix-converter-based DVR. Equations for each method are derived. Furthermore some parameters to characterize the behavior of each method for this special kind of DVR are defined. Finally, comparisons are done in various

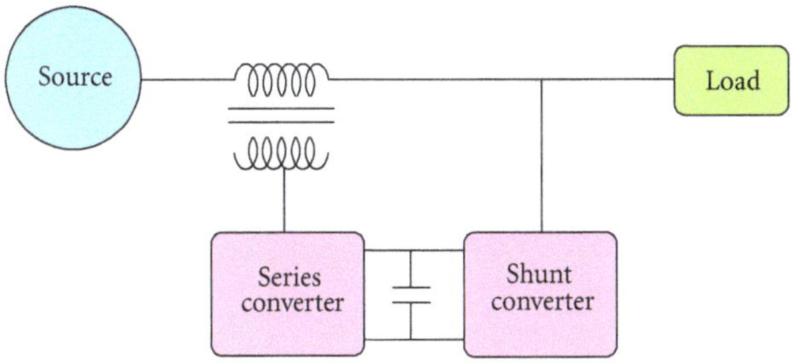

FIGURE 1: A DVR with no energy storage system and load-side-connected converter.

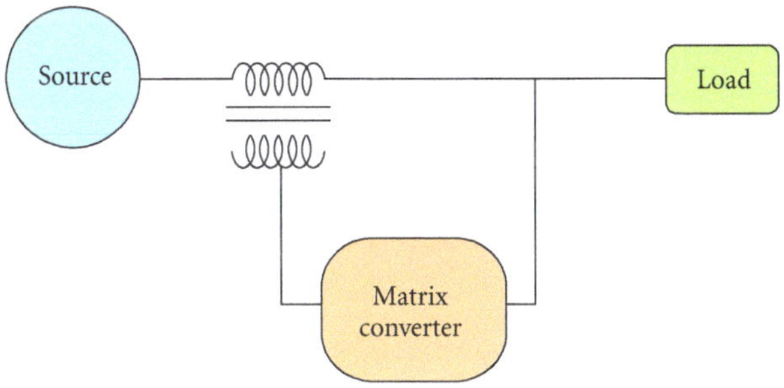

FIGURE 2: A DVR based on matrix converter.

aspects to show the characteristics of each compensating algorithm. Moreover, there are some studies on the DVR performance limitation in this topology.

In this paper, the structure of a DVR with matrix converter is introduced in Section 2. Different Compensation strategies in DVR are studied in Section 3. Comparison between the compensation methods with different aspects are done in Section 4, followed by the limitation of this DVR topology in Section 5. Finally, Section 6 concludes the discussion.

2. DVR Topology

DVR controls the load voltage by injecting an appropriate voltage during disturbances. It restores the load voltage to its rated value so that sensitive loads could operate in their normal conditions during disturbance. Injecting voltage may need active or reactive power and DVR should be able to provide that. Generally there are two types of DVR: the one with no energy storage system, and the one with energy storage system. Each type is divided to two categories: a supply-side-connected converter, and a load-side-connected converter. All of these topologies are compared in [3]. Results show that the no energy storage DVR topology with load-side-connected converter has better advantages over other topologies. This kind of structure for DVR is shown in Figure 1.

By replacing the two converters of this DVR with a matrix converter, the DC-link is removed. This topology is shown in Figure 2. The structure of a three phase to three-phase matrix converter is shown in Figure 3.

With appropriate switching algorithm, matrix converter is able to provide the required output voltage from its input

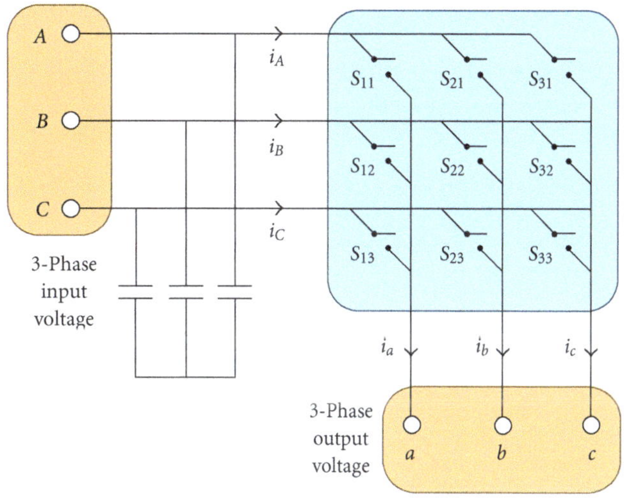

FIGURE 3: Matrix converter structure.

terminals. There are various modulation techniques to control matrix converter such as Alesina-Venturini modulation, Optimum Alesian-Venturini modulation, and Space Vector modulations (SVM). The last method is evaluated as the best one which is studied in [5].

3. Compensation Strategies of DVR

Load compensation could be done through active or reactive power injection by DVR. There are three compensation techniques based on using active, reactive, or both power by DVR. These methods are recognized as prefault compensation method, in-phase compensation method, and energy-optimized compensation method. These strategies are introduced in this section followed by derivation of their respective equations. Some simulations would show the characteristics of each method individually.

3.1. Prefault Compensation Method. This method is mainly used for loads that are sensitive to phase jump such as thyristor controlled drives. In other words, both magnitude and phase of the load voltage are compensated in this strategy such that the load voltage during disturbances is as same as the one before disturbances. Figure 4 shows this kind of compensation in per-unit plane.

In Figure 4, $I_{\text{load}}^{\text{Prefault}} = 1\angle 0$ is the load current before disturbance that is considered as the reference vector. The load is selected as a general type so the load voltage could be defined as $V_{\text{load}}^{\text{Prefault}} = 1\angle\varphi$. The disturbance might change the source voltage magnitude from 1 to k with δ degree phase jump. Therefore the source voltage after disturbance can be defined as $V_s^{\text{Postfault}} = k\angle(\varphi+\delta)$. DVR should inject a voltage with appropriate magnitude and phase to compensate this disturbance. The DVR injected voltage is defined as $V_{\text{DVR}} = x\angle\beta$. Also the load voltage and current after compensation are the same as their values before disturbance which means $V_{\text{load}}^{\text{Postfault}} = V_{\text{load}}^{\text{Prefault}}$ and $I_{\text{load}}^{\text{Postfault}} = I_{\text{load}}^{\text{Prefault}}$. Moreover the resultant phase jump between load voltage before and after disturbance, γ, is zero in this compensating method. By these

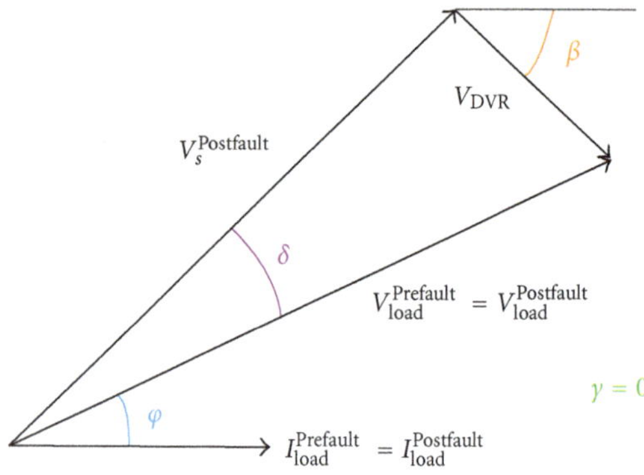

FIGURE 4: Prefault compensation strategy.

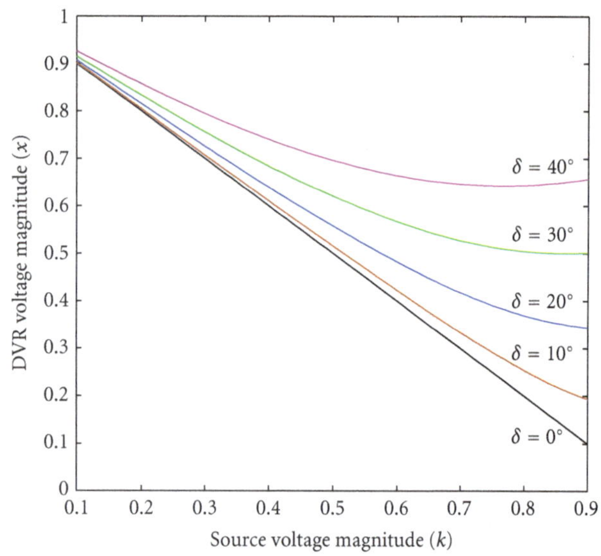

FIGURE 5: Prefault compensation strategy: the magnitude of the injected voltage versus the magnitude of the source voltage.

definitions, the DVR-injected voltage can be calculated as follows:

$$\begin{aligned} V_{\text{DVR}} &= V_{\text{load}}^{\text{Postfault}} - V_s^{\text{Postfault}} \Longrightarrow \\ V_{\text{DVR}} &= 1\measuredangle\varphi - k\measuredangle(\varphi+\delta) \Longrightarrow \\ V_{\text{DVR}} &= [\cos(\varphi) - k\cos(\varphi+\delta)] \\ &\quad + j[\sin(\varphi) - k\sin(\varphi+\delta)] \Longrightarrow \end{aligned} \tag{1}$$

$$x = \sqrt{1 + k^2 - 2k\cos(\delta)}, \tag{2}$$

$$\beta = \tan^{-1}\left(\frac{\sin(\varphi) - k\sin(\varphi+\delta)}{\cos(\varphi) - k\cos(\varphi+\delta)}\right). \tag{3}$$

Equation (2) shows that the magnitude of the injected voltage only depends on the degree of distortion in the input voltage waveform. Figure 5 shows this parameter versus the magnitude of the input voltage for different values of δ.

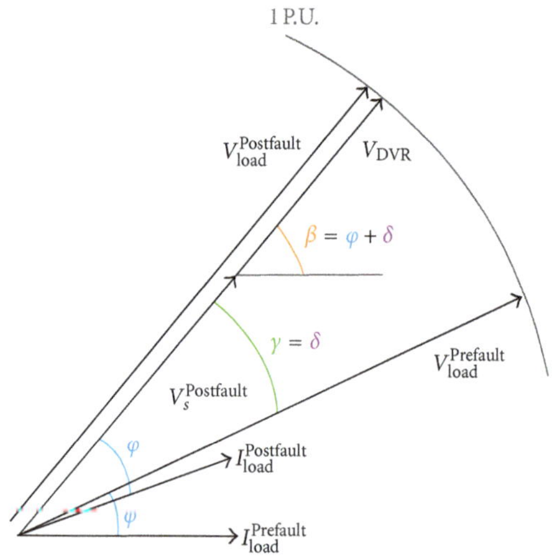

FIGURE 6: In-phase compensation strategy.

3.2. In-Phase Compensation Method. In this method, DVR only compensates the load voltage magnitude so the phase jump would be remained uncompensated. Figure 6 shows this control strategy. It is obvious that there is a phase difference between the load voltages before and after compensation. By using the same definitions in the previous section, the DVR-injected voltage is as follows:

$$V_{\text{DVR}} = 1\measuredangle(\varphi+\delta) - k\measuredangle(\varphi+\delta) = (1-k)\measuredangle(\varphi+\delta) \Longrightarrow \tag{4}$$

$$x = 1 - k, \tag{5}$$

$$\beta = \varphi + \delta. \tag{6}$$

Therefore, the load voltage after compensation can be written as $V_{\text{load}}^{\text{Postfault}} = 1\measuredangle(\varphi+\delta)$, so that the load current would be $I_{\text{load}}^{\text{Postfault}} = 1\measuredangle\delta$. Moreover, there is a phase jump equal to $\gamma = \delta$ in the load voltage.

The magnitude of the injected voltage for any disturbance's phase jump and any load power factor is shown in Figure 7 that satisfies (5).

3.3. Energy-Optimized Compensation Method. It should be noticed that both previous strategies need a storage system to operate. However in energy-optimized method, the capacity of this storage system is minimized. Furthermore in shallow sags, the compensation could be done only with reactive power, and the amount of active power injection is zero. This strategy is based on injecting voltage in such a way that the injected voltage vector would be at almost 90° to the resultant load current vector. This will minimize the injection of active power and even it may be zero. Figure 8 shows the basics of this strategy.

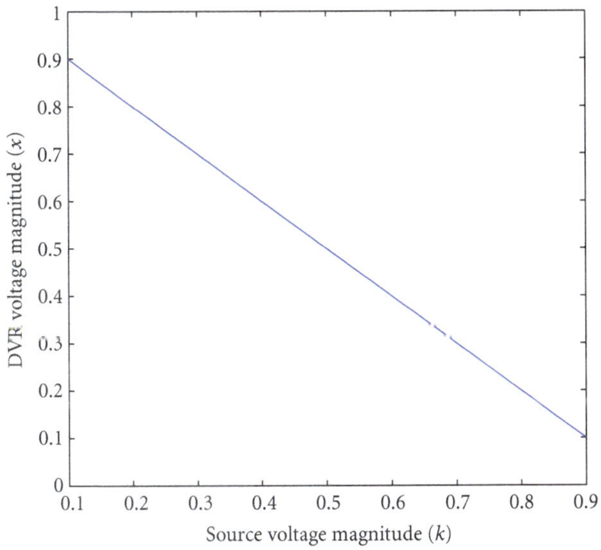

FIGURE 7: In-phase compensation strategy: the magnitude of the injected voltage versus the magnitude of the source voltage.

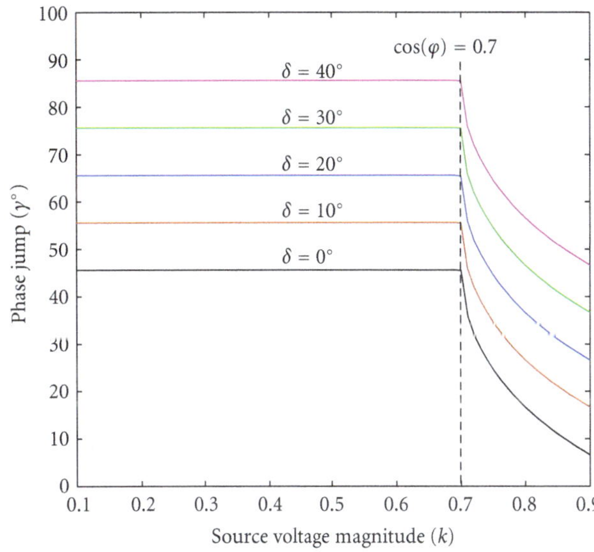

FIGURE 9: Energy-optimized compensation strategy: the phase jump of the load voltage versus the magnitude of the source voltage for PF = 0.7.

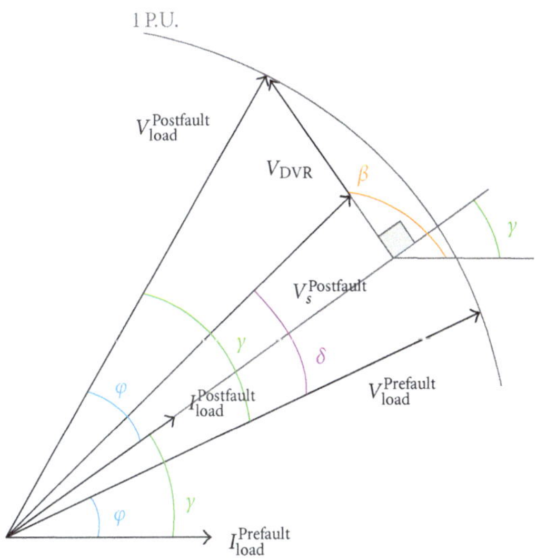

FIGURE 8: Energy-optimized compensation strategy.

With the similar definitions like previous sections for voltage and currents, the injected voltage can be calculated as follows:

$$\begin{aligned} V_{DVR} &= V_{load}^{Postfault} - V_{s}^{Postfault} \Longrightarrow \\ V_{DVR} &= 1\measuredangle(\varphi+\gamma) - k\measuredangle(\varphi+\delta) \Longrightarrow \\ V_{DVR} &= [\cos(\varphi+\gamma) - k\cos(\varphi+\delta)] \\ &\quad + j[\sin(\varphi+\gamma) - k\sin(\varphi+\delta)] \Longrightarrow \end{aligned} \tag{7}$$

$$x = \sqrt{1 + k^2 - 2k\cos(\delta-\gamma)}, \tag{8}$$

$$\beta = \tan^{-1}\left(\frac{\sin(\varphi+\gamma) - k\sin(\varphi+\delta)}{\cos(\varphi+\gamma) - k\cos(\varphi+\delta)}\right). \tag{9}$$

According to Figure 8 the injection voltage phase angle is $\beta = \pi/2 + \gamma$. By replacing this value in (9), the load phase jump can be calculated after some manipulations:

$$\begin{aligned} \tan^{-1}\left(\frac{\sin(\varphi+\gamma) - k\sin(\varphi+\delta)}{\cos(\varphi+\gamma) - k\cos(\varphi+\delta)}\right) &= \frac{\pi}{2} + \gamma \Longrightarrow \\ \frac{\sin(\varphi+\gamma) - k\sin(\varphi+\delta)}{\cos(\varphi+\gamma) - k\cos(\varphi+\delta)} &= -\cot\gamma \Longrightarrow \end{aligned} \tag{10}$$

$$\gamma = \varphi + \delta - \cos^{-1}\left(\frac{1}{k}\cos(\varphi)\right). \tag{11}$$

To ensure that (11) is feasible, $(1/k)\cos(\varphi) \le 1$ has to be established. So the source amplitude must be more than the load power factor or $k \ge \cos(\varphi)$. If this situation does not yield, γ should be calculated through (12):

$$\gamma = \varphi + \delta. \tag{12}$$

Figures 9 and 10 show the load phase jump and the injected voltage magnitude for different values of δ. The load power factor is chosen as 0.7.

In Figures 11 and 12, $\delta = 30°$ is constant and the load phase jump and the injected voltage magnitude is plotted for different load power factor. There are some notices in these figures. It should be noted that the injected voltage magnitude is independent of δ. Moreover the break-points in these figures are relevant to the load power factor. In other words, the load can be compensated only through the injection of reactive power for $k \ge \cos(\varphi)$; otherwise there would be an active power injection too.

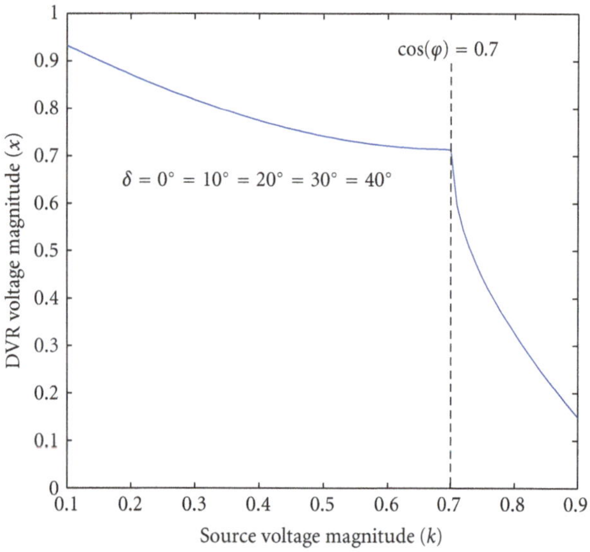

FIGURE 10: Energy-optimized compensation strategy: the magnitude of the injected voltage versus the magnitude of the source voltage for PF = 0.7.

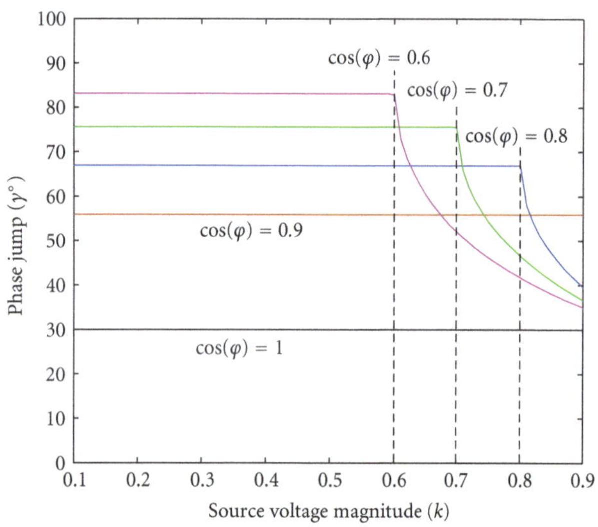

FIGURE 11: Energy-optimized compensation strategy: the phase jump of the load voltage versus the magnitude of the source voltage for $\delta = 30°$.

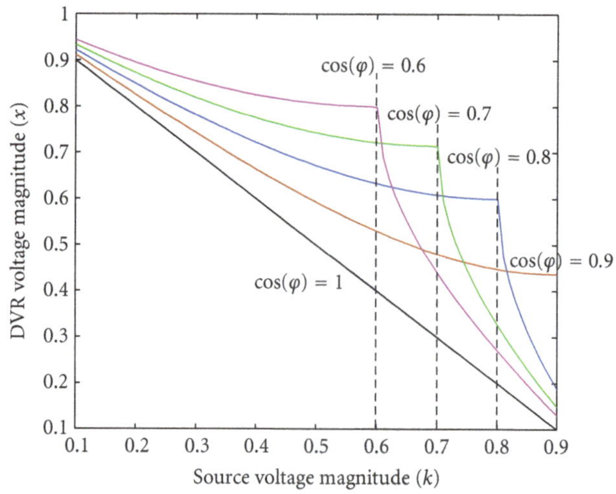

FIGURE 12: Energy-optimized compensation strategy: the magnitude of the injected voltage versus the magnitude of the source voltage for $\delta = 30°$.

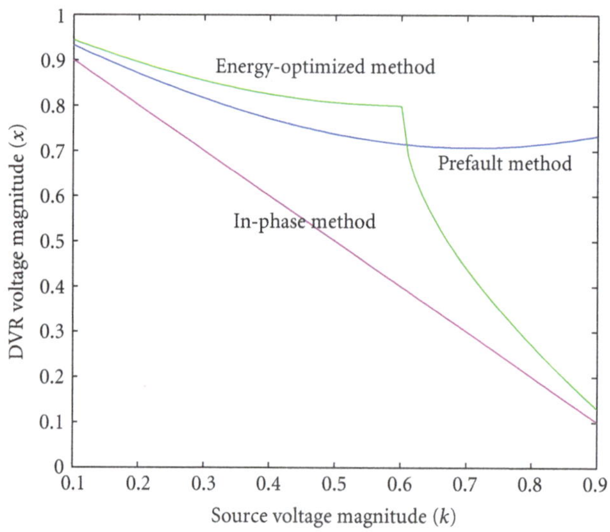

FIGURE 13: Comparison between different control strategies: the magnitude of the injected voltage versus the magnitude of the source voltage.

4. Comparison between Compensation Strategies

In this section, all compensation strategies are compared with each other. There are different criteria for comparison. Suppose that a disturbance with $\delta = 45°$ occurs and the load power factor is $\cos(\varphi) = 0.6$.

4.1. Injection Voltage Magnitude. In Figure 13 the magnitude of the injected voltage versus the magnitude of the source voltage is shown for all compensation methods. It could be observed that the in-phase method has the least injection voltage magnitude. Furthermore for shallow disturbances, energy optimized method has less injection voltage magnitude than the prefault method. For large disturbances, the prefault compensation strategy has the most magnitude of the DVR injected voltage. This means that in this compensation method, DVR components should have a higher voltage rating compared with the other methods.

4.2. Load Voltage Phase Jump. Figure 14 shows the phase jump of the load voltage for the source voltage magnitude between 0.1 to 0.9 per unit. This figure shows that the energy optimized method has the most load phase jump angle followed by the in-phase method. This parameter can be a criterion of the load voltage distortion especially at the start and the end of the disturbance occurring time. The prefault method compensates the load voltage to its prefault value so it makes no phase jump in the load voltage.

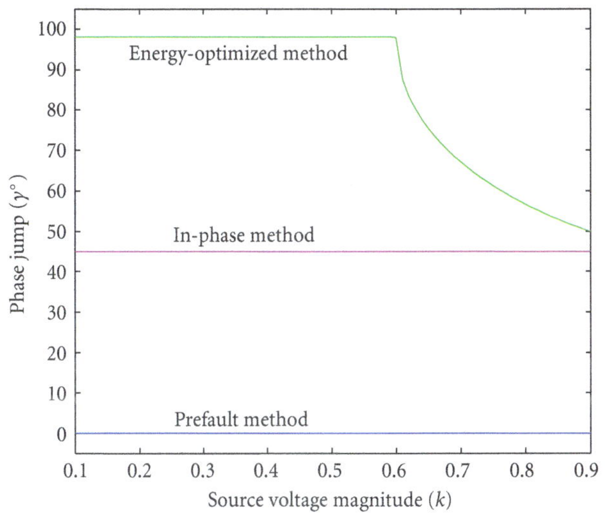

FIGURE 14: Comparison between different control strategies: the phase jump of the load voltage versus the magnitude of the source voltage.

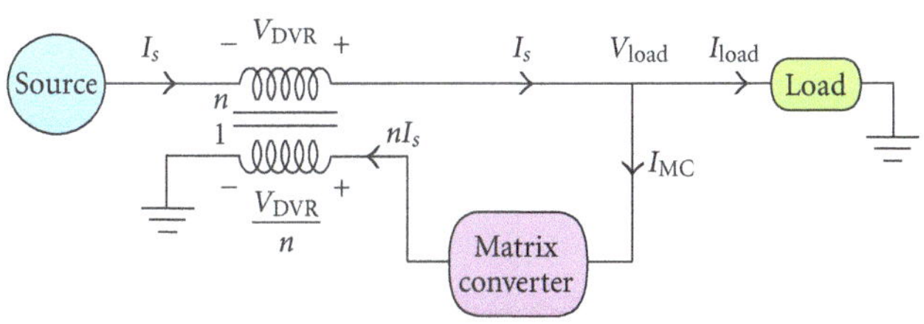

FIGURE 15: Schematics of a DVR based on matrix converter.

4.3. Source Current Magnitude. According to Figure 15, the matrix-converter-based DVR has no energy storage and the required power for compensation is taken from the grid itself. So the source current can be a comparative characteristic of different compensation techniques for this kind of DVR.

As it has been mentioned, matrix converter is able to control its input power factor. The matrix converter input voltage is the load voltage which is $V_{\mathrm{load}}^{\mathrm{Postfault}} = 1\measuredangle(\varphi+\gamma)$ after compensation. By considering a unity power factor for the input of matrix converter, the current of matrix converter would be $I_{\mathrm{MC}}^{\mathrm{Postfault}} = y\measuredangle(\varphi+\gamma)$. Also the source current is defined as $I_s^{\mathrm{Postfault}} = z\measuredangle(\zeta)$. Because there is no energy storage in the structure, the input and output active power of the matrix converter should be equal. Therefore

$$\begin{aligned}
&P_{\mathrm{MC}_1}^{\mathrm{Postfault}} = P_{\mathrm{MC}_2}^{\mathrm{Postfault}} \Longrightarrow \\
&\mathrm{Re}\left\{\left[V_{\mathrm{load}}^{\mathrm{Postfault}}\right]\left[I_{\mathrm{MC}}^{\mathrm{Postfault}}\right]^*\right\} \\
&\quad = \mathrm{Re}\left\{\left[\frac{1}{n}V_{\mathrm{DVR}}\right]\left[nI_s^{\mathrm{Postfault}}\right]^*\right\} \Longrightarrow \qquad (13)\\
&\mathrm{Re}\left\{\left[1\measuredangle(\varphi+\gamma)\right]\left[y\measuredangle(\varphi+\gamma)\right]^*\right\} \\
&\quad = \mathrm{Re}\left\{\left[x\measuredangle\beta\right]\left[z\measuredangle\zeta\right]^*\right\} \Longrightarrow \\
&\quad y = xz\cos(\zeta-\beta). \qquad (14)
\end{aligned}$$

Now by using KCL we have

$$\begin{aligned}
&I_{\mathrm{load}}^{\mathrm{Postfault}} = I_s^{\mathrm{Postfault}} - I_{\mathrm{MC}}^{\mathrm{Postfault}} = z\measuredangle\zeta - y\measuredangle(\varphi+\gamma) \Longrightarrow \\
&1\measuredangle\gamma = \left[z\cos(\zeta) - y\cos(\varphi+\gamma)\right] \qquad (15)\\
&\qquad + j\left[z\sin(\zeta) - y\sin(\varphi+\gamma)\right] \Longrightarrow \\
&\sqrt{z^2+y^2-2zy\cos(\zeta-\varphi-\gamma)} = 1, \qquad (16)\\
&\tan^{-1}\left(\frac{z\sin(\zeta)-y\sin(\varphi+\gamma)}{z\cos(\zeta)-y\cos(\varphi+\gamma)}\right) = \gamma. \qquad (17)
\end{aligned}$$

After some manipulation, (17) yields to

$$\sin(\zeta-\gamma) = \frac{y\sin(\varphi+\gamma) - y\tan(\gamma)\cos(\varphi+\gamma)}{z\sqrt{1+\tan^2(\gamma)}}. \qquad (18)$$

By inserting (14) in to (18) we have

$$\zeta = \tan^{-1}\left(\frac{\sin(\gamma)\sqrt{1+\tan^2(\gamma)} + x\cos(\beta)\mathcal{A}}{\cos(\gamma)\sqrt{1+\tan^2(\gamma)} + x\sin(\beta)\mathcal{A}}\right), \qquad (19)$$

where $\mathcal{A}$ denotes $[\sin(\varphi+\gamma) - \tan(\gamma)\mathrm{csos}(\varphi+\gamma)]$ and by inserting (14) in to (16):

$$z = \frac{1}{\sqrt{1+x^2\cos^2(\zeta-\beta) - 2x\cos(\zeta-\beta)\cos(\zeta-\varphi-\gamma)}}. \qquad (20)$$

Equations (19) and (20) can be applied to all of the compensation strategies. It should be noted that for the prefault method $\gamma = 0$, for the in-phase method $\gamma = \delta$, and for the energy optimized method γ can be calculated by (11) or (12).

Figure 16 shows the magnitude of the input source current versus the magnitude of the source voltage for different control strategies. It shows that the energy optimized method and the prefault method has the most and the least magnitude of the input current, respectively. The higher value of the input current makes a more severe disturbance for the upstream loads because it makes more voltage drop. Moreover, the DVR components should have a higher current rating.

4.4. Matrix Converter Current Magnitude and Capacity. According to Figure 15, the postfault matrix converter current is $I_{\mathrm{MC}}^{\mathrm{Postfault}} = y\measuredangle(\varphi+\gamma)$. The capacity of the matrix converter can be calculated as follows:

$$\left|S_{\mathrm{MC}}\right| = \left|V_{\mathrm{load}}^{\mathrm{Postfault}}\right|\left|I_{\mathrm{MC}}^{\mathrm{Postfault}}\right| = y. \qquad (21)$$

Figure 17 shows the magnitude of this current (or the matrix converter capacity) versus the magnitude of the source voltage. It can be seen that the energy optimized method needs least matrix converter rating to perform.

4.5. Voltage Transfer Ratio of Injection Transformer. According to Figure 15, the input and output voltage of matrix converter are related to each other:

$$\left|\frac{V_{\mathrm{DVR}}}{n}\right| = q\left|V_s^{\mathrm{Postfault}}\right| \Longrightarrow n = \frac{x}{qk}, \qquad (22)$$

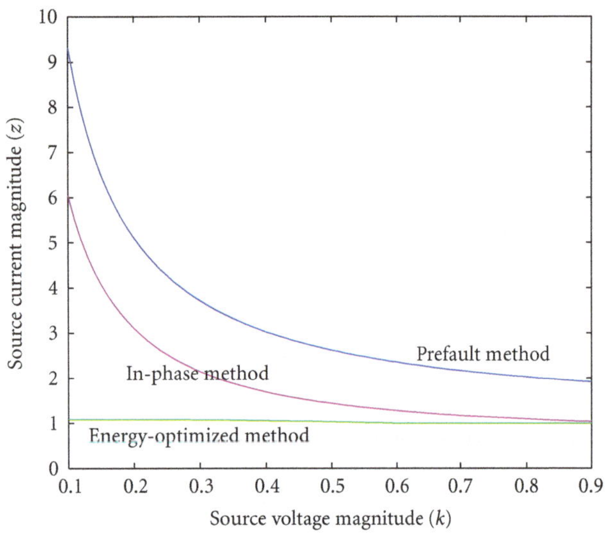

Figure 16: Comparison between different control strategies: the magnitude of the input source current versus the magnitude of the source voltage.

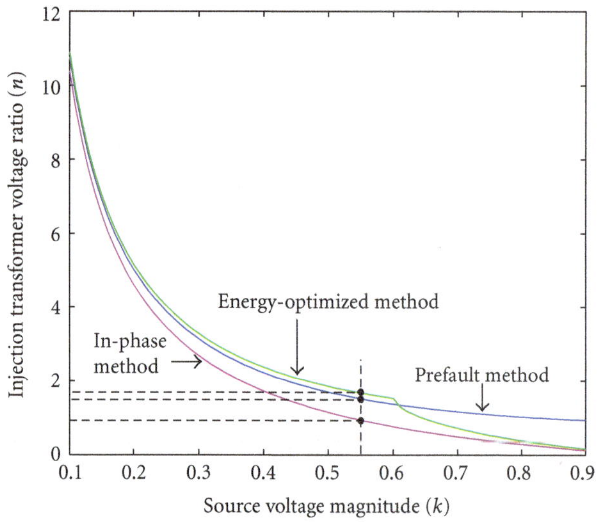

Figure 18: Comparison between different control strategies: the injection transformer voltage transfer ratio versus the magnitude of the source voltage.

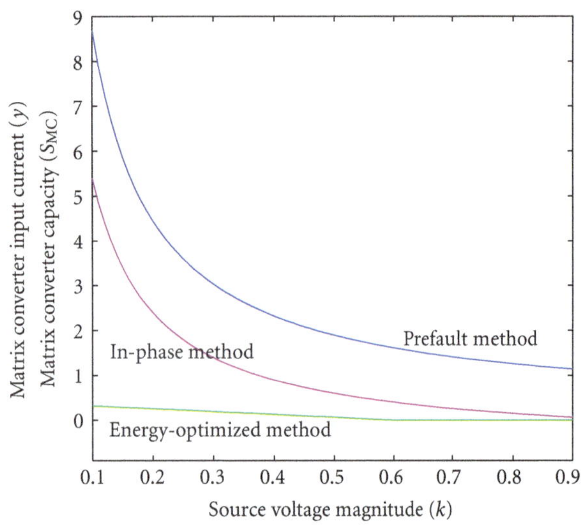

Figure 17: Comparison between different control strategies: Matrix converter input current (or capacity) versus the magnitude of the source voltage.

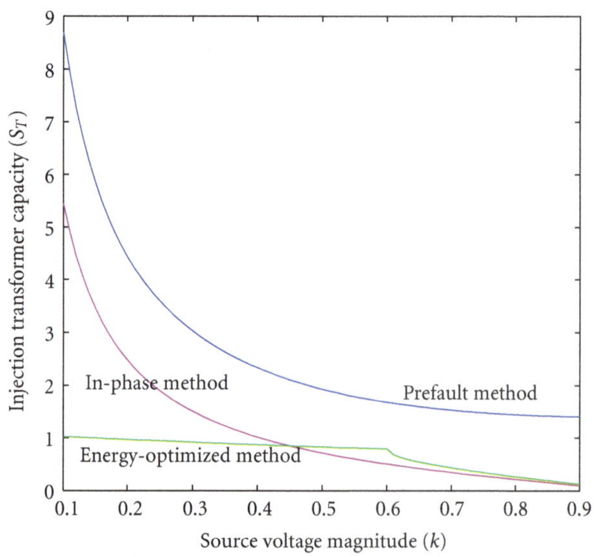

Figure 19: Comparison between different control strategies: the injection transformer capacity versus the magnitude of the source voltage.

where q is the matrix converter transfer ratio. The maximum value of q is 0.866 [5]. Therefore the minimum value of the transformer voltage ratio can be calculated for every compensation algorithm. The result is plotted in Figure 18. For example the minimum value of n to compensate a disturbance with $k = 0.55$ are 1.52, 0.94, and 1.68 for prefault, in-phase, and energy-optimized method, respectively.

4.6. Injection Transformer Capacity. The injection transformer capacity can be calculated as follows:

$$|S_T| = |V_{\mathrm{DVR}}| \left| I_s^{\mathrm{Postfault}} \right| = xz \tag{23}$$

and Figure 19 shows this parameter characteristic for all compensation strategies.

5. DVR Performance Limitation

In the matrix-converter-based DVR, the compensation limit depends on the thevenin impedance of the network. For instance, in a deep Sag condition DVR consumes a large source current to compensate this disturbance because in this topology there is no energy storage device. This large current produces a large voltage drop on the network impedances. As it was mentioned before, the maximum voltage transfer ratio of a matrix converter is 0.866, which is less than unity.

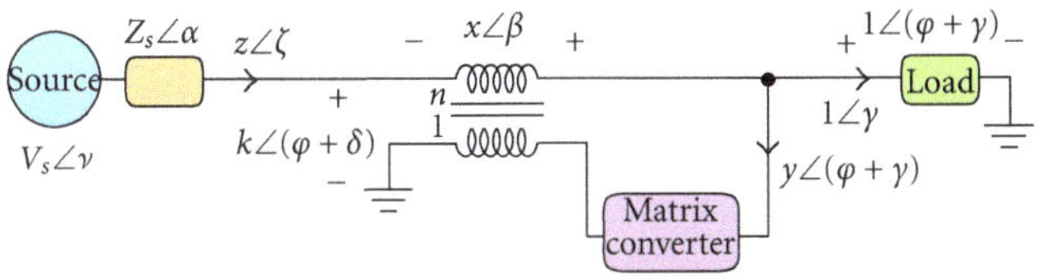

Figure 20: Effect of network thevenin impedance on DVR compensation limits.

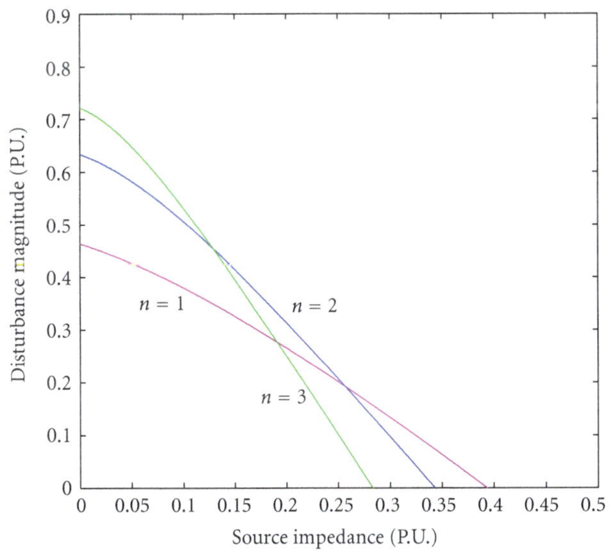

Figure 21: Maximum compensation capability versus the network thevenin impedance for different voltage ratio of injection transformer.

Therefore the maximum compensation capability of the DVR can be calculated. Equations (8) and (22) yield to:

$$k = \frac{\cos(\delta-\gamma) - \sqrt{\cos^2(\delta-\gamma) + n^2q^2 - 1}}{1 - n^2q^2}. \quad (24)$$

According to Figure 20, V_s and δ can be calculated as follows:

$$V_s\angle\nu - (Z_s\angle\alpha)(z\angle\zeta) = k\angle(\varphi+\delta) \Longrightarrow \quad (25)$$

$$\delta = -\varphi + \tan^{-1}\left(\frac{V_s\sin(\nu) - Z_s z\sin(\alpha+\zeta)}{V_s\cos(\nu) - Z_s z\cos(\alpha+\zeta)}\right), \quad (26)$$

$$V_s = \sqrt{K^2 + Z_s^2 z^2 + 2kZ_s z\cos(\varphi+\delta-\alpha-\zeta)}. \quad (27)$$

Consider that $\gamma = 0$ for the Prefault compensation method. Equations (9), (19), (20), (22), (24), (26), and (27) could be solved by numerical techniques. Figure 21 shows the maximum disturbance magnitude that can be compensated versus the amount of the network thevenin impedance for three transformer voltage ratio ($\cos(\varphi) = 0.86$, $\alpha = 85^\circ$, $\nu = 30^\circ$). It shows that for small-grid impedances a larger transformer voltage ratio is needed to compensate voltage disturbances. However, this situation is different when the network impedance is large such as weak distribution networks.

6. Conclusion

In this paper, different compensation strategies to provide the injection voltage for a matrix-converter-based DVR are analyzed. Different comparison criteria are studied. Results show that if the load is sensitive to voltage phase jump, prefault method should be used. However, this method draws more current from the source which needs DVR components with higher current rating. If the load is insensitive to phase jump, either in-phase or energy optimized methods can be used to compensate it. The in-phase method has the least injection voltage magnitude that makes the components cheaper. Moreover for similar disturbance compensation, the in-phase strategy needs less transformer voltage ratio compared to other methods. The energy-optimized method needs the least input current magnitude to compensate disturbances. This will create less voltage drop for upstream loads and also the possibility to design a DVR with less current rating components. At the end the DVR limits are calculated for this topology. Results show that for a strong network a higher transfer ratio of the injection transformer brings the ability to compensate deeper sags. However for a weak network, that has larger short circuit impedance, a lower transfer ratio makes DVR compensating more severe disturbances.

Acknowledgment

It is a pleasure to thank those who made this work possible, especially *Shiraz Electrical Distribution Company* that supported this paper.

References

[1] D. D. Sabin, "An assessment of distribution system power quality," Tech. Rep. TR-106294-V2, Electric Power Research Institute, Palo Alto, Calif, USA, 1996.
[2] IEEE STD 1159–1995, "IEEE Recommended Practice for Monitoring Electric Power Quality".
[3] J. G. Nielsen and F. Blaabjerg, "A detailed comparison of system topologies for dynamic voltage restorers," *IEEE Transactions on Industry Applications*, vol. 41, no. 5, pp. 1272–1280, 2005.
[4] M. Venturini and A. Alesina, "Generalised transformer: a new bidirectional, sinusoidal waveform frequency converter with continuously adjustable input power factor," in *Proceedinds of the IEEE Power Electronics Specialists Conference (PESC '80)*, pp. 242–252, Atlanta, Ga, USA, June 1980.
[5] D. Casadei, G. Serra, A. Tani, and L. Zarri, "Matrix converter modulation strategies: a new general approach based on space-vector representation of the switch state," *IEEE Transactions on Industrial Electronics*, vol. 49, no. 2, pp. 370–381, 2002.
[6] B. Wang and G. Venkataramanan, "Dynamic voltage restorer utilizing a matrix converter and flywheel energy storage," in *Proceedings of the IEEE Industry Applications Conference 42nd IAS Annual Meeting*, pp. 208–215, September 2007.
[7] E. Babaei and M. F. Kangarlu, "A new topology for dynamic voltage restorers without dc link," in *Proceedinds of the IEEE Symposium on Industrial Electronics & Applications (ISIEA '09)*, pp. 1016–1021, October 2009.
[8] J. M. Lozano, J. M. Ramirez, and R. E. Correa, "A novel dynamic voltage restorer based on matrix converters," in *Proceedings*

of the IEEE Proceedings Modern Electric Power Systems (MEPS '10), Wroclaw, Poland, 2010, paper 15.2.

[9] M. Jazayeri and H. Abdollahzadeh, "A novel DVR control system design for compensating all types of voltage sags based on pre-fault method," *European Journal of Scientific Research*, vol. 33, no. 1, pp. 70–85, 2009.

[10] S. Zhe, G. ChunLin, X. YongHai, X. XiangNing, L. Yingying, and T. Shun, "A new analysis method for compensation strategy of DVR and minimum energy control," in *Proceedings of the 5th IEEE Conference on Industrial Electronics and Applications (ICIEA '10)*, pp. 1321–1325, June 2010.

[11] J. G. Nielsen, F. Blaabjerg, and N. Mohan, "Control strategies for dynamic voltage restorer compensating voltage sags with phase jump," in *Proceedings of the16th Annual IEEE Applied Power Electronics Conference and Exposition (APEC '01)*, vol. 2, pp. 1267–1273, March 2001.

[12] R. Strzelecki and G. Benysek, "Control strategies and comparison of the dynamic voltage restorer," in *Proceedings of the 6th International Conference on Power Quality and Supply Reliability*, pp. 79–82, August 2008.

11

Modeling, Control, and Simulation of a Solar Hydrogen/Fuel Cell Hybrid Energy System for Grid-Connected Applications

Tourkia Lajnef, Slim Abid, and Anis Ammous

Power Electronic Group (PEG), ENIS, BP W, 3038 Sfax, Tunisia

Correspondence should be addressed to Tourkia Lajnef; lajnefturkia@yahoo.fr

Academic Editor: Jose Pomilio

Different energy sources and converters need to be integrated with each other for extended usage of alternative energy, in order to meet sustained load demands during various weather conditions. The objective of this paper is to associate photovoltaic generators, fuel cells, and electrolysers. Here, to sustain the power demand and solve the energy storage problem, electrical energy can be stored in the form of hydrogen. By using an electrolyser, hydrogen can be generated and stored for future use. The hydrogen produced by the electrolyser using PV power is used in the FC system and acts as an energy buffer. Thus, the effects of reduction and even the absence of the available power from the PV system can be easily tackled. Modeling and simulations are performed using MATLAB/Simulink and SimPowerSystems packages and results are presented to verify the effectiveness of the proposed system.

1. Introduction

At present, most of energy demand in the world relies on fossil fuels such as petroleum, coal, and natural gas that are being exhausted very fast. One of the major severe problems of global warming is one of these fuels combustion products, carbon dioxide; these are resulting in great danger for life on our planet [1].

Fossil fuels can have as an alternative some renewable energy sources like solar, wind, biomass, and so; among them on the photovoltaic (PV) generator which converts the solar radiation into electricity, largely used in low power applications. The photovoltaic generator is chosen for its positive points including being carbon free and inexhaustible; moreover, it does not cause noise for it is without moving parts and with size-independent electric conversion efficiency [2].

Nevertheless, the power generated by a PV system is influenced by weather conditions; for example, at night or in cloudy periods, it would not generate any power or application. In addition, it is difficult to store the power generated by a PV system for future use. The best method to overcome this problem is to integrate the PV generator with other power sources such as an electrolyser, hydrogen storage tank, FC system, or battery due to their good features such as high efficiency response, modular production, and fuel flexibility [3, 4]. Its coordination with a PV system could be successful for both grid-connected and stand-alone power applications. Thanks to the rapid response capability of the fuel cell power system, the photovoltaic fuel cell hybrid system can be able to overcome the inconvenience of the intermittent power generation. Furthermore, unlike a secondary battery, the FC does not only store energy but also produce electricity for unlimited time to support the PV power generator. Hence, the coordination between the FC power system and the photovoltaic generator becomes necessary in order to smooth out the PV power fluctuations.

This paper focuses on developing a simulation model to design and size the hybrid system for a variety of loading and meteorological conditions. This simulation model is performed using Matlab and SimPowerSystems and results are presented to verify the effectiveness of the proposed system.

2. Modeling

2.1. A Dynamic Model of PV Generator. PV arrays are built up with combined series/parallel combinations of PV solar cells, which allow extracting the characteristic parameters of the one-diode equivalent model for a single solar cell.

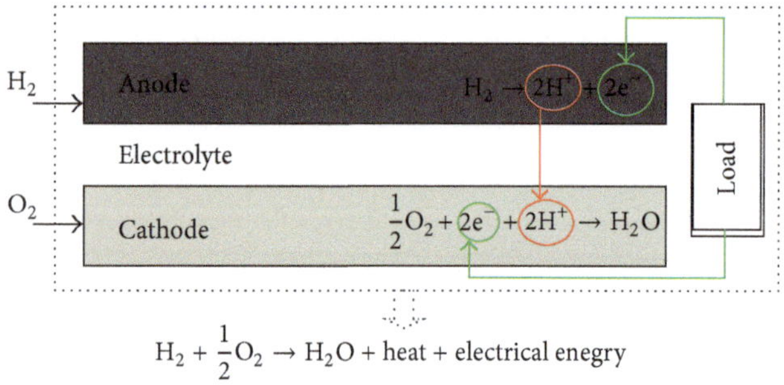

FIGURE 1: Cross-sectional view of a PEMFC.

The PV cell output voltage is a function of the photocurrent that is mainly determined by load current depending on the solar irradiation level during the operation [2]:

$$V_{PV} = \frac{N_s kT}{q} \ln\left[\frac{I_{sc} - I_{PV} + I_0 N_p}{N_p I_0}\right] - \frac{N_s}{N_p} R_s I_{PV}. \quad (1)$$

The parameters used in the mathematical model of the PV system are shown in the abbreviations section.

2.2. A Dynamic Model of PEMFC. The PEMFC is an electrochemical device which allows the electric energy conversion of the chemical energy contained in a reaction between a fuel, the hydrogen, and an oxidizer, the oxygen. The temperature effects have been taken into account in the typical range of low temperature PEM (60–100°C) and a thermal behaviour submodel has been introduced.

A bias voltage is applied across the electrochemical cell in order to induce electrochemical reactions at both electrodes. Water is introduced at the anode and dissociated into oxygen, protons, and electrons. The protons are driven by an electric field through the PEM to the cathode where they combine with the electrons arriving from the external circuit to form hydrogen gas. The assessment of these two half reactions produces water, heat, and electricity as Figure 1 shows.

Many models have been proposed to simulate the fuel cell in the literature [4, 5]. This model is built by utilizing the relationship between the output voltage and potential pressure of hydrogen, oxygen, and water. Figure 2 shows the detailed PEMFC model, which is then embedded into the SimPowerSystems of MATLAB-controlled voltage source and integrated into the overall system.

The proportional relationship of the molar flow of gas through a valve with its partial pressure can be expressed as

$$\frac{q_{H_2}}{P_{H_2}} = \frac{K_{an}}{\sqrt{M_{H_2}}} = K_{H_2}. \quad (2)$$

For hydrogen molar flow, the derivative of the partial pressure can be calculated using the perfect gas equation as follows [5]

$$\frac{d}{dt} P_{H_2} = \frac{RT}{V_{an}} \left(q_{H_2}^{in} - q_{H_2}^{out} - q_{H_2}^{r}\right). \quad (3)$$

The relationship between the hydrogen flow and the FC system current can be written as [5]

$$q_{H_2}^{r} = \frac{N_0 I}{2F} = 2K_r I. \quad (4)$$

Using (4) and (3), the hydrogen partial pressure can be rewritten in the (s) domain as [5]

$$P_{H_2} = \frac{1/K_{H_2}}{1 + \tau_{H_2} s} \left(q_{H_2}^{in} - 2K_r I\right), \quad (5)$$

where

$$\tau_{H_2} = \frac{V_{an}}{K_{H_2} RT}. \quad (6)$$

Typically, the fuel cell output voltage is obtained from the sum of three effects [5, 6], the Nernst potential, the cathode and anode activation overvoltage, and the ohmic overvoltage due to internal resistance:

$$V_{cell} = E + \eta_{act} + \eta_{ohmic}, \quad (7)$$

where

$$\eta_{act} = -B \ln(CI), \qquad \eta_{ohmic} = -R^{int} I. \quad (8)$$

Now, the Nernst voltage in terms of gas molarities may be expressed as [5]

$$E = N_0 \left[E_0 + \frac{RT}{2F} \log\left[P_{H_2} P_{O_2}^{0.5}\right]\right]. \quad (9)$$

The FC system consumes hydrogen obtained from the on-board high-pressure hydrogen tanks according to the power demand. A feedback control strategy is used to control hydrogen flow rate according to the output power of the FC system. This feedback control is achieved where FC current how the output is taken back into the input while converting the hydrogen into molar form [5, 7].

The quality of hydrogen found from the hydrogen tank is given by:

$$q_{H_2}^{req} = \frac{N_0 N_s I}{2FU}. \quad (10)$$

2.3. An Electrolyser Model. Electrolysis of water is the dissociation of water molecules into hydrogen and oxygen gas. The electrochemical reaction of water electrolysis is given by

$$H_2O_{(liquid)} + \text{electrical energy} \Longrightarrow H_{2(gas)} + \frac{1}{2} O_2. \quad (11)$$

The rate of hydrogen reacting is directly proportional to the electrical current in the equivalent electrolysis circuit [8, 9], given by

$$n_{H_2} = \frac{\eta_F n_c i_e}{2F}. \quad (12)$$

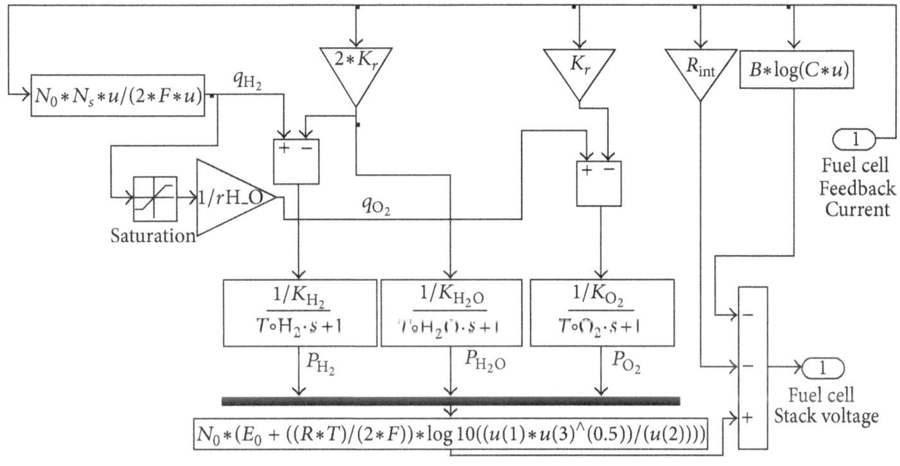

FIGURE 2: Dynamic model of the FC system.

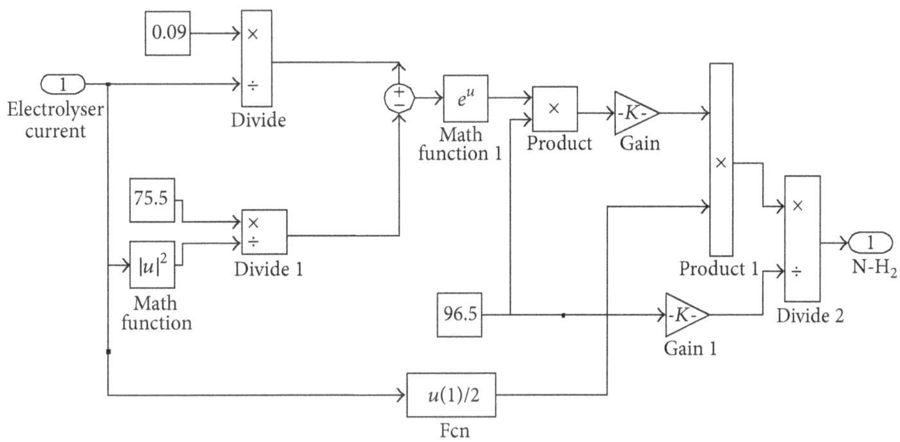

FIGURE 3: The Simulink diagram of the electrolyser model.

The relation between the real hydrogen flow rate and the theoretical one is defined as the Faraday's efficiency. In general, it is assumed to be more than 99%. The Faraday efficiency is expressed by [9, 10]

$$\eta_F = 96.5e^{(0.09/i_e - 75.5/i_e^2)}. \tag{13}$$

Figure 3 shows the simulation model of the electrolyser implemented in MATLAB/Simulink.

2.4. A Hydrogen Storage System. The amount of hydrogen required by the PEMFC is sent directly from the electrolyzer system based on the relationship between the output power and the hydrogen requirement of the PEMFC system. The remaining amount of hydrogen is sent to the storage tank [11]. The parameters used in the hydrogen storage system are listed in the abbreviations section.

In this study, the dynamic of the storage is obtained as follows [10, 11]:

$$\widehat{P_b - P_{bi}} = Z\frac{N_{H_2} R T_b}{M_{H_2} V_b}. \tag{14}$$

Neither the compression dynamics nor the compression energy requirements are accounted for in our calculations. All auxiliary power requirements such as pumps, valves were ignored in the dynamic model. The Simulink version of the hydrogen storage model is depicted in Figure 4.

2.5. An Averaged Model of the Converters. The simulation of power electronic systems behavior using semiconductor refined models gives an accurate, but unaffordable results. The structure of the switch cell is given in Figure 5; it has two basic switching cells (P-cell and N-cell).

Each cell consists of one controlled switch and one diode. The two active switches are directly controlled by external control signals. The diodes are indirectly controlled by the state of the controlled switches and the circuit conditions. The load is presented by an inductor L and a voltage source (V_s). The DC loop inductance is modeled by an inductor L_{st}. Depending on the sign of the load current $I_L(t)$, only two devices (one controlled device and one diode) operate simultaneously [12–15].

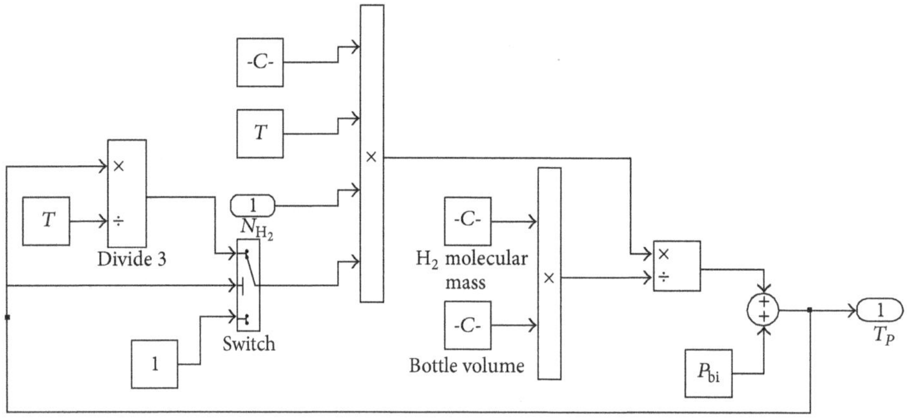

FIGURE 4: The Simulink model of the hydrogen storage system.

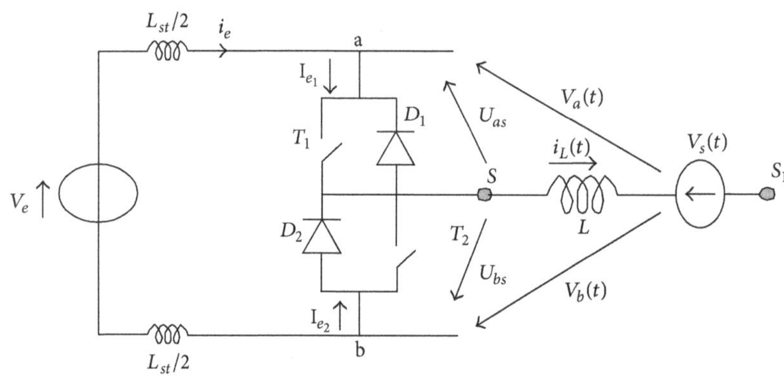

FIGURE 5: (a) The PWM switch, (b) the equivalent circuit of the PWM switch averaged model.

The proposed averaged model of the PWM-switch is presented in Figure 5(b). This model contains a controlled voltage source (V_1) and a controlled current source (I_1). The PWM-switch is the only nonlinear element which is supposed to be responsible for the nonlinear behavior of the converter, Considering T_s as the switching period of the controlled switches and (d) the duty ratio which is the ratio of the on-time value (T_{on}) of the upper controlled switch (T_1) and the switching period T_s.

In Figure 5(b), the current source (I_1) and the voltage source (V_1) are given by

$$V_1 = \langle V_{as} \rangle,$$
$$I_1 = \langle i_{e_2} \rangle, \tag{15}$$

where $\langle U_{as} \rangle$ and $\langle i_{e_2} \rangle$ are the time averaged values of the instantaneous terminal waveforms $U_{as}(t)$ and $i_{e_2}(t)$, respectively, over one cycle T_s.

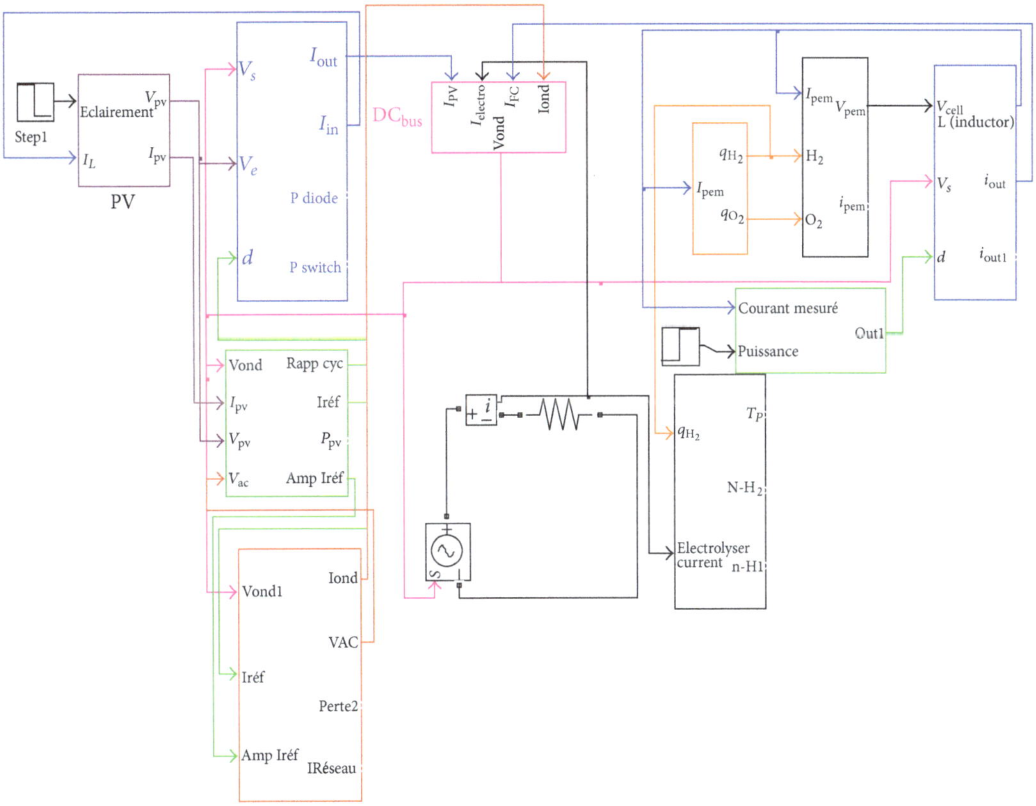

FIGURE 6: Schematic drawing of the PVFC hybrid system.

In this paper we have chosen MOSFET's devices for the controlled switch and the PIN devices for diodes. We notice that the load current i_L is considered constant and equal to the averaged value of the real current $i_L(t)$ in the load over the switching period T_s. During devices turn-on and turn-off phases, the different voltage magnitudes are given by

$$V_x = -\left(L_{st}\left(\frac{dI_F}{dt} + \frac{dI_R}{dt}\right) - V_d\right),$$
$$V_{RM} = -\left(V_e + L_{st}\frac{dI_R}{dt} - V_t\right), \quad (16)$$
$$V_{st} = L_{st}\frac{dI_F}{dt}, \qquad V_q = V_e + V_d - L_{st}\frac{dI_F}{dt}.$$

The averaged values of the voltage across the device of the voltage source V_1 and current I_1 are obtained by integrating the voltage and the current evolutions over one cycle T_s.

2.6. The PV System Model and Integration to the Overall Hybrid Model. As shown in Figure 6, the majority of the PVFC system comprises a solar-cell model, a PEM fuel cell generator, and a water electrolyser. The parameters of the PV systems are listed in Table 1. The PEMFC system parameters are given in Table 2.

DC-DC converters are widely used in PV generating systems as an interface between the PV generator and the water electrolyser. An optimized controller for the DC-DC must find the optimum duty cycle which leads the PV generator as close as possible to its MPPT and ensure that the working point at the water electrolyser is a safety point. Since the fuel cell stack operates at a low DC voltage range (102 V in this paper), the DC-DC converter must boost the DC voltage and invert it to the AC grid frequency (230 V/50 Hz here) for grid-connected operations. To keep the DC buses fixed at 400 V, we chose to use a hysteresis regulator.

TABLE 1: PV system model parameters.

PV system model parameters	Value
The number of parallel cells per strings (N_p)	36
The number of series cells per strings (N_s)	1
The number of parallel modules	6
Ideality or completion factor (a)	1.9
Boltzmann's constant (k)	1.3805e − 23 [j/°K]
PV cell temperature (T)	298 [°K]
Electron charge (q)	1.6e − 19C
Short-circuit cell current (representing insolation level) (I_{sc}) (A)	2.926
PV cell reverse saturation current (I_0) (A)	0.00005
Series resistance of PV cell (R_s) (Ω)	0.0277

Hence, the fuel cell control problem is translated into an output current control requirement, to be realized by the DC/DC converter, in order to ensure optimal operation for

TABLE 2: FC system model parameters.

FC system model parameters	Value
Activation voltage constant (B)	0.04777 [A^{-1}]
Activation voltage constant (C)	0.0136
Conversion factor (CV)	2
Faraday's constant (F)	96484600 [C/kmol]
Hydrogen time constant (τ_{H_2})	3.37 [s]
Hydrogen valve constant (K_{H_2})	$4.22 * 10^{-5}$ [kmol/(atms)]
Hydrogen-oxygen flow ratio (r_{H-O})	1.168
K_r constant = $N_0/4 * F$	$2.2802 * 10^{-7}$ [kmol/(atms)]
Number of cells (N_0)	88
Number of stacks (N_s)	1
Oxygen time constant (τ_{o_2})	6.74 [s]
Oxygen valve constant (K_{o_2})	$2.11 * 10^{-5}$ [kmol/(atms)]
FC system internal resistance (R_{int})	0.00303 (Ω)
FC absolute temperature (T)	343 [K]
Universal gas constant (R)	8314.47 [j/(kmol K)]
Utilization factor (U)	0.8
Water time constant (τ_{H_2O})	18.418 [s]
Water valve constant (K_{H_2O})	$7.716 * 10^{-6}$ [kmol/(atms)]

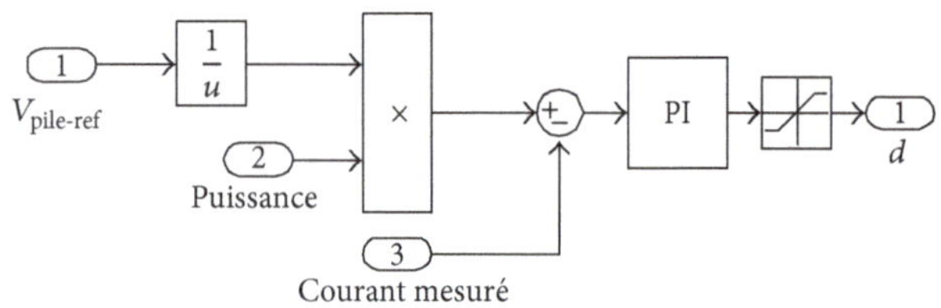

FIGURE 7: Inner control loop insuring fast dynamics of PEMFC.

a given fuel flow rate. Under these conditions, the cell output power is directly related to its fuel consumption at the selected optimum operating point of the V-I characteristic.

Operating the fuel cell at different output power levels requires suitable variation of its input flow rate, to be realized by the overall control system of the cell. The power demand requirement of the fuel cell is translated into a current demand input by dividing with stack output voltage:

$$I_{\text{demand}} = \frac{P_{\text{demand}}}{V_{\text{fc}}}. \tag{17}$$

2.7. Control Strategy of the Hybrid Power System. The proposed control strategy is based on a power decoupling strategy in the frequency domain of the power source. This energy-management strategy fulfills the fast energy demands if the load respects the integrity of each source. The inner control loop is tuned to drive the PEMFC current. V_{Bus} can be considered constant during current regulation. With this assumption, a classical proportional-integral (PI) controller is proposed (Figure 7).

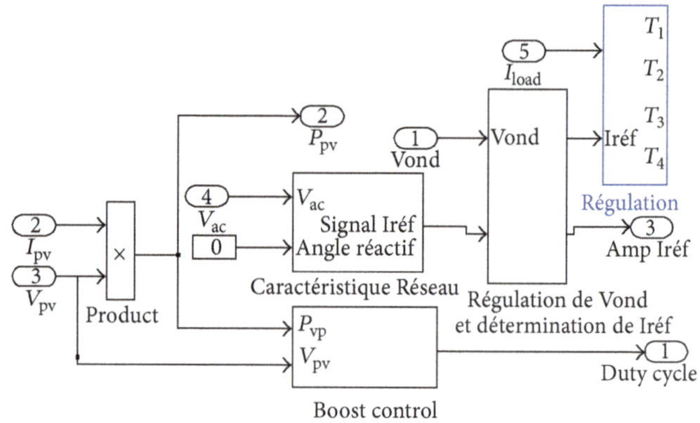

FIGURE 8: Current regulation implementation.

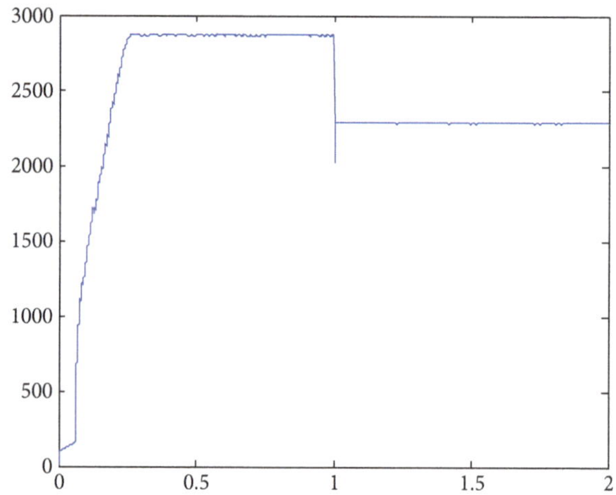

FIGURE 9: Power drawn from solar system.

The implementation structure of the loop regulation with hysteresis is given in Figure 8 scheme. This cascaded control structure shows that it could satisfy the different requirements of the hybrid system both from the point of view of the feed load and from the point of view of the component limits.

3. Results and Discussion

In this section we present simulation results for the coupling between the PV/FC and the PEM electrolyser through the DC-DC converter controller. The PV generator is a 2880 W power plant at 1000 (w/m^2) solar radiation. From Figures 9 and 10, we observe that the current and the available power from the PV generator decreases because of radiation variation.

Initially, the total power generated by the PV generator is sent to the electrolyser through a DC-DC converter to generate hydrogen. The hydrogen produced by the electrolyser causes the pressure of the storage tank to vary as shown in Figure 11. These results are obtained for two values of solar radiation 1000 (w/m^2) and 800 (w/m^2). We notice that the final value of pressure is used as initial condition in the following simulation.

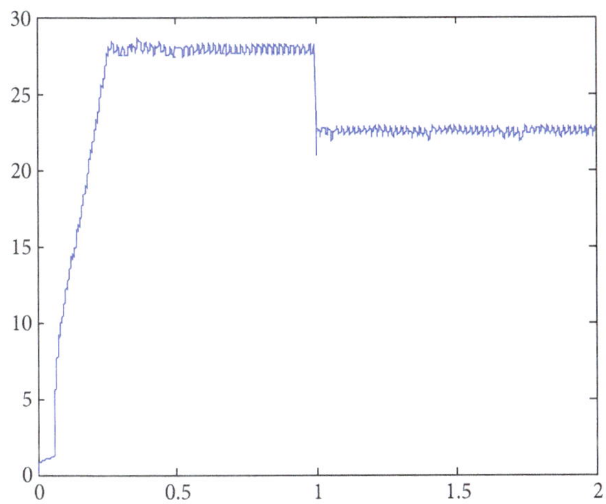

FIGURE 10: The current variation for the PV solar system.

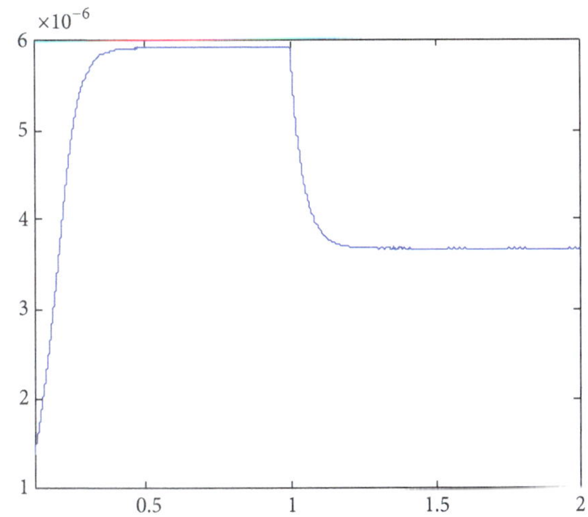

FIGURE 11: Amount of hydrogen produced by electrolyser.

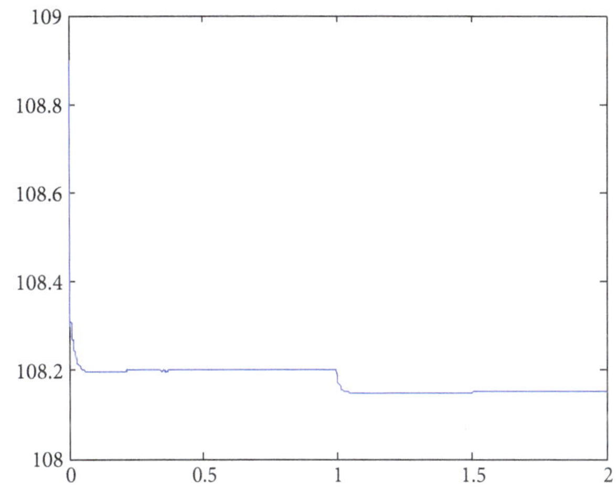

FIGURE 12: Voltage of the fuel cell.

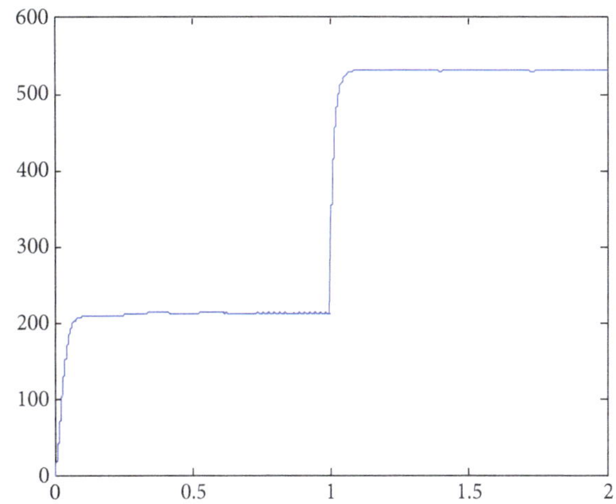

FIGURE 13: Power produced by the FC system.

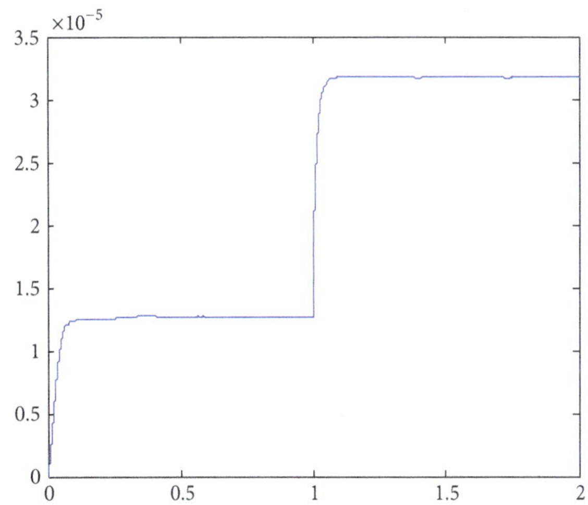

FIGURE 14: The molar flow of hydrogen consumed by FC.

After this, the total power generated by the PV system is sent to grid via a DC/AC converter. At 1000 (w/m^2) solar radiation, the PV generator used in this study is capable of delivering 2880 W while the FC delivers 200 W, but at 800 (w/m^2) solar radiation (at 1 s) the power produced by the PV system tends to 2300 W. So the generated power by the PV is less than the demand; power will be supplied from the FC system. The power produced by the FC system tends to 500 W at 1 s of time simulation.

The internal voltage of the FC system decreases when the FC output power increases. This relation between power and the voltage of the FC system authenticates the reliability of the FC model. The transient response of the FC system voltage to the load changes varies according to the amount of power supplied by the FC system as shown in Figure 12. The power produced by the FC system is given in Figure 13.

The amount of hydrogen moles consumed by the FC system is proportional to the power drawn from the FC system. The hydrogen flow to the FC system per second is depicted in Figure 14.

Figure 15 shows the hydrogen storage tank pressure variation corresponding to the amount of hydrogen extracted

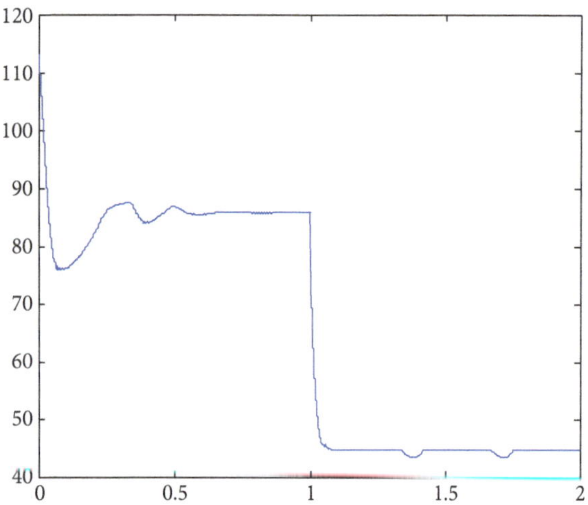

FIGURE 15: The pressure variation of the hydrogen storage tank.

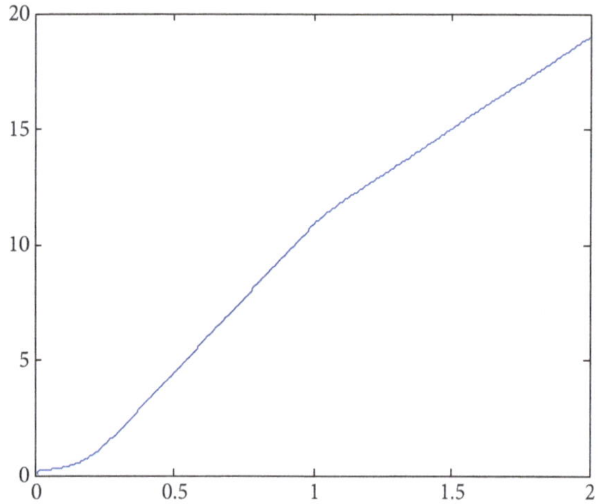

FIGURE 16: The pressure variation of storage hydrogen.

from the storage tank. It is evident that the hydrogen storage tank pressure decreases with time as more and more hydrogen is extracted from the storage tank. The pressure variation of storage hydrogen is as illustrated in Figure 16.

4. Conclusion

In this paper, a PV/FC generator and PEM electrolyser have been described for a PV/FC system intended for grid-connected operations. Special attention has been paid to the modeling of temperature dependence, concentration over potential, and limiting current in the PEM electrolyser model. Then, the power conditioning system, including the DC/DC and DC/AC converters, is presented and typical waveforms are shown from its simulation in MATLAB/Simulink.

Abbreviations

a: Ideality or completion factor
I_0: PV cell reverse saturation current [A]
I_{PV}: PV cell output current [A]
I_{sc}: Short-circuit cell current (representing insolation level [A])
k: Boltzmann's constant [j/°K]
M_v: Voltage factor
N_p: The number of parallel strings
N_s: The number of series cells per string
q: Electron charge [C]
R_s: Series resistance of PV cell [Ω]
T: PV cell temperature [°K]
V_{MP}: PV cell voltage corresponding to maximum power [V]
V_{oc}: Open-circuit voltage [V]
V_{PV}: Terminal voltage for PV cell [V]
M_{H_2}: Molar mass of hydrogen [kg kmol^{-1}]
N_{H_2}: Hydrogen moles per second delivered to the storage tank [kmol/s]
P_{b}: Pressure of tank [pascal]
P_{bi}: Initial pressure of the storage tank [pascal]
R: Universal (Rydberg) gas constant [j/(kmol K)]
T_b: Operating temperature [°K]
V_b: Volume of the tank [m^3]
Z: Compressibility factor as a function of pressure.

References

[1] K.-S. Ro and S. Rahman, "Two-loop controller for maximizing performance of a grid-connected photovoltaic-fuel cell hybrid power plant," *IEEE Transactions on Energy Conversion*, vol. 13, no. 3, pp. 276–281, 1998.

[2] Y. Sukamongkol, S. Chungpaibulpatana, and W. Ongsakul, "A simulation model for predicting the performance of a solar photovoltaic system with alternating current loads," *Renewable Energy*, vol. 27, no. 2, pp. 237–258, 2002.

[3] J. J. Hwang, D. Y. Wang, N. C. Shih, D. Y. Lai, and C. K. Chen, "Development of fuel-cell-powered electric bicycle," *Journal of Power Sources*, vol. 133, no. 2, pp. 223–228, 2004.

[4] J. J. Hwang, W. R. Chang, F. B. Weng, A. Su, and C. K. Chen, "Development of a small vehicular PEM fuel cell system," *International Journal of Hydrogen Energy*, vol. 33, no. 14, pp. 3801–3807, 2008.

[5] M. Y. El-Sharkh, A. Rahman, M. S. Alam, P. C. Byrne, A. A. Sakla, and T. Thomas, "A dynamic model for a stand-alone PEM fuel cell power plant for residential applications," *Journal of Power Sources*, vol. 138, no. 1-2, pp. 199–204, 2004.

[6] J. Padullés, G. W. Ault, and J. R. McDonald, "Integrated SOFC plant dynamic model for power systems simulation," *Journal of Power Sources*, vol. 86, no. 1, pp. 495–500, 2000.

[7] J. Hamelin, K. Agbossou, A. Laperrière, F. Laurencelle, and T. K. Bose, "Dynamic behavior of a PEM fuel cell stack for stationary applications," *International Journal of Hydrogen Energy*, vol. 26, no. 6, pp. 625–629, 2001.

[8] M. J. Khan and M. T. Iqbal, "Dynamic modeling and simulation of a small wind-fuel cell hybrid energy system," *Renewable Energy*, vol. 30, no. 3, pp. 421–439, 2005.

[9] O. Ulleberg, *Stand-alone power systems for the future: optimal design, operation and control of solar-hydrogen energy systems [Ph.D. thesis]*, Norwegian University of Science and Technology, 1998.

[10] K. Sapru, N. T. Stetson, S. R. Ovshinsky et al., "Development of a small scale hydrogen production-storage system for hydrogen applications," in *Proceedings of the 32nd Intersociety Energy Conversion Engineering Conference*, pp. 1947–1952, August 1997.

[11] H. Görgün, "Dynamic modelling of a proton exchange membrane (PEM) electrolyzer," *International Journal of Hydrogen Energy*, vol. 31, no. 1, pp. 29–38, 2006.

[12] H. J. N. Spruijt, D. M. O'Sullivan, and J. B. Klaassens, "PWM-switch modeling of DC-DC converters," *IEEE Transactions on Power Electronics*, vol. 10, no. 6, pp. 659–664, 1995.

[13] J. Chen, R. W. Erickson, and D. Maksimovic', "Averaged switch modeling of boundary conduction mode DC-to-DC converters," in *Proceedings of the 27th Annual Conference of the IEEE Industrial Electronics Society (IECON '01)*, vol. 2, pp. 844–849, Denver, Colo, USA, November 2001.

[14] A. Ammous, K. Ammous, M. Ayedi, Y. Ounajjar, and F. Sellami, "An advanced PWM-switch model including semiconductor device nonlinearities," *IEEE Transactions on Power Electronics*, vol. 18, no. 5, pp. 1230–1237, 2003.

[15] S. Abis, K. Ammous, H. Morel, and A. Ammous, "Advanced averaged model of PWM-switch operating in CCM and DCM conduction modes," *International Review of Electrical Engineering*, vol. 2, no. 4, pp. 544–556, 2007.

12

Simple Hybrid Model for Efficiency Optimization of Induction Motor Drives with Its Experimental Validation

Branko Blanuša[1] and Bojan Knezevic[2]

[1] *Faculty of Electrical Engineering, University of Banja Luka, Patre 5, 78000 Banja Luka, Bosnia and Herzegovina*
[2] *Faculty of Mechanical Engineering, University of Banja Luka, Bulevar Stepe Stepanovica 75, 78000 Banja Luka, Bosnia and Herzegovina*

Correspondence should be addressed to Branko Blanuša; bbranko@etfbl.net

Academic Editor: Jose Pomilio

New hybrid model for efficiency optimization of induction motor drives (IMD) is presented in this paper. It combines two strategies for efficiency optimization: loss model control and search control. Search control technique is used in a steady state of drive and loss model during transient processes. As a result, power and energy losses are reduced, especially when load torque is significant less related to its rated value. Also, this hybrid method gives fast convergence to operating point of minimal power losses and shows negligible sensitivity to motor parameter changes regarding other published optimization strategies. This model is implemented in vector control induction motor drive. Simulations and experimental tests are performed. Results are presented in this paper.

1. Introduction

Induction motor is a widely used electrical motor and a great energy consumer. The vast majority of induction motor drives are used for heating, ventilation, and air conditioning (HVAC). These applications require only low dynamic performance, and in most cases only voltage source inverter is inserted between grid and induction motor as cheapest solution. The classical way to control these drives is constant V/f ratio, and simple methods for efficiency optimization can be applied [1]. From the other side there are many applications where, like electrical vehicles, electric energy has to be consumed in the best possible way and use of induction motors. These applications require an energy optimized control strategy [2].

One interesting algorithm which can be applied in a drive controller is algorithm for efficiency optimization.

In a conventional setting, the field excitation is kept constant at rated value throughout its entire load range. If machine is underloaded, this would result in overexcitation and unnecessary copper losses. Thus in cases where a motor drive has to operate in wider load range, the minimization of losses has great significance. It is known that efficiency improvement of IMD can be implemented via motor flux level and this method has been proven to be particularly effective at light loads and in a steady state of drive. Also flux reduction at light loads gives less acoustic noise derived from both converter and machine. From the other side low flux makes motor more sensitive to load disturbances and degrades dynamic performances [3].

Drive loss model is used for optimal drive control in loss model control (LMC) [3–7]. These algorithms are fast because the optimal control is calculated directly from the loss model. But power loss modeling and calculation of the optimal control can be very complex. Often the loss model is not accurate enough.

Search strategy methods have an important advantage compared to other strategies [8–11]. It is completely insensitive to parameter changes, while effects of the parameter variations caused by temperature and saturation are very expressed in two other strategies. The online efficiency optimization control on the basis of search, where the flux is decremented in steps until the measured input power settles down to the lowest value, is very attractive. Algorithm is applicable universally to any motor. Besides all good characteristics of search strategy methods, there is an outstanding problem in its use. For many applications flux convergence to its optimal value is too slowly. Also, flux is never reached its optimal value then in small steps oscillates around it.

For electrical drives that work in periodic cycles, it is possible to calculate the optimal trajectory of magnetization flux, using optimal control theory, so that power losses in one working cycle are minimal [12]. These methods give good results if the working conditions do not change.

Hybrid method combines good characteristics of two optimization strategies SC and LMC [3, 13–15]. It was enhanced attention as interesting solution for efficiency optimization of controlled electrical drives. During transient process LMC is used, so fast flux changes and good dynamic performances are kept. Search control is used for efficiency optimization in a steady state of drive. Loss model of IM in d-q rotational system and procedure for parameter identification in a loss model based on Moore-Penrose pseudoinversion is given in Section 2. New hybrid model is presented in Section 3. Qualitative analyses of this method with simulation and experimental results are given in Section 4. At the end, obtained results are presented in Conclusion.

2. Power Loss Modeling

The process of energy conversion within motor drive converter and motor leads to the power losses in the motor windings and magnetic circuit as well as conduction and commutation losses in the inverter [6].

The losses in the motor consist of hysteresis losses and eddy current losses in a magnetic circuit (iron losses), losses in the stator and rotor windings (copper loss), and stray losses. In nominal operating conditions the iron losses are typically 2-3 times smaller than the copper losses, but at low loads, these losses are dominant. These losses consist of hysteresis and eddy current losses. Eddy current losses are proportional to the square of supply frequency, and hysteresis losses are proportional to supply frequency. Both components of iron losses are dependent of stator flux level, so next expression is suitable to represent these losses:

$$P_{\gamma\mathrm{Fe}} = c_{\mathrm{eddy}}\psi_{sd}^2\omega_s^2 + c_{\mathrm{hys}}\psi_{sd}^2\omega_s, \tag{1}$$

where c_{eddy} is eddy current and c_{hys} is hysteresis loss coefficients.

Copper losses are appeared as a result of the passing the electric current through the stator and rotor windings. These losses are proportional to the square of current through stator and rotor windings, and they are given by

$$p_{\gamma\mathrm{Cu}} = R_s i_s^2 + R_r i_{sq}^2. \tag{2}$$

The total additional losses typically do not exceed 5% when the drive works with light loads. This case is the most important for the power loss minimization algorithms, so stray losses are not considered as a separate loss component.

Losses in the drive converter consist of the losses in the rectifier and the conductive and switching losses in the inverter. The losses in the rectifier are independent of the magnetizing flux and not specifically taken into account. Only conductive losses in the inverter are dependent on the magnetizing flux, and these can be presented in the next form:

$$P_{\mathrm{INV}} = R_{\mathrm{INV}} i_s^2 = R_{\mathrm{INV}}\left(i_{sd}^2 + i_{sq}^2\right). \tag{3}$$

Based on the previous considerations, the losses in the induction motor drive, dependent on the magnetizing flux, can be expressed as follows:

$$\begin{aligned} P_\gamma = {} & \left(R_{\mathrm{INV}} + R_s\right) i_{sd}^2 + \left(R_{\mathrm{INV}} + R_s + R_r\right) i_{sq}^2 \\ & + c_{\mathrm{eddy}}\omega_s^2\psi_{sd}^2 + c_{\mathrm{hys}}\omega_s\psi_{sd}^2. \end{aligned} \tag{4}$$

Take into account expression for output power:

$$P_{\mathrm{out}} = d\omega\psi_{sd} i_{sq}, \tag{5}$$

where P_{out} is output power and d is constant which depends on the characteristics of Park's rotating transformation (in this case it is 1.5), and the relation

$$P_{\mathrm{in}} = P_\gamma + P_{\mathrm{out}} \tag{6}$$

expression for input power can be written in the next form:

$$P_{\mathrm{in}} = a i_{sd}^2 + b i_{sq}^2 + c_1\omega_s^2\psi_{sd}^2 + c_1\omega_s\psi_{sd}^2 + d\omega\psi_{sd} i_{sq}, \tag{7}$$

where $a = R_s + R_{\mathrm{INV}}$, $b = R_s + R_{\mathrm{INV}} + R_r$, $c_1 = c_{\mathrm{eddy}}$, $c_2 = c_{\mathrm{hys}}$, and d is positive constant.

Two typical cases are differed:

(1) linear dependence of magnetizing flux from the magnetizing current,

(2) nonlinear dependence of magnetizing flux from the magnetizing current.

In the algorithms for loss minimization, magnetizing flux is less than or equal to the nominal value, so it is used in the linear part of magnetization characteristics.

Starting from the expressions (4) and taking into account expression (5), the power losses can be expressed as a function of i_{sd}, T_{em}, and ω_s as follows:

$$P_\gamma\left(i_{sd}, T_{\mathrm{em}}, \omega_s\right) = \left(a + c_1 L_m^2\omega_s^2 + c_2 L_m^2\omega_s\right) i_d^2 + \frac{bT_{\mathrm{em}}^2}{\left(dL_m i_{sd}\right)^2}. \tag{8}$$

Slip angular speed can be defined as follows:

$$\omega_{sl} = \omega_s - \omega \approx \frac{i_{sq}}{T_r i_{sd}}, \tag{9}$$

where T_r is time rotor constant.

Based on expression (8) power losses can be given as function of magnetizing current i_{sd} and operating conditions $(\omega, T_{\mathrm{em}})$:

$$\begin{aligned} P_\gamma\left(i_{sd}, T_{\mathrm{em}}, \omega\right) = {} & \left(a + c_1 L_m^2\omega^2 + c_2 L_m^2\omega\right) i_{sd}^2 \\ & + \frac{\left(2\omega L_m + c_2 L_m\right) T_{\mathrm{em}}}{dT_r} \\ & + \frac{T_{\mathrm{em}}^2}{d^2}\left(\frac{b}{L_m^2} + \frac{c_1}{T_r^2}\right)\frac{1}{i_{sd}^2}. \end{aligned} \tag{10}$$

Putting $k_1 = a + c_1 L_m^2 \omega^2 + c_2 L_m^2 \omega$, $k_2 = (T_{em}^2/d^2)(b/L_m^2 + c_1/T_r^2)$, and $k_3 = (2\omega L_m + c_2 L_m) T_{em}/dT_r$ (10) can be written as follows:

$$P_\gamma (i_d, T_{em}, \omega) = k_1 i_{sd}^2 + \frac{k_2}{i_{sd}^2} + k_3. \quad (11)$$

Parameter k_3 is a function of T_r which is time variant especially due to temperature changes. Time rotor constant is continuously updated in the algorithm for parameter identification, so the parameter k_3 too.

First and second derivations of P_γ in a function of i_{sd} are

$$\frac{\partial P_\gamma}{\partial i_d} = 2k_1 i_{sd} - 2\frac{k_2}{i_{sd}^3},$$

$$\frac{\partial^2 P_\gamma}{\partial^2 i_{sd}} = 2k_1 + 6\frac{k_2}{i_{sd}^4}, \quad k_1 > 0,\ k_2 > 0,\ \frac{\partial^2 P_\gamma}{\partial^2 i_{sd}} > 0, \quad (12)$$

$$\text{for } i_{sd\,\min} \le i_{sd} \le i_{sd\,\max}.$$

Value $i_{sd\,\max}$ is chosen to be I_{dn}, and $i_{sd\,\min}$ is determined based on characteristics of drive and requirements for electromagnetic torque reserve. In this case $i_{sd\,\min} = 0.5 I_{dn}$, where I_{dn} is nominal value of i_{sd} current. On the basis of expressions in (12), it can be concluded that in the steady state of drive and known operating conditions exists accurate value of i_{sd} which gives minimal losses as follows:

$$i_{sdLMC}^* = \sqrt[4]{\frac{k_2}{k_1}}. \quad (13)$$

Based on the specified optimal value of d component stator current vector and command of electromagnetic torque, q component of the stator current is determined as follows:

$$i_{sqLMC}^* = \frac{T_{em}}{dL_m i_{sdLMC}^*}. \quad (14)$$

2.1. Parameter Identification in a Loss Model. This procedure of parameter identification is a mainly based on the procedures described in [6, 7]. In order to determine the parameters in the model losses, a, b, c_1, and c_2, it is necessary to measure the input power and know the exact value of the output power. The exact value of the output power is needed to determine correct information of the losses in the drive and to avoid coupling between load pulsation and work of optimization controller. Also, the samples of the values that define input powers v_{DC}, i_{DC} ($P_{in} = v_{DC} i_{DC}$) are available in the controller, and these values are used for protect and control functions (dynamic braking, the soft-start operation, PWM compensation, etc.).

The proposed procedure of parameter identification is shown in Figure 1. The samples that exist in the expression (7) are memorized at each period. Usually the T_S is 100–200 μs. Due to high-frequency components that do not contribute to identification of the relevant parameters $W = [a\ b\ c_1\ c_2\ d]^T$, input signal and the input power are averaged in the period $T = QT_S$:

$$\begin{aligned}\int_{nT}^{(n+1)T} P_{in}(t)\,dt &= a\int_{nT}^{(n+1)T} i_d^2(t)\,dt + b\int_{nT}^{(n+1)T} i_q^2(t)\,dt \\ &+ c_1\int_{nT}^{(n+1)T} \left[\psi_d^2(t)\,\omega_s^2(t)\,dt\right] \\ &+ c_2\int_{nT}^{(n+1)T} \left[\psi_d^2(t)\,\omega_s(t)\,dt\right] \\ &+ d\int_{nT}^{(n+1)T} \left[\omega(t)\,i_q(t)\,\psi_{sd}(t)\,dt\right] \Longrightarrow \\ Y_N &= aA_N + bB_N + c_1 C_{N1} + c_2 C_{N2} + dD_N. \end{aligned} \quad (15)$$

Values A_N, B_N, C_{N1}, C_{N2}, and D_N, $N = 1, \ldots, M$ successively form the columns $P(:,1), P(:,2), P(:,3), P(:,4)$, and $P(:,5)$ of matrix $P_{M\times5}$.

Vector Y is formed from M averaged values of input power Y_N, $N = 1, \ldots, M$. Vector W_g is calculated by applying Moore-Penrose pseudoinversion (Figure 1) as follows:

$$W_g = [a_g\ b_g\ c_{g1}\ c_{g2}\ d_g]^T = (P^T P)^{-1} PY. \quad (16)$$

That is W_g is approximate solution of the matrix equations $PW = Y$, which gives a minimum value $\|PW - Y\|$ or minimum mean-square error.

The choice of Q is of great importance for the process of parameters identification (see (15)) because the input power fluctuations should be sufficient in the period $T = QT_S$. Otherwise $P^T P$ matrix will be singular or aspire singularity, that is, $\det(P^T P)$ will be close to 0. The process of determining the parameters in the loss model is repeated during the operation of electric drive.

3. Hybrid Model for Efficiency Optimization

Electrical drive with block for efficiency optimization is shown in Figure 2. Electric drive is supplied from the primary power network 3×380 V. This voltage is rectified, and the voltage and current in DC link are measured. The drive inverter is current regulated (CR) voltage three-phase inverter. Reference and measured d and q components of stator current are kept in the current controllers. These controllers are realized in rotational d, q coordinate system as a linear PI controller. Outputs of controller are d and q component of stator voltage. These voltages are transformed (B-transformation) in three-phase reference voltages v_a^*, v_b^*, and v_c^*. These voltages are scaled and lead to PWM modulator, where control signals of inverter switches are generated. This drive works as speed controlled drive. Speed reference and measured speed are led into speed controller. Speed controller is realized as PI controller in incremental form, with proportional coefficient in a feedback local branch. The output of speed controller is reference value of electromagnetic torque. Reference value of stator current d component is determined in a block for efficiency optimization and q component of stator current

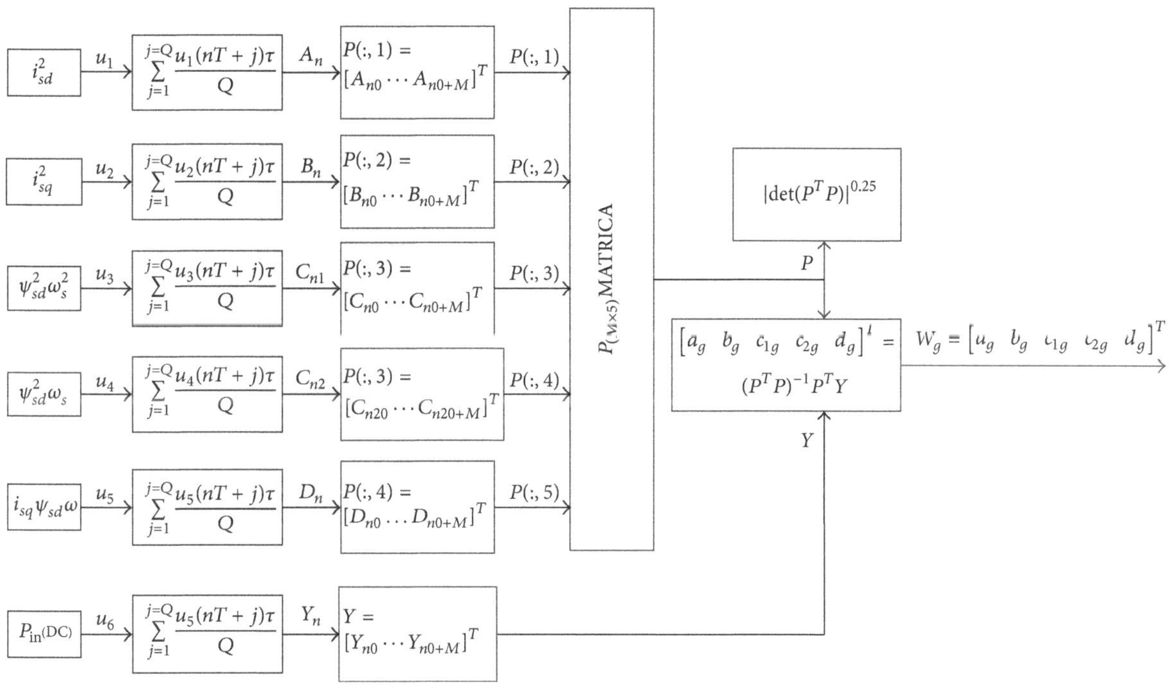

FIGURE 1: A method for determining the parameters a, b, c_1, c_2 and, d from the input signal.

vector on the basis of d stator current vector component and electromagnetic torque reference (see (14)). Position of rotor flux vector is determined in indirect vector control block (IVC).

Hybrid model for efficiency optimization consists from 3 blocks, LMC, SC, and steady state control (SSC) block. LMC is used during transient states caused of external speed or torque demand [3]. Optimal control (i^*_{sdLMC}, i^*_{sqLMC}) is calculated directly from loss model for a given operational conditions what obtains power loss optimization and good dynamic performances. SC is used in a steady state for a constant output power.

On the basis of speed reference and measured speed, SSC block defines its output and controls switches (Figure 2). If transient states is detected, LMC is active, and its outputs are forwarded to indirect vector control (IVC) block and current regulators. When steady state is detected in SSC block, last value of magnetizing current during transient state is used as starting point for search algorithm.

3.1. Loss Model Controller. Optimal control calculation in LMC for a given operational conditions is described in Section 2.

Expressions for d and q component of stator current are defined by (8) and (10).

This method is sensitive to parameter changes due to temperature changes and magnetic circuit saturation, what consequently leads to error in a current references calculation. So, algorithm for parameter identification is always active, and parameters in the loss model are continuously updated (Figure 2).

3.2. Search Controller. Search algorithm is used in steady state, which is detected in the SSC block. Error that exists between the current reference i_d that is generated in the LMC model and in the search model appears as a consequence of inverter and stray losses which are not included in the model. The applied search algorithm is simple. Since the current i_{sd} is very close to the value which gives minimal losses, small step of magnetization current $\Delta i_{sd} = 0.01 I_{dn}$ is chosen.

For two successive values of the i_{sd} current, power losses are determined. Sign of Δi_{sd} is maintained if power losses are reduced. Otherwise, the sign of Δi_{sd} is opposite in the next step:

$$i_{sd}(n) = i_{sd}(n-1) - \operatorname{sgn}\left(\Delta P_\gamma (n-1)\right) \Delta i_{sd}. \qquad (17)$$

When the two values of magnetization current i_{sd1} and i_{sd2} were found, so the sign of power loss is changed between these values, and new reference of i_{sd} current is specified as:

$$i^*_{sdSC} = \frac{i_{sd1} + i_{sd2}}{2}. \qquad (18)$$

In this way, there are no oscillations of i_{sd} current, air gap flux, and electromagnetic torque, which are characteristics of the search algorithm.

4. Simulation and Experimental Results

4.1. Simulation Results. Hybrid method for efficiency optimization of IMD has been studied through computer simulation in MATLAB-Simulink.

Speed reference and load torque are shown in Figure 3. The steep change of load torque appears with the aim

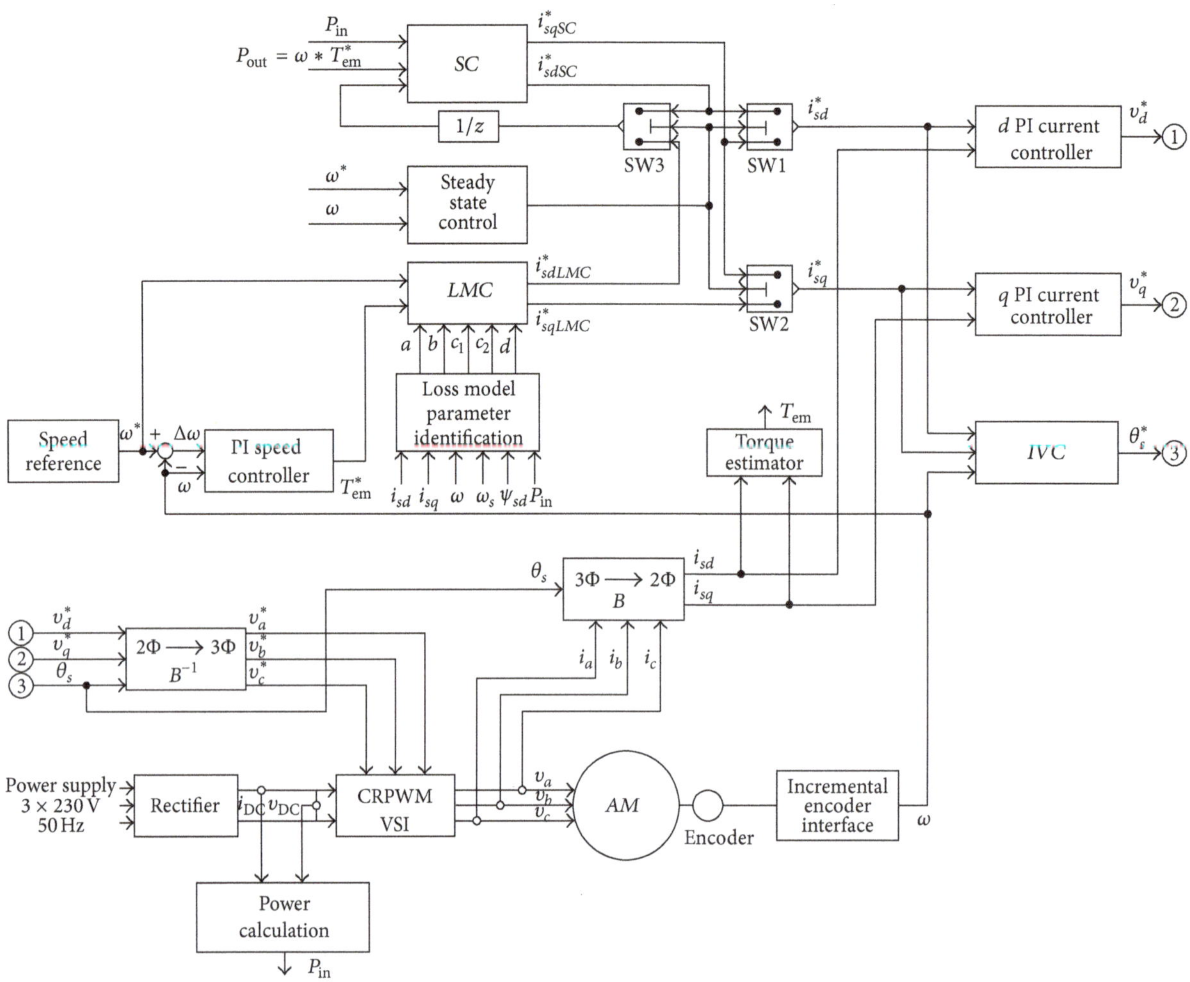

FIGURE 2: Overall proposed block diagram of efficiency optimization controller in IMD.

of testing the drive behavior in the dynamic mode and its robustness for a sudden load perturbations. Graphs of magnetization current (i_{sd}) and active component (i_{sq}) of stator current vector for a given operating conditions and applied hybrid method are shown in Figure 4.

Graphs of power loss for nominal flux and hybrid method are given in Figure 5. It is obvious that the use of hybrid methods gives significantly less power losses than when the flux is maintained at nominal value for a light load of drive.

4.2. Experimental Results. The experimental tests have been performed on the setup which consists of the following:

(i) induction motor (3 phases, Δ380 V/Y220 V, 3.7/2.12 A, $\cos\phi = 0.71$, 1410 r/min, 50 Hz),

(ii) incremental encoder connected with the motor shaft,

(iii) three-phase drive converter (DC/AC converter and DC link),

(iv) PC and dSPACE1102 controller board with TMS320C31 floating point processor and peripherals,

(v) interface between controller board and drive converter.

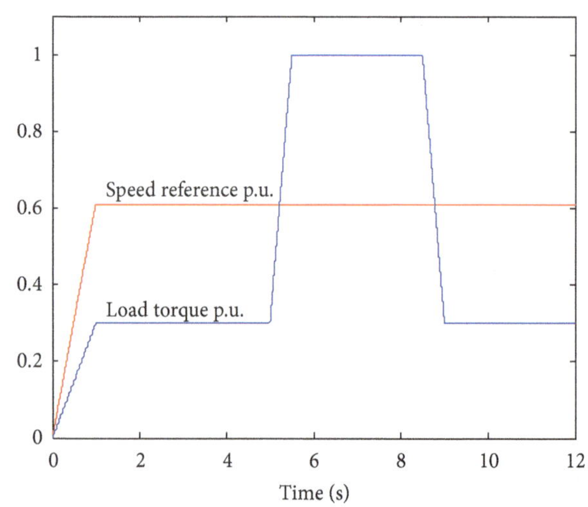

FIGURE 3: Speed reference and load torque [p.u.].

Parameters of used induction motor are given in the appendix of the paper.

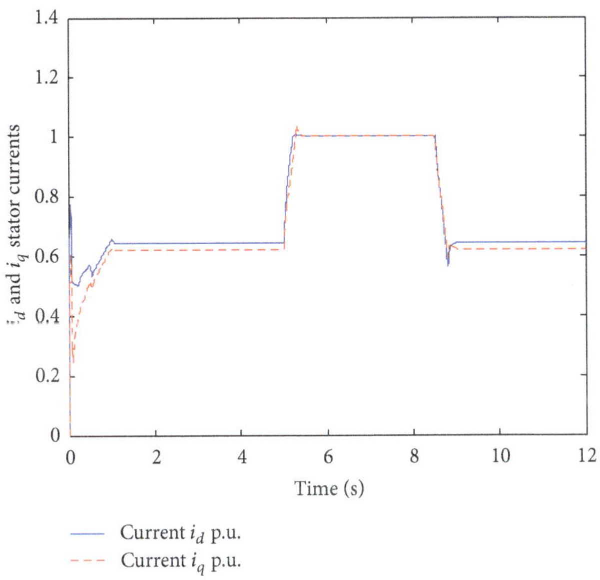

FIGURE 4: Graphs of d and q component of stator current for a given operating conditions.

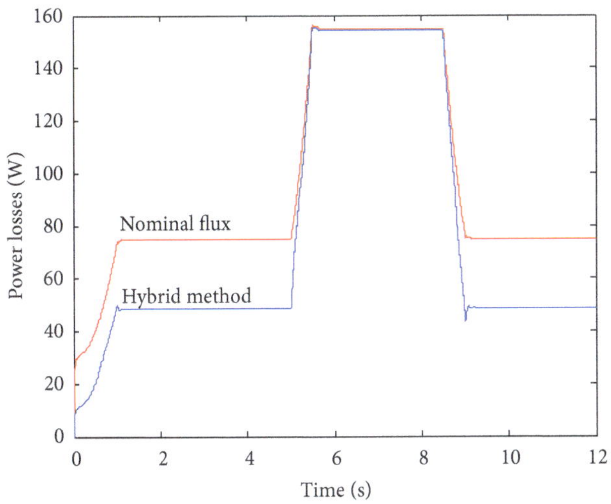

FIGURE 5: Graph of power loss for nominal flux and applied hybrid method.

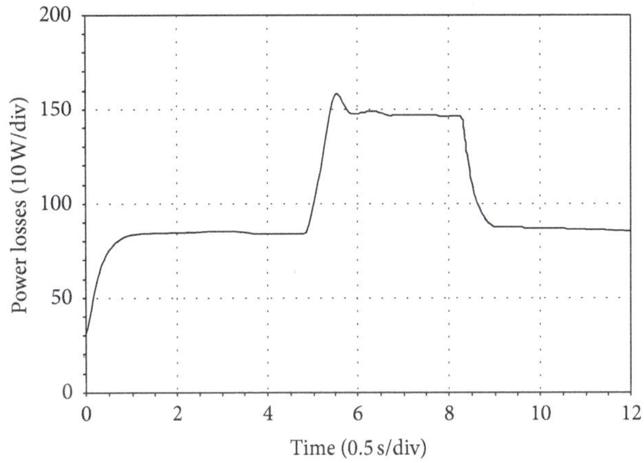

FIGURE 6: Power losses for nominal magnetization flux and operational conditions shown in Figure 3.

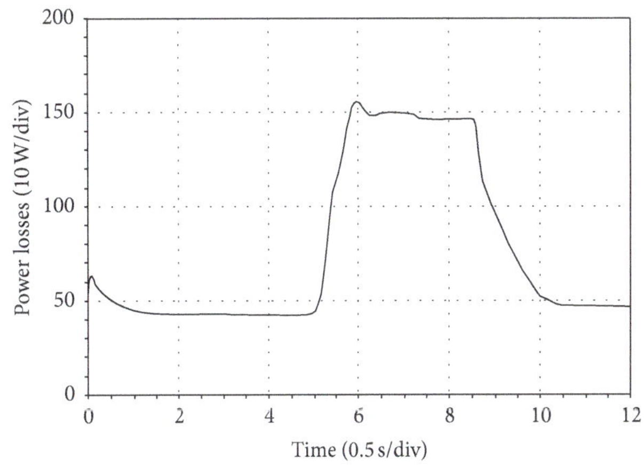

FIGURE 7: Power losses when the hybrid method is applied and operational conditions shown in Figure 3.

The algorithm observed in this paper used the MATLAB-Simulink software, dSPACE real-time interface, and C language. Handling real-time applications is done in ControlDesk.

All experimental tests and simulations have been done in the same operating conditions of the drive, and some comparisons between algorithms for efficiency optimization are made through the experimental tests. Graphs of power losses for nominal flux and when the hybrid method is applied are shown in Figures 6 and 7, respectively. Graphs of magnetization current for a given operational conditions are shown in Figure 8, and speed response is shown in Figure 9.

Based on graph in Figure 6, it can be concluded that the speed drop for a step change of load is small and speed response is strictly aperiodic.

5. Conclusion

Algorithms for efficiency optimization of induction motor drive are briefly described. Detailed theoretical analysis of power losses in induction motor is presented. Algorithm for parameter identification in the loss model based on Moore-Penrose pseudoinversion is presented. Also, hybrid algorithm for efficiency optimization has been applied. According to the theoretical analysis, performed simulations, and experimental tests we have arrived to the following conclusions.

If load torque has a value close to nominal or higher, magnetizing flux is also nominal regardless of whether an algorithm for efficiency optimization is applied or not.

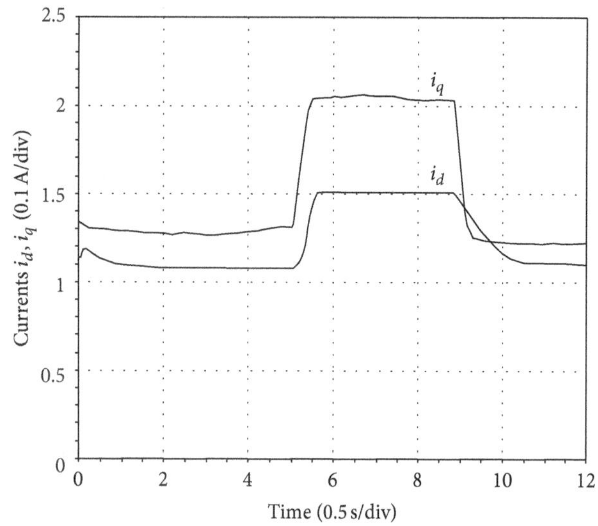

FIGURE 8: Magnetization current i_d and active current when the hybrid method is applied and operational conditions shown in Figure 3.

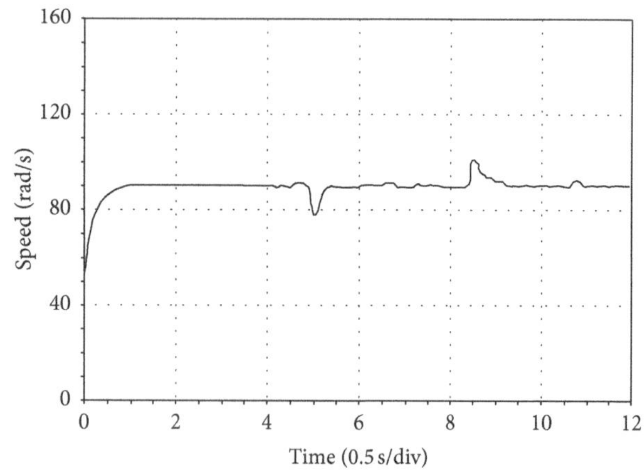

FIGURE 9: Speed response when the hybrid method is applied and operational conditions shown in Figure 3.

For a light load, hybrid method for efficiency optimization gives significant power loss reduction (Figures 5, 6, and 7).

There is no oscillation of magnetization flux which is characteristic of the search algorithms (Figures 4 and 8).

Also, this hybrid method shows good dynamic performances and no sensitivity to parameter changes (Figure 9).

Implementation of presented algorithm is simple, and it can be universally applied to any electrical motor. Changes are only related to different models of power losses.

Appendix

Parameters of used induction motor

3 phase $\Delta 220/Y380$ V (supply voltage)

3,7/2,12 A (nominal stator current)

0.75 kW (nominal mechanical power)

$\cos\phi = 0,71$ (power factor)

1410 r/min (nominal mechanical speed)

$R_s = 10.4\ \Omega$ (stator resistance)

$R_r = 11.6\ \Omega$ (rotor resistance)

$L_{\gamma s} = 22$ mH (stator leakage inductance)

$L_{\gamma r} = 22$ mH (stator leakage inductance)

$L_{mn} = 0,557$ H (magnetization inductance)

$J_m = 0,0072\ \text{kgm}^2$ (inertia moment)

$I_{dn} = 1.501$ A (nominal value of i_{sd} current)

$I_{qn} = 2.093$ A (nominal value of i_{sq} current).

Nomenclature

R_s, R_r: Resistance of stator and rotor winding
L_s, L_r: Self-inductance of stator and rotor
L_m: Magnetizing inductance
σ: Leakage factor
L'_s: Stator transient inductance
P: Number of pole pairs
ϑ_s, ω_s: Rotor flux angle and angular speed
ϑ, ω: Rotor angle and angular speed
ω_s: slip speed
T_{em}: Elektromagnetic torque
ψ_{sd}: Magnetizing flux
i_{sd}, i_{sq}: d and q components of stator current vector.

References

[1] F. Abrahamsen, F. Blaabjerg, J. K. Pedersen, P. Z. Grabowski, and P. Thgersen, "On the energy optimized control of standard and high-efficiency induction motors in CT and HVAC applications," *IEEE Transactions on Industry Applications*, vol. 34, no. 4, pp. 822–831, 1998.

[2] M. Chis, S. Jayaram, and K. Rajashekara, "Neural network-based efficiency optimization of EV drive," in *Proceedings of the Canadian Conference on Electrical and Computer Engineering (CCECE '97)*, pp. 454–457, May 1997.

[3] E. S. Sergaki and G. S. Stavrakakis, "Online search based fuzzy optimum efficiency operation in steady and transient states for DC and AC vector controlled motors," in *Proceedings of the International Conference on Electrical Machines (ICEM '08)*, Vilamoura, Portugal, September 2008.

[4] F. Abrahamsen, J. K. Pedersen, and F. Blaabjerg, "State-of-Art of optimal efficiency control of low cost induction motor drives," in *Proceedings of Conference on Fuzzy Systems (PESC '96)*, pp. 920–924, 1996.

[5] M. N. Uddin and S. W. Nam, "Development of a nonlinear and model-based online loss minimization control of an IM drive," *IEEE Transactions on Energy Conversion*, vol. 23, no. 4, pp. 1015–1024, 2008.

[6] B. Blanusa, P. Matić, Z. Ivanovic, and S. N. Vukosavic, "An improved loss model based algorithm for efficiency optimization of the induction motor drive," *Electronics*, vol. 10, no. 1, pp. 49–52, 2006.

[7] S. N. Vukosavic and E. Levi, "Robust DSP-based efficiency optimization of a variable speed induction motor drive," *IEEE Transactions on Industrial Electronics*, vol. 50, no. 3, pp. 560–570, 2003.

[8] G. C. D. Sousa, B. K. Bose, and J. G. Cleland, "Fuzzy logic based on-line efficiency optimization control of an indirect vector-controlled induction motor drive," *IEEE Transactions on Industrial Electronics*, vol. 42, no. 2, pp. 192–198, 1995.

[9] D. de Almeida Souza, W. C. P. de Aragao Filho, and G. C. D. Sousa, "Adaptive fuzzy controller for efficiency optimization of induction motors," *IEEE Transactions on Industrial Electronics*, vol. 54, no. 4, pp. 2157–2164, 2007.

[10] S. Ghozzi, K. Jelassi, and X. Roboam, "Energy optimization of induction motor drives," in *Proceedings of the IEEE International Conference on Industrial Technology (ICIT '04)*, pp. 602–610, December 2004.

[11] Z. Liwei, L. Jun, W. Xuhui, and T. Q. Zheng, "Systematic design of fuzzy logic based hybrid on-line minimum input power search control strategy for efficiency optimization of IM," in *Proceedings of the CES/IEEE 5th International Power Electronics and Motion Control Conference (IPEMC '06)*, pp. 1012–1016, August 2006.

[12] B. Blanusa and S. N. Vukosavic, "Efficiency optimized control for closed-cycle operations of high performance induction motor drive," *Journal of Electrical Engineering*, vol. 8, pp. 81–88, 2008.

[13] C. Chakraborty and Y. Hori, "Fast efficiency optimization techniques for the indirect vector-controlled induction motor drives," *IEEE Transactions on Industry Applications*, vol. 39, no. 4, pp. 1070–1076, 2003.

[14] B. Blanuša, P. Matić, and B. Dokić, "New hybrid model for efficiency optimization of induction motor drives," in *Proceedings of 52nd International Symposium ELMAR-2010*, pp. 313–317, 2010.

[15] P. Was, *Electrical Machines and Drives*, Oxford University Press, 1992.

FPGA-Based Fixed Point Implementation of a Real-Time Induction Motor Emulator

L. Charaabi

L.S.E., Ecole Nationale d'Ingénieur de Tunis, BP 37, 1002 Le Belvédère, Tunisia

Correspondence should be addressed to L. Charaabi, lotfi.charaabi@enit.rnu.tn

Academic Editor: Jose Pomilio

This paper investigates the numerical issue of a discrete-time induction-motor emulator implementation. The stability analysis of the finite-word-length implementation shows a coupling between required word length and the sample rate. We propose specific guidelines to analyze this coupling and to estimate the required data word length for both signals and coefficients of the model. To respect algorithm requirements, an FPGA-based implementation was used for architecture development. The direct torque control is implemented to verify in real time the AC-motor emulator prototype.

1. Introduction

Control algorithms of electrical drives are usually tested by the attachment of electrical motors. Experimental testing requires direct measuring by employing measurement instrumentation and sensors that would be complex, impractical, noise sensitive, and expensive. Traditionally these tests are performed on motor and inverter models under software simulation environments which are, in most cases, non-real-time and unable to exactly replicate real operational conditions.

In order to provide a real-time verification of the implemented control algorithm and increase the realism of simulation, the control algorithm is tested while connected to a real-time emulator for the plant (induction motor and inverter) [1]. Due to their high capacities of executing in real-time complex algorithm, FPGA technology is a good candidate for this kind of RT emulator [2].

For reasons of cost, simplicity, speeds and memory space, the real-time emulator of the induction motor is performed with the Euler's shift discretization method and quantized with fixed point format [3].

The robustness of the discretized algorithm is a critical issue in fixed-point format implementation. It is well known that, stable, closed-loop system may become unstable when the algorithm is implemented using a fixed-point processor due to finite-word-length effects. The fractional part precisions in fixed point are chosen to guarantee a minimum signal-to-noise ratio for finite-word-length quantization effects. The integer part is computed using norms [4].

In an FPGA implementation, it is possible to use separate fixed-point format for each coefficient and signal in the algorithm. Hence, the use of FPGA allows maintaining a higher computational precision at critical points.

In this paper, authors propose a real-time induction motor emulator designed in fixed point format for an FPGA implementation. The starting point is a continuous-time model. Discrete-time models are derived using the traditional shift form approximation. A study in terms of stability and finite-world-length effects is shown and a design technique for choosing the coefficient and signal bit widths is given [5, 6].

To demonstrate the accurateness of the real-time induction motor emulator, a particular example is developed: the direct torque control (DTC) of an induction motor [7].

The paper is organized as follows. The second section reminds the algorithmic study of a discrete-time induction motor model-based shift form approximation. Section 3 gives an efficient methodology for choosing the finite fixed-point word length. Section 4 shows the FPGA-based architecture development and gives emulation results of the induction motor emulator with the DTC algorithm.

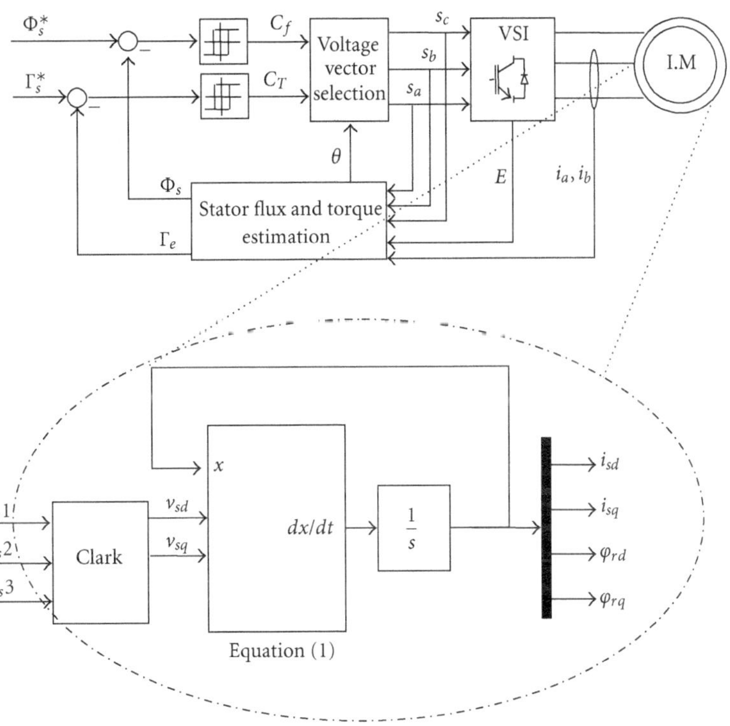

FIGURE 1: Block diagram of the DTC technique.

2. Discrete-Time Model Based on Shift Form Approximation

The block diagram of the complete system is outlined in Figure 1 including the control scheme, the induction motor, and the power chain.

As it can be seen, the DTC algorithm is decomposed in three specific blocks.

(i) Transformation block, which is composed by two subalgorithms based on the direct Concordia transformation algorithm for the current and voltage three phases models.

(ii) Estimation block, which is composed by the calculation subalgorithms of flux ($\phi_{s(d,q)}, \phi_s$), torque (Γ_e), and flux sector position. The discrete-time model of the flux subalgorithm is based on the shift form approximation:

$$\phi_{s(d,q)}(k+1) = \phi_{s(d,q)}(k) + T_{vs(d,q)} - Ti_{s(d,q)}. \quad (1)$$

(iii) command block, which is composed by a look-up table (LUT) corresponding to the switching function of the power inverter.

In order to provide a real-time verification of the implemented control algorithm, the DTC is tested while connected to a real-time emulator for the induction motor and inverter.

The induction machine is a nonlinear high-order system and for this reason complicated models must be used to control it. The dynamic behaviour of the induction motors can be described by a set of differential equations in a rotating reference frame with an angular velocity of ω_λ. Moreover it is assumed that a stationary reference frame is fixed to the stator axis. The proposed model also assumes the classical decoupling between mechanical and electrical modes. It yields that under these assumptions, the state model of the induction motor (IM) is [8]

$$\begin{aligned} \frac{d}{dt}y(t) &= A(\omega)y(t) + Bu(t), \\ y(t) &= Cx(t) + Du(t), \end{aligned} \quad (2)$$

where, $x = y = \begin{bmatrix} i_{sd} \\ i_{sq} \\ \phi_{rd} \\ \phi_{rq} \end{bmatrix}$, $u = \begin{bmatrix} u_{sd} \\ u_{sq} \end{bmatrix}$ matrices A, B, C, and D are 4-by-4, 4-by-2, 4-by-4, and 4-by-2, respectively:

$$A(\omega) = \begin{bmatrix} -\gamma & 0 & \frac{M_{sr}}{\sigma L_s L_r T_r} & \frac{M_{sr}}{\sigma L_s L_r}\omega \\ 0 & -\gamma & -\frac{M_{sr}}{\sigma L_s L_r}\omega & \frac{M_{sr}}{\sigma L_s L_r T_r} \\ \frac{M_{sr}}{T_r} & 0 & -\frac{1}{T_r} & -\omega \\ 0 & \frac{M_{sr}}{T_r} & \omega & -\frac{1}{T_r} \end{bmatrix},$$

$$B = \begin{bmatrix} \frac{1}{\sigma L_s} & 0 \\ 0 & \frac{1}{\sigma L_s} \\ 0 & 0 \\ 0 & 0 \end{bmatrix}, \quad C = \begin{bmatrix} 1 & 0 & 0 & 0 \\ 0 & 1 & 0 & 0 \\ 0 & 0 & 1 & 0 \\ 0 & 0 & 0 & 1 \end{bmatrix}, \quad D = \begin{bmatrix} 0 & 0 \\ 0 & 0 \\ 0 & 0 \\ 0 & 0 \end{bmatrix}. \tag{3}$$

$T_r = L_r/R_r$ is the rotor time constant, σ is leakage coefficient with $\sigma = 1 - M_{sr}^2/L_sL_r$, $\gamma = (R_s + M_{sr}/L_rT_r)/\sigma L_s$, and ω is the angular velocity.

By means of (2), the general equation of the electromagnetic torque is transformed to an advantageous form:

$$\Gamma_{em} = p\frac{M_{sr}}{L_r}\left(\phi_{rd}i_{sq} - \phi_{rq}i_{sd}\right). \tag{4}$$

The purpose of this section is to develop a computationally efficient discrete-time approximation of the continuous motor model operating in real time.

The most used discretization method is based on Forward shift approximation [8]. The shift form approximation is given by

$$\frac{d}{dt} = \frac{(z-1)}{T}, \tag{5}$$

where T is the sampling period.

Introducing (5) in (2) yields

$$\begin{aligned} x_z(k+1) &= A_z x_z(k) + B_z u(k), \\ y_z(k) &= C x_z(k) + D u(k), \end{aligned} \tag{6}$$

where

$$x_z(k) = y_z(k) = \begin{bmatrix} i_{sd}(k) \\ i_{sq}(k) \\ \phi_{rd}(k) \\ \phi_{rq}(k) \end{bmatrix}, \quad A_z = (I + AT), \quad B_z = BT. \tag{7}$$

The data flow graph (DFG) corresponding to the induction motor model is presented in Figure 2. Coefficients a_i depend on the parameters of the machine and are obtained for a per unit model.

The following lines show the different coefficients used for the per unit (PU) DTC algorithm

$$A = \begin{bmatrix} -a_1 & 0 & a_2 & a_3\omega \\ 0 & -a_1 & -a_3\omega & a_2 \\ a_5 & 0 & a_6 & -\omega \\ 0 & a_5 & \omega & -a_6 \end{bmatrix}, \tag{8}$$

where, $a_1 = \gamma$, $a_2 = (M_{sr}/\sigma L_s L_r T_r)(\phi_u/I_u)$, $a_3 = (M_{sr}/\sigma L_s L_r)(v_u/I_u)$, $a_4 = 1/\sigma L_s$, $a_5 = (M_{sr}/T_r)(I_u/\phi_u)$ and $a_6 = 1/T_r$.

The base values are determined from the maximal values by using the following equations, where I_m, v_m, and f_m are, respectively, the phase maximal current, the phase to neutral maximal voltage, and the maximal frequency of the induction motor:

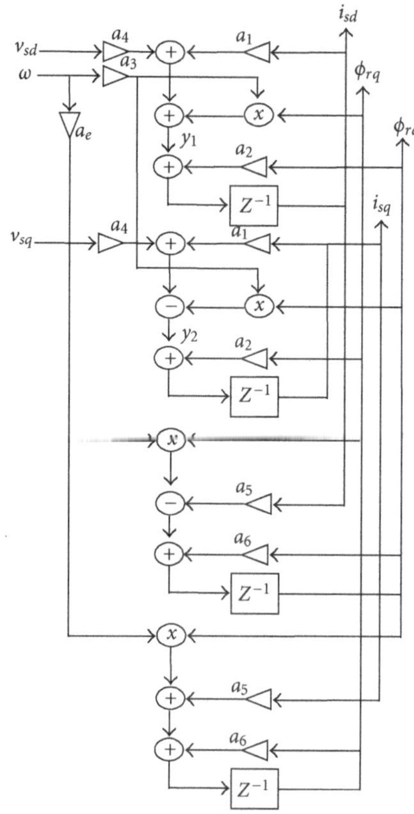

FIGURE 2: The DFG of induction motor model.

$$I_u = \sqrt{2}I_m, \qquad v_u = \sqrt{2}v_m, \qquad \phi_u = \frac{v_m}{2\pi f_m}. \tag{9}$$

The solution of the continuous system (2) is

$$x(t) = e^{At}x(0) + \int_0^t e^{A(t-\tau)}Bu(\tau)d\tau. \tag{10}$$

And if we consider the solution at the instant kT, we obtain

$$x(kT) = \left(e^{AT}\right)^k x(0) + \int_0^{kT} e^{A(kT-\tau)}Bu(\tau)d\tau. \tag{11}$$

The solution at the instant kT to the system (6) is the following:

$$x_z(kT) = (I + AT)^k x_z(0) + T\sum_{l=0}^{k-1}(I + AT)^{k-l-1}Bu(lT). \tag{12}$$

Therefore, the matrices that rule the continuous and discrete time systems are, respectively, e^{AT} and $(I + AT)$, and, consequently, the continuous and discrete time systems will be described by the eigenvalues or the poles of these matrices.

The poles of the matrix e^{AT} are $\lambda = e^{\beta T}$, β being a pole of the matrix A of the continuous time system; the poles of the matrix $(I + AT)$ are $(\lambda = 1 + \beta T)$.

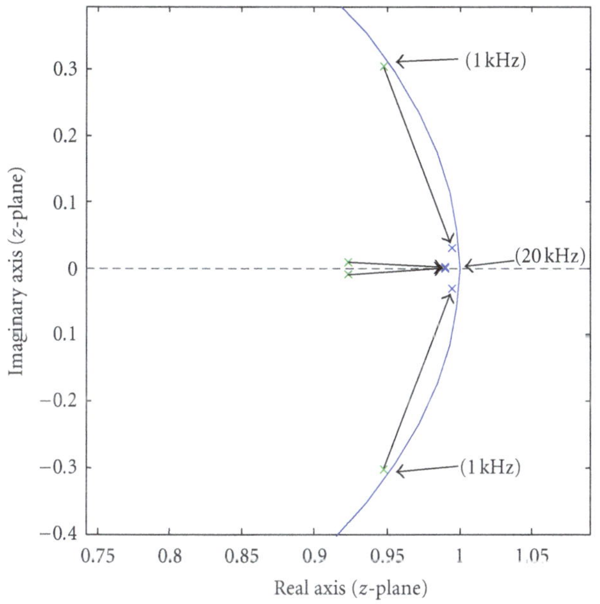

FIGURE 3: Z-domain pole location of the motor model as function of the sampling frequency.

In this study, a low power induction motor (1.5 kW) has been considered. The corresponding continuous-time model has four poles for electromagnetic model as illustrated below:

$$\beta_1 = (-46,86 + 0,5j)\omega + \frac{\sqrt{11336 - \omega^2 + 46i\omega}}{2},$$
$$\beta_3 = \overline{\beta_1},$$
$$\beta_2 = (-46,86 + 0,5j)\omega - \frac{\sqrt{11336 - \omega^2 + 46i\omega}}{2},$$
$$\beta_4 = \overline{\beta_2}. \quad (13)$$

Let us suppose the following quadratic error function:

$$\varepsilon = \left(e^{\beta T} - 1 - \beta T\right)\left(e^{\beta T} - 1 - \beta T\right)^*. \quad (14)$$

Therefore, for sampling periods T tending to zero the difference between $e^{\beta T}$ and $(1 + \beta T)$ becomes very small for each β_i pole of A, and, consequently, it will optimize the resemblance between the time-domain response of the continuous-time and the discrete-time system.

Discrete-time systems resulting from the shift form approximation are sometime unstable. It is possible, however, to select the sampling rate such that the discrete-time system is always stable when the corresponding continuous-time system is stable. For the shift form realization, the stability domain is located inside the unit circle. Therefore, a z-domain pole is stable if its module is less than one:

$$\left|\beta_i T + 1\right| < 1. \quad (15)$$

Therefore, the most critical z-domain pole corresponds to the s-domain pole having the greatest module. The pole β_1 has the greatest module as function of ω.

The locations of poles of the motor for different sampling rate values are shown in Figure 3. It can be seen that for high sampling rate, the poles become very close to one. This can be problematic, particularly when finite precision formats are considered.

3. Efficient Methodology for Choosing the Word Length of Coefficients and Variables

The following methodology is developed for choosing each coefficient and variable bit width.

3.1. Representation of Coefficients. When the model is implemented in fixed-point format, each coefficient a_k must be given a finite-precision representation composed by an integer and fractional part. Each part is estimated separately.

The integer part position for a coefficient a_k must be estimated by taking the base 2 logarithm of the maximum coefficient:

$$B_{\text{int}} = \left\lfloor \log_2 |a_k| \right\rfloor + 1, \quad (16)$$

where $\lfloor x \rfloor$ rounds x to the nearest integer less than or equal to x.

The fractional part must be determined by the maximum allowable perturbation of the coefficient from its ideal infinite-precision value. This can be achieved by computing the quantized parameter deviation or sensitivity and the stability margin. The angular velocity is selected to be the worst-case value. By means of Figure 3, for pole λ_1 and its conjugate, ω is fixed to the maximal value and for pole λ_2 and its conjugate ω is equal to zero.

Let $\{\hat{a}_k\}$ be the quantized parameter set consisting of all quantized coefficients of the state matrix A_z. When the model is implemented in fixed-point format, each coefficient of the $\{a_k\}$ set is perturbed:

$$\{\Delta a_k = \hat{a}_k - a_k\}. \quad (17)$$

Due to the finite-word-length effects, each Δa_k perturbation is bounded by

$$|\Delta a_k| \le 2^{-B_F}, \quad (18)$$

where B_F is the number of bits of the fractional part.

Due to the perturbation $\{\Delta a_k\}$ each pole $\lambda_i[(a_k)]$ is moved to $\hat{\lambda}_i = \lambda_i[A_z(a_k + \Delta a_k)]$ and it follows from a first-order approximation that the deviation can be computed with

$$\Delta\lambda_i = \hat{\lambda}_i - \lambda_i = \sum_{k=1}^{N} \frac{\partial \lambda_i}{\partial a_k}\Delta a_k, \quad (19)$$

where N is the number of modified parameters ($\Delta a_k \ne 0$).

As an example, for poles λ_2 and λ_4, the worst-case deviation is obtained with $\omega = 0$. Then, the expression of deviation

$$\Delta\lambda_2 = \Delta\lambda_4 = \frac{\partial \lambda_2}{\partial a_1}\Delta a_1 + \frac{\partial \lambda_2}{\partial a_2}\Delta a_2 + \frac{\partial \lambda_2}{\partial a_5}\Delta a_5 + \frac{\partial \lambda_2}{\partial a_6}\Delta a_6 \quad (20)$$

with, $\partial\lambda_2/\partial a_1 = -\partial\lambda_2/\partial a_6 = -(T/2)(1 + (a_1 - a_6)/\sqrt{(a_1 - a_6)^2 + 4a_5a_2})$ and $\partial\lambda_2/\partial a_2 = (a_2/a_5)(\partial\lambda_2/\partial a_5) = -(T/2)(2a_5/\sqrt{(a_1 - a_6)^2 + 4a_5a_2})$.

The quantized pole $\widehat{\lambda}_i$ may be outside the unit circle, especially with high sampling rate; thus, it is critical to know when the finite-word-length error leads to instability. This means to determine how close $\lambda_i[(a_k)]$ are to the unit circle. Let us consider the following related measure for each pole:

$$\left| \frac{\sum_{k=1}^{N}(\partial\lambda_i/\partial a_k)\Delta a_k}{1 - |\lambda_i(a_k)|} \right| < \varepsilon, \tag{21}$$

where $1 - |\lambda_i(a_k)|$ is called the stability margin of the ith pole and ε is a prescribed maximum allowable percent change in pole location relative to the stability margin.

Assuming that all coefficients have the same word-length fractional part, then $\Delta a_k = \Delta a \ \forall k \in \{1, \ldots, N\}$; it follows from inequality (21) that

$$|\Delta a| < \frac{||1 - |\lambda_i(a_k)||}{\left|\sum_{k=1}^{N}(\partial\lambda_i/\partial a_k)\right|}\varepsilon. \tag{22}$$

From inequalities (22) and (18), the required fractional part word length can be derived as

$$B_F = \left\lfloor \log_2\left(\frac{\left|\sum_{k=1}^{N}(\partial\lambda_i/\partial a_k)\right|}{||1 - |\lambda_i(a_k)||} \frac{1}{\varepsilon} \right) \right\rfloor + 1. \tag{23}$$

For the studied motor, setting $\varepsilon = 5\%$ and $T = 100\,\mu s$, the conditions (23) yield 19 bits for fractional part word length for A matrix coefficients.

3.2. Representation of Signals. The next step in the fixed-point motor model implementation is to estimate the format for each signal. An incremental methodology is used to determine the appropriate representation for the state variables. As shown in Figure 4, the first step is to find transfer function from each input and the appropriate signal in the DFG. Then the dynamic range (maximum amplitude) of each signal is carried out by using l_∞ and l_1 norms, and, thus, the integer word length range is identified. Finally, the fractional part word-length is estimated by analysis of the signal-to-quantization-noise ratio (SQNR) for each signal.

The dynamic range (maximum amplitude) of each signal of the DFG allows identifying their integer range. In practice the maximal current supported by the motor as well as the maximal produced torque and fluxes is known. Therefore, for the per-unit motor model, these dynamic ranges are bounded by 1. Notice that the voltage inputs and the electrical velocity are also normalized. Thus, these dynamic ranges are all bounded by 1:

$$\begin{aligned} \|i_{sd}\|_\infty &= \left\|i_{sq}\right\|_\infty = \|\Gamma_{em}\|_\infty = \left\|v_{sq}\right\|_\infty = \|v_{sd}\|_\infty \\ &= \|\omega\|_\infty = \|\phi_{rd}\|_\infty = \left\|\phi_{rq}\right\|_\infty = 1, \end{aligned} \tag{24}$$

where $\|\cdot\|_\infty$ denotes the the l_∞ norm, defined by

$$\|x\|_\infty = \max_k |x(k)|. \tag{25}$$

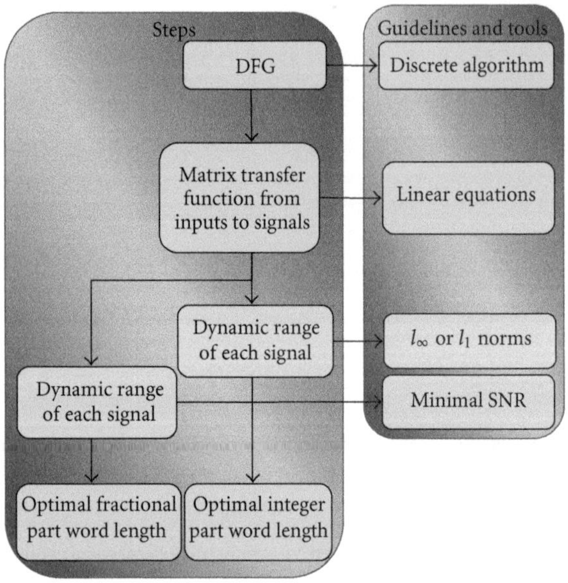

FIGURE 4: Incremental methodology for signal representation.

TABLE 1: Range bounds for the motor model signals.

Signal	Range bound	Integer size
$v_{sd}, v_{sq}, \Gamma_{em}, \omega, i_{sd}, i_{sq}, \phi_{rd}, \phi_{rq}$	1	1
y_1, y_2	1.03	2

Now considering internal y_1 and y_2 signals of the DFG, one can see that

$$\begin{aligned} y_1(z) &= a_4 v_{sd}(z) + a_3\phi_{rd}(z)\omega(z) + a_1 i_{sd}, \\ y_2(z) &= a_4 v_{sq}(z) + a_3\phi_{rq}(z)\omega(z) + a_1 i_{sq}. \end{aligned} \tag{26}$$

Therefore, signals y_1 and y_2 are bounded by the same quantity:

$$\|y_1\|_\infty = \|y_2\|_\infty \le |a_4| + |a_1| + |a_2|. \tag{27}$$

The integer part word length of all the signals is then estimated by taking the base 2 logarithm of the range bound. Table 1 illustrates the signal range bounds and the corresponding integer part word length.

The fractional part word length is estimated by analyzing the signal-to-quantization-noise ratio (SQNR) for each signal [9]. The SQNR is measured in decibels as

$$\text{SQNR} = 10\log_{10}\left(\frac{\sigma^2_{\text{signal}}}{\sigma^2_{\text{noise}}}\right). \tag{28}$$

There are two possible sources of quantization noise in the induction motor emulator implementation, namely:

(i) the noise introduced by the limited resolution of a possible control technique, producing an input noise variance σ_i^2,

(ii) the truncation noise which is introduced when products or sums are quantized. This noise propagates through the emulator and appears at outputs.

It is assumed that the quantization noise has the following properties.

(i) Each quantization noise source is a stationary white noise process.

(ii) Each noise source is uncorrelated with all other noise sources and the input of the system.

(iii) The error resulting from quantization can then be modeled as a random variable uniformly distributed over the appropriate error range. Therefore, the noise variance due to truncation is

$$\sigma_t^2 = \frac{2^{-2B}}{12}, \quad (29)$$

where B is the fractional part word length.

To investigate the noise propagation, a noise quantization model of the induction motor emulator is developed in Figure 5. The addition is performed without truncation. The quantization is performed after each multiplication; the quantization noise resulting from N multiplications is therefore N-times higher than the one generated by a single multiplication. It yields that the variances of noise e_{t1} and e_{t2} are

$$\sigma_{t1}^2 = \sigma_{t4}^2 = 4\frac{2^{-2B_a}}{12}, \qquad \sigma_{t2}^2 = 3\frac{2^{-2B_b}}{12}. \quad (30)$$

The output noise variances can be estimated using the matrix transfer function from the input vector $u(k) = [e_{t1}(k), e_{t1}(k), e_{t2}(k), e_{t2}(k)]$ to the output vector $y(k) = [f_i(k), f_i(k), f_\phi(k), f_\phi(k)]$. Using matrix notation in (5), the transfer function matrix response is

$$H_{v\to\phi,i}(z) = D' + C(zI - A_z)^{-1}B' = \begin{bmatrix} H_{11}(z) & H_{12}(z) & H_{13}(z) & H_{14}(z) \\ -H_{12}(z) & H_{11}(z) & -H_{14}(z) & H_{13}(z) \\ H_{31}(z) & H_{32}(z) & H_{33}(z) & H_{34}(z) \\ -H_{32}(z) & H_{31}(z) & -H_{34}(z) & H_{33}(z) \end{bmatrix}, \quad (31)$$

where

$$B' = \begin{bmatrix} a_4 & 0 & 0 & 0 \\ 0 & a_4 & 0 & 0 \\ 0 & 0 & 1 & 0 \\ 0 & 0 & 0 & 1 \end{bmatrix}, \qquad D' = \begin{bmatrix} 0 & 0 & 0 & 0 \\ 0 & 0 & 0 & 0 \\ 0 & 0 & 0 & 0 \\ 0 & 0 & 0 & 0 \end{bmatrix}. \quad (32)$$

Let $h_{x\to y}(k)$ denote the impulse response from input x to output y for a linear system. Assuming that the input signal has a maximal value $x_{\max}$, the output y must be bounded [10] so that

$$\|y\|_\infty \le x_{\max}\left\|h_{x\to y}\right\|_1, \quad (33)$$

where $\|\cdot\|_1$ denotes the l_1 norm, defined by $\|x\|_1 = \sum_{k=0}^{\infty}|x(k)|$.

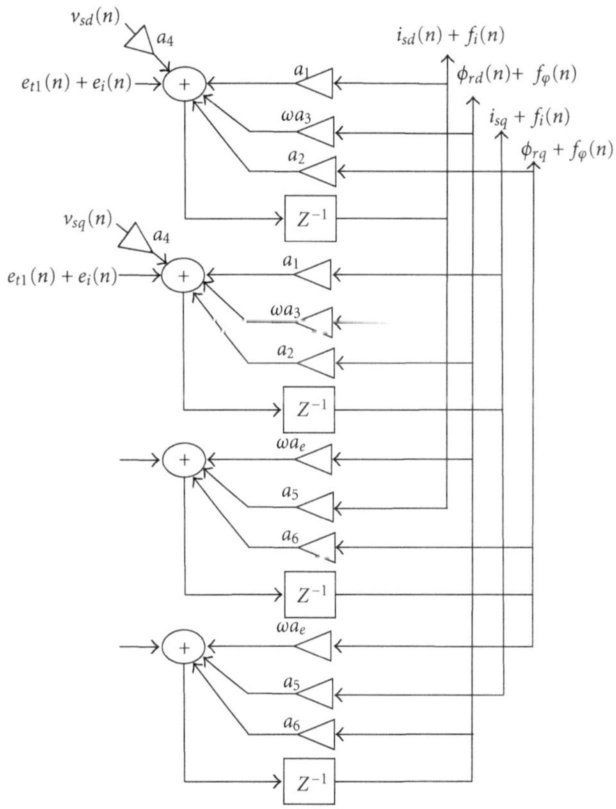

FIGURE 5: An additive noise model for the truncation and control technique noise through the induction motor model.

Using (31) and (33), the variances of resulting output noises are governed by these bounds [11]:

$$\begin{aligned} \left\|\sigma_{fi}^2\right\|_\infty &\le (\sigma_{t1}^2 + \sigma_i^2)(\|H_{11}(z)\|_1 + \|H_{12}(z)\|_1) \\ &\quad + (\sigma_{t2}^2)(\|H_{13}(z)\|_1 + \|H_{14}(z)\|_1), \\ \left\|\sigma_{f\phi}^2\right\|_\infty &\le (\sigma_{t1}^2 + \sigma_i^2)(\|H_{31}(z)\|_1 + \|H_{32}(z)\|_1) \\ &\quad + (\sigma_{t2}^2)(\|H_{33}(z)\|_1 + \|H_{34}(z)\|_1). \end{aligned} \quad (34)$$

Each coefficient $\|H_{ij}(z)\|_1$ is computed to have a maximal value as function of electrical velocity ω:

$$\left\|H_{ij}(z)\right\|_1 = \max_{\omega}\left\|H_{ij}(z,\omega)\right\|_1. \quad (35)$$

The l_1 norms of discrete-time impulse responses corresponding to these transfer functions are computed approximately by finite sums. Figure 6 shows the impulse response for each coefficient of the matrix $H_{v\to\phi,i}(z)$.

$$\left\|\sigma_{fi}^2\right\|_\infty \le 9697\frac{2^{-2B}}{12} + 1246(\sigma_i^2), \quad (36)$$

$$\left\|\sigma_{f\phi}^2\right\|_\infty \le 2275\frac{2^{-2B}}{12} + 115(\sigma_i^2). \quad (37)$$

Assuming that all signals have the same word length, the first inequality gives the worst-case variance error and

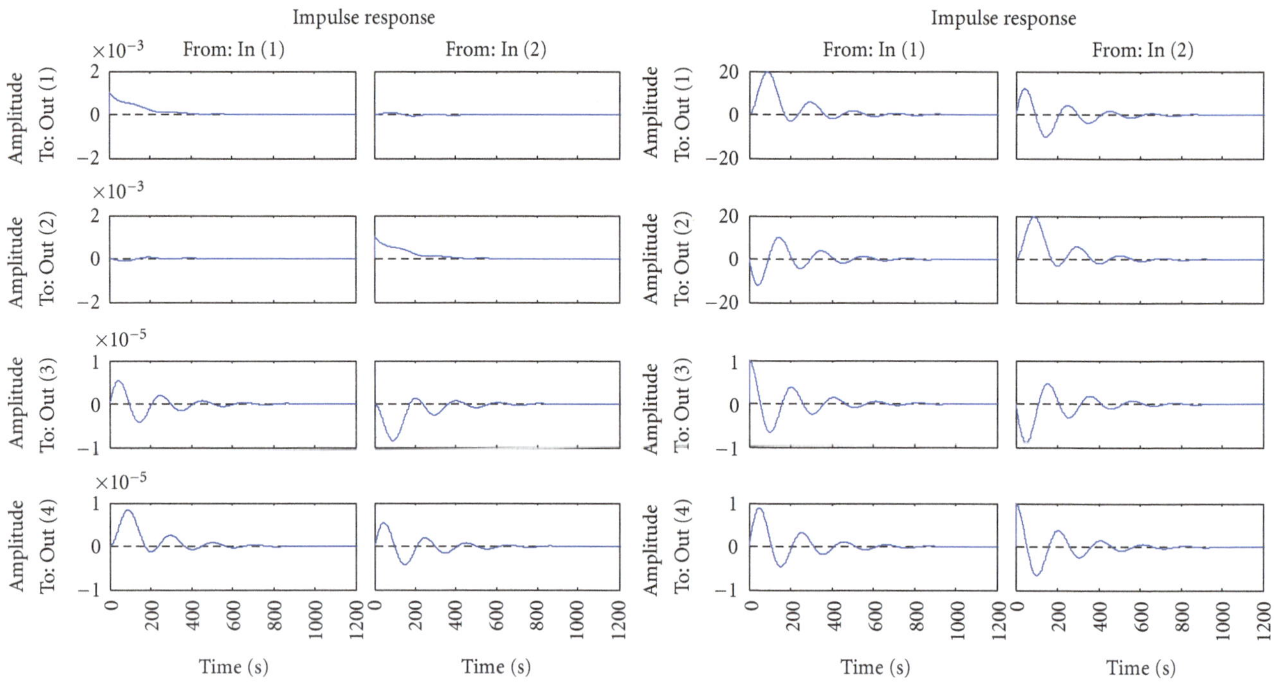

FIGURE 6: Impulse response of the matrix $H_{v \to \phi, i}$ with sample period $T = 100\,\mu$s.

therefore the fractional part word length will be selected taking into account this condition.

The variances of the sinusoidal input and output signals of the induction motor emulator are expressed in terms of the root mean square. Hence (18) gives

$$\sigma^2_{\text{signal}} = \text{signal}^2_{rms} = \frac{1}{2}. \tag{38}$$

Assume that

$$\sigma_i^2 = \frac{2^{-2B_i}}{12}, \tag{39}$$

where B_i is the fractional part word length of the input voltage. The resolution is strongly dependent on the PWM one.

Substituting (38) and (28) in (36) gives this condition:

$$\text{SQNR}_{fi} \geq -10 \log_{10}\left(1628.2^{-2B} + 208.2^{-2B_i}\right). \tag{40}$$

Figure 7 shows the SQNR as function of the fractional part word length for different B_i bits. With a 10-bit B_i wordlength, we obtain SQNR = 75 db for 18-bit B wordlength.

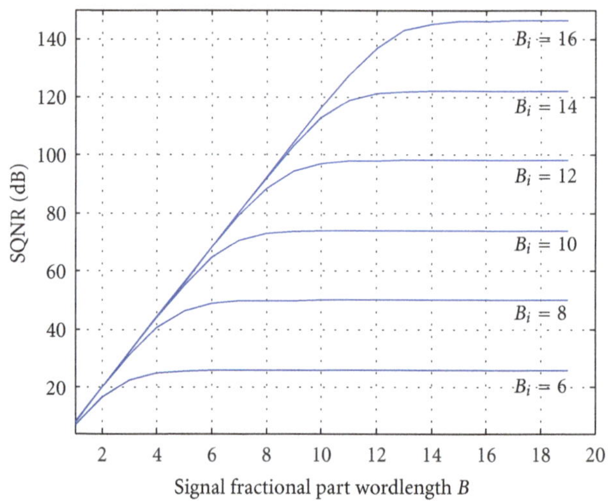

FIGURE 7: SQNR as function of the signal fractional part word length.

The emulator module requires important hardware resource. Therefore, the corresponding architecture is optimized in terms of consumed resources by the AAA methodology [12] which leads to only 3 multipliers.

As in [13], the proposed architecture consists in a control unit and a data path. The control unit generates a suitable timing schedule to control the data path and other local sequencers. Figure 8 depicts the proposed system architecture.

4. FPGA-Based Induction Motor Emulator

4.1. System Architecture. The induction motor emulator is implemented using modular and standard design principles on a Xilinx Development kit, which contain a FPGA from SPARTAN III family, the XC3s200.

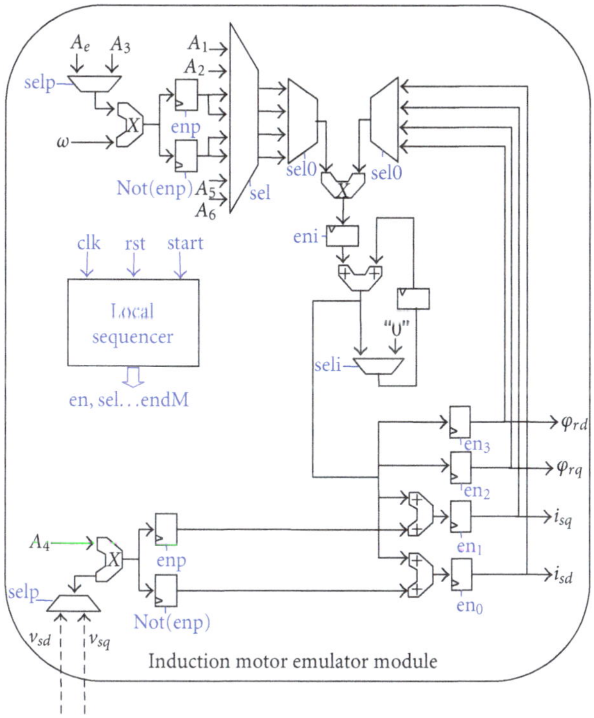

FIGURE 8: Proposed system architecture.

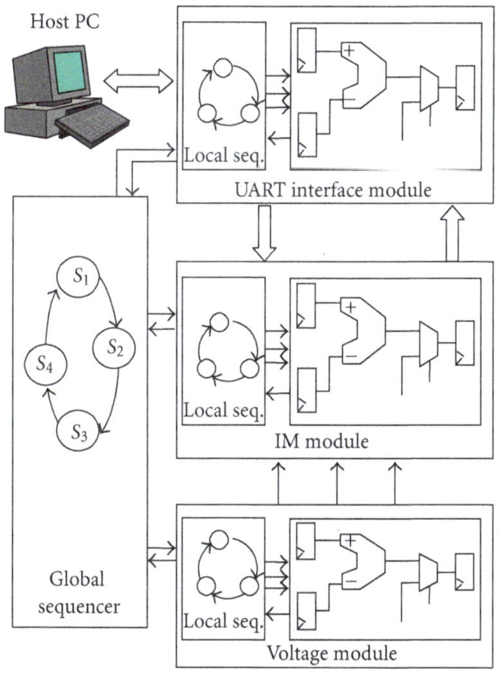

FIGURE 9: Proposed system architecture.

After the specification of the algorithm with data wordlength and sampling frequency, we propose to develop a modular architecture for the induction motor model to be implemented through an FPGA device.

4.2. System Architecture under Test in Open Loop. Figure 9 depicts the proposed system architecture. It includes a set of dedicated modules to guarantee the real-time hardware in the loop (HIL) simulation.

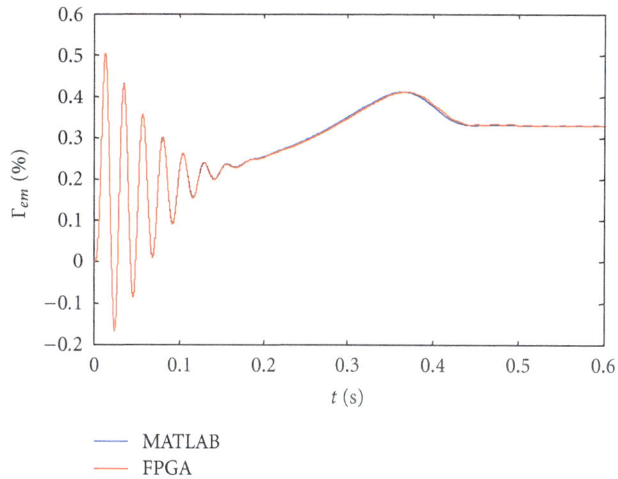

FIGURE 10: Simulated torque and torque produced by the implemented model on FPGA during a start-up.

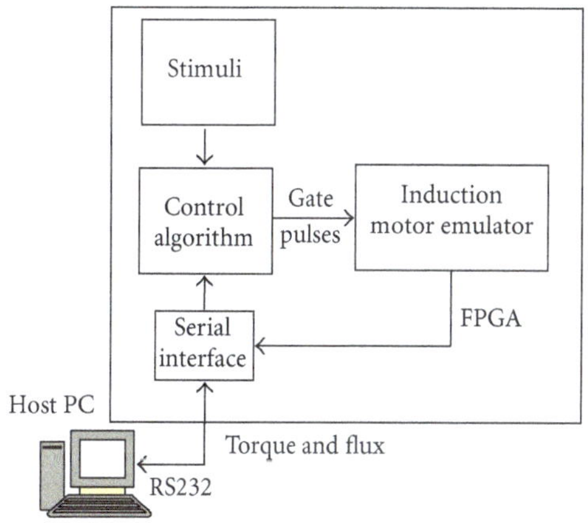

FIGURE 11: Real-time test bed.

The functional blocks of the global architecture are listed below.

(i) The UART (Universal Asynchronous Receiver Transmitter) module that provides a serial communication between the host PC and the implemented architecture. The UART allows both reconfigurations in real time of the induction motor model by tuning its coefficients and data acquisition to be visualized on the PC. To solve the dialogs between the UART implemented on the FPGA board and the PC, MATLAB's serial port interface is used.

(ii) The input voltage stimuli module. It generates three digital values, which represent the three phases voltage. One can control the output waves amplitude and frequency in real time via the UART module. This module is used to test the induction motor (IM) module.

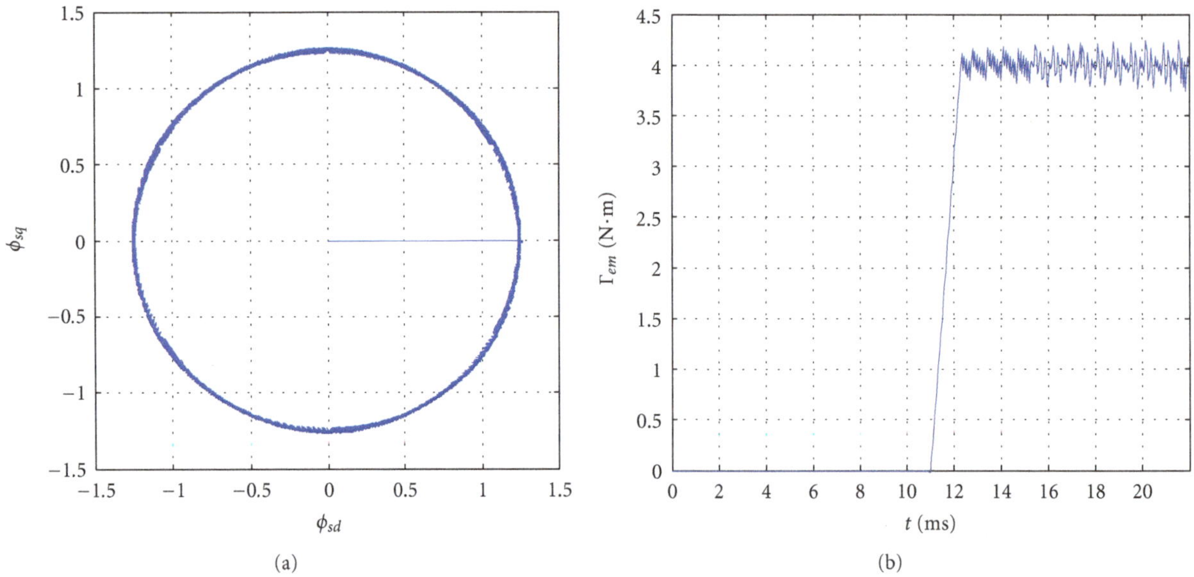

Figure 12: Real-time simulation results, obtained with DTC algorithm and the IM emulator: (a) steady-state stator flux vector locus, (b) estimated torque response to a step control from 0 N·m to 4 N·m. $T = 100\,\mu$s, $\omega = 500$ rpm and $\Phi_s^* = 1.15$ wb.

(iii) The IM module that requires the most important hardware resources. It implements the induction motor model.

(iv) The global sequencer module. It represents the control unit that generates a suitable timing schedule to control data path and other local sequencers.

The real-time simulation results presented in Figure 10 with 10 kHz sampling frequency show a slight difference between the torque produced by the implemented motor model architecture and the simulated torque under MATLAB environment under continuous-time conditions. This difference is too small to affect the performance of the real-time model.

4.3. System Architecture under Test with DTC Algorithm. The DTC is a well-known induction motor control strategy [7]. This control algorithm is selected to be tested with the real time induction motor model. The implemented architecture of the DTC control algorithm has been tested with the induction motor model architecture in real time. Figure 11 illustrates the proposed real time test bed. It includes a set of dedicated modules to guarantee a real-time simulation.

(i) The UART (Universal Asynchronous Receiver Transmitter) module that provides a serial communication between the host PC and the implemented architecture.

(ii) The stimuli module. It generates the references for the DTC control algorithm.

The torque and the stator flux are collected from the serial interface and visualized under Matlab-Simulink environment. Figure 12 illustrates the real-time emulation results.

5. Conclusions

This work has presented an optimized fixed-point format induction motor model intended for real-time simulation and emulation. With the analysis of the coupling between the sample rate and the data word length, this work has provided a theoretical guideline to find the optimal hardware implementation. The proposed architecture of the model has been successfully verified by the development and implementation of a real-time test bed that contains the AC-motor model and a DTC control algorithm.

The methodology for estimating the appropriate data word length may be applied to other various discrete-time system realizations.

Acknowledgment

This paper was supported by the Tunisian Ministry of High Education and Research: UR-LSE-ENIT-03/UR/ES05.

References

[1] S. Lentijo, A. Monti, E. Santi, C. Welch, and R. Dougal, "A new testing tool for power electronic digital control," in *Proceedings of the IEEE 34th Annual Power Electronics Specialists Conference*, vol. 1, pp. 81–87, Acapulco, Mexico, June 2003.

[2] T. Grandpierre, C. Lavarenne, and Y. Sorel, "Optimized rapid prototyping for real-time embedded heterogeneous multiprocessors," in *Proceedings of the 1999 7th International Conference on Hardware/Software Codesign (CODES '99)*, pp. 74–78, Rome, Italy, May 1999.

[3] J. Wu, S. Chen, G. Li, and J. Chu, "Optimal finite-precision state-estimate feedback controller realizations of discrete-time systems," *IEEE Transactions on Automatic Control*, vol. 45, no. 8, pp. 1550–1554, 2000.

[4] D. Menard and O. Sentieys, "Automatic evaluation of the accuracy of fixed-point algorithms," in *Proceedings of the IEEE/ACM Conference on Design, Automation and Test in Europe (DATE '02)*, Paris, France, March 2002.
[5] H.-M. Cheng and G. T.-C. Chiu, "Coupling between sample rate and required wordlength for finite precision controller implementation with delta transform," in *Proceedings of the American Control Conference (ACC '07)*, pp. 3588–3593, New York, NY, USA, July 2007.
[6] R. S. H. Istepanian, G. Li, J. Wu, and J. Chu, "Analysis of sensitivity measures of finite-precision digital controller structures with closed loop stability bounds," *IEEE Proceedings of control Theory Application*, vol. 145, no. 5, pp. 472–478, 1998.
[7] I. Takahachi and T. Nogushi, "A new quick response and high efficiency control strategy of an induction motor," *IEEE Transactions on Industry Applications*, vol. IA-22, no. 5, pp. 820–827, 1986.
[8] O. Vainio, S. J. Ovaska, and J. J. Pasanen, "A digital signal processing approach to real-time AC motor modeling," *IEEE Transactions on Industrial Electronics*, vol. 39, no. 1, pp. 36–45, 1992.
[9] L. Charaabi, E. Monmasson, and I. Slama-Belkhodja, "FPGA-based real-time emulation of induction motor using fixed point representation," in *Proceedings of the 34th Annual Conference of the IEEE Industrial Electronics Society (IECON '08)*, pp. 2393–2398, November 2008.
[10] Z. Fang, J. E. Carletta, and R. J. Veillette, "A methodology for FPGA-based control implementation," *IEEE Transactions on Control Systems Technology*, vol. 13, no. 6, pp. 977–987, 2005.
[11] M. H. Hayes, *Schaum's Outline of Theory and Problems of Digital Signal Processing*, McGraw-Hill, New York, NY, USA, 1998.
[12] T. Grandpierre, C. Lavarenne, and Y. Sorel, "Optimized rapid prototyping for real-time embedded heterogeneous multiprocessors," in *Proceedings of the 1999 7th International Conference on Hardware/Software Codesign (CODES'99)*, pp. 74–78, May 1999.
[13] E. Monmasson and M. N. Cirstea, "FPGA design methodology for industrial control systems-a review," *IEEE Transactions on Industrial Electronics*, vol. 54, no. 4, pp. 1824–1842, 2007.

14

Magnetic-Field Immunity Examination and Evaluation of Transcutaneous Energy-Transmission System for a Totally Implantable Artificial Heart

Takahiko Yamamoto and Kohji Koshiji

Department of Electrical Engineering, Faculty of Science and Technology, Tokyo University of Science, 2641, Yamazaki, Noda, Chiba 278-8510, Japan

Correspondence should be addressed to Takahiko Yamamoto, yamamoto@rs.noda.tus.ac.jp

Academic Editor: Shuo Wang

Transcutaneous energy transmission (TET) is the most promising noninvasive method for supplying driving energy to a totally implantable artificial heart. Induction-heating (IH) cookers generate a magnetic flux, and if a cooker is operated near a transcutaneous transformer, the magnetic flux generated will link with its external and internal coils. This will affect the performance of the TET and the artificial heart system. In this paper, we present the design and development of a coil to be used for a magnetic immunity test, and we detail the investigation of the magnetic immunity of a transcutaneous transformer. The experimental coil, with five turns like a solenoid, was able to generate a uniform magnetic field in the necessary bandwidth. A magnetic-field immunity examination of the TET system was performed using this coil, and the system was confirmed to have sufficient immunity to the magnetic field generated as a result of the conventional operation of induction-heating cooker.

1. Introduction

Immediate energy transmission using a cable is an easy method for supplying driving energy to a totally implantable artificial heart [1–5]. However, this method increases the risk of infectious disease and decreases the patient's quality of life (QOL). Transcutaneous energy transmission (TET) is the most promising non-invasive method of energy transfer [3–5].

If an artificial heart stops because of malfunctioning of the TET system, the consequences are fatal. Therefore, according to the Pharmaceutical Affairs Act in Japan, an evaluation of electromagnetic compatibility (EMC) is indispensable [5]. The EMC requirements for an artificial heart system state that it is necessary to reduce the electromagnetic interference (EMI) and electromagnetic susceptibility (EMS) while maintaining performance. However, even if the examination satisfies the Pharmaceutical Affairs Act, the safety of the system is not necessarily guaranteed from the viewpoint of EMC. Recently, there have been some reports on the malfunctioning of medical devices such as pacemakers that was caused by magnetic-field leakage from induction-heating (IH) cookers [6]. However, EMS evaluation that assumes such an event is not currently obligatory for medical equipment.

A transcutaneous transformer transmits driving energy to an artificial heart implanted inside the body by using electromagnetic induction between two coils inside and outside the body. IH cookers generate a magnetic flux, and if a cooker is operated near a transcutaneous transformer, the magnetic flux generated will link with the transformer's external and internal coils [7]. This can affect the performance of the TET system and the artificial heart system. Hence, it is necessary to investigate the magnetic-field immunity of the TET system. In this paper, we report on the design and development of a coil for the magnetic immunity test, and we detail the investigation of the immunity of transcutaneous transformers.

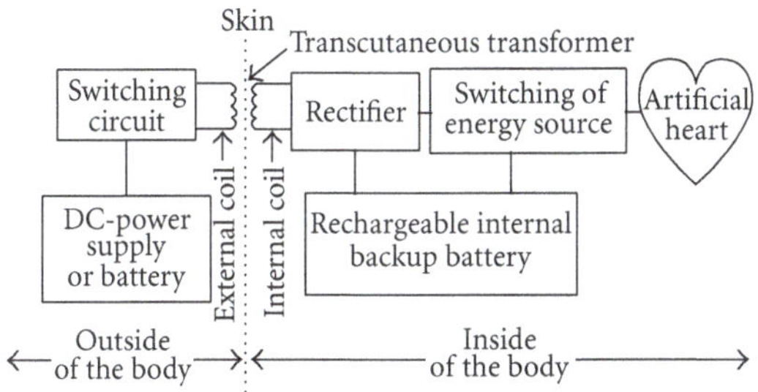

FIGURE 1: Block diagram of TET system.

2. TET System

The TET system allows the noninvasive transmission of energy to the inside of the body. Figure 1 displays a block diagram of the TET system. Outside the body, DC electric power that is supplied by a stabilized DC power supply or an external rechargeable battery is converted into high-frequency (300 kHz) AC electric power by a push-pull-type inverter circuit. The AC electric power is transmitted to the inside of the body through the transcutaneous transformer, as shown in Figure 2(a). The transcutaneous transformer consists of an external coil and an internal coil, of nine turns each, using Litz wire. Two primary coils outside the body are closely wound in the toroidal-type ferrite. To detach a coil outside the body easily, the ferritic core combined two ferrites of the character type C. The half of the annulus ring of the coil that is inside the body (nine turns) was buried under the hypodermis, and the remaining half was wrapped with the skin and thrust out like an arch (Figure 2(b)). Figure 2(c) shows an installation of the passing skin transformer on a goat.

Inside the body, AC electric power is converted into DC power in order to drive the artificial heart actuator and charge the rechargeable internal backup battery.

3. Design of the Coil for the Magnetic Immunity Test

3.1. Deciding the Magnetic-Field Immunity Examination Frequency. The magnetic-field leakage to the outside of the IH cooker was measured, and the bandwidth at which the magnetic-field immunity test should be carried out was decided on the basis of this result. The magnetic field around the IH cooker (near field) when heating a water-filled pan with the maximum output (1.4 kW) of the cooker was measured using a near-magnetic-field probe (Hewlett Packard, 11941A, CA, USA) and a spectrum analyzer (Anritsu, MS2661C, Kanagawa, Japan). The distance from the IH cooker was 35 cm because it was assumed that the patient was working on the counter in the vicinity of the IH cooker.

Figure 3 shows the results of the magnetic-field leakage from the IH cooker. The exciting frequency was 27 kHz; the magnetic-field leakage at this frequency was 128 dBμA/m, and it decreased as the frequency increased. The maximum frequency in the leakage field from the IH cooker was 1 MHz.

(a) Composition of transcutaneous transformer

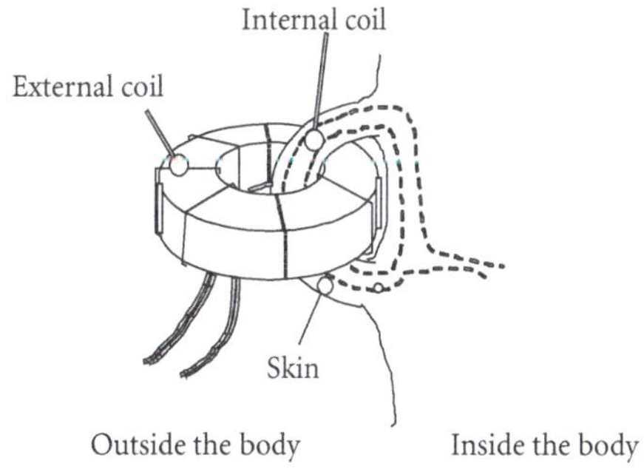

(b) Illustration of internal coil embedding and the transformer wearing

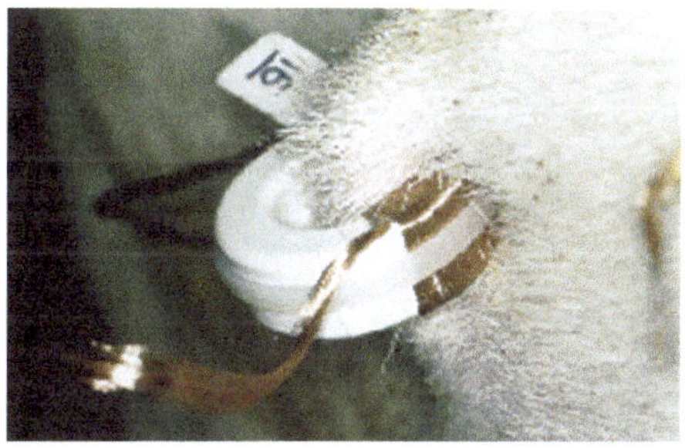

(c) Goat wearing transcutaneous transformer

FIGURE 2: Transcutaneous transformer.

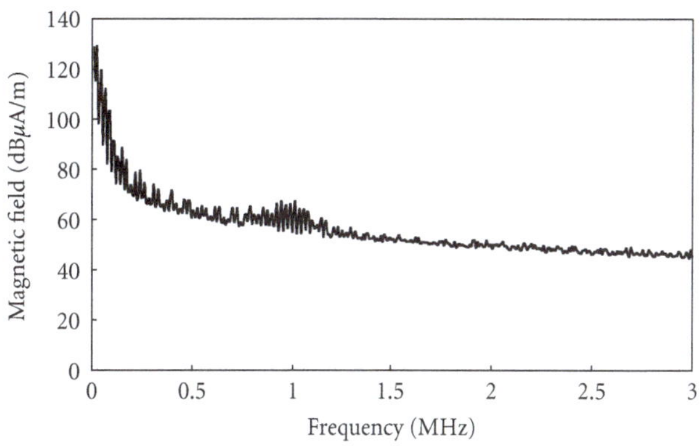

FIGURE 3: Magnetic field from IH cooker (near field).

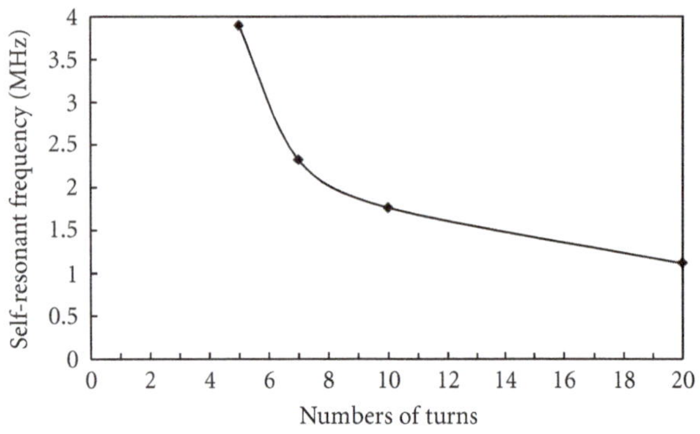

Figure 4: Self-resonant frequency of experimental coil (measured).

Therefore, we chose a test frequency range from 1 kHz to 3 MHz.

3.2. Fabrication of Test Coil for Trial Purposes. A coil is used to generate the magnetic field in the examination of the magnetic-field immunity. In this test, the coil must be able to generate a uniform magnetic field within the evaluation bandwidth. According to IEC 61000-4-3, the immunity standard for radiated radiofrequencies, the uniform field area of the immunity test must be within the range from 0 to + 6 dB. Therefore, in this paper, the field is considered to be uniform when the fluctuation band of the generated magnetic field is within 6 dB.

A Helmholtz coil can generate a uniform magnetic field. However, the stray capacitance between the winding lines is large because the wire of the coil is concentrated. The self-resonant frequency of the coil decreases, and there is a possibility that the magnetic field cannot be generated in the necessary bandwidth. Therefore, we selected a solenoid coil that increased the self-resonant frequency of the coil for the examination and that was able to generate a uniform magnetic field.

The required power supply, which was generated by increasing the number of coil turns, is low, and therefore, the immunity test can be performed easily and at a low cost. However, the self-resonant frequency decreases with an increasing number of turns. Therefore, the optimal number of coil turns is defined as the maximum number of turns for which the self-resonant frequency of the coil increases more than the bandwidth of the test.

A solenoid coil was fabricated for a cubic with a length of 1 m for the TET system. Figure 4 shows the relation between the number of turns and the self-resonant frequency. The self-resonant frequency of the coil with five turns was 3.79 MHz, and it rose more than 3 MHz of the maximum frequency of the test. It was thus decided that five turns was the appropriate number of turns for the magnetic-field immunity test coil.

Next, the uniform magnetic field of the prototype coil was evaluated. The inside of the coil side was assumed to be the x- and y-axes, and the vertical direction on the loop side was assumed to be the z-axis. The magnetic field generated in the coil by using a signal generator (ROHDE and SCHWARZ, SMY01, Munich, Germany) and a power amplifier (NF Corporation 4025, Saitama, Japan) when the frequency was assumed to be 30 kHz (the excitation frequency of the IH cooker) was measured. The magnetic field was measured by using the near-field probe and the spectrum analyzer. Figure 5 shows the magnetic-field strength distribution in the xy-plane on the top of the solenoid coil (z = 100 cm). The uniformity in the magnetic field in the area of a square with sides of length 0.8 m on the loop side was 3.3 dB, which complied with the IEC61000-4-3 standard. The magnetic-field strength distribution in the axial direction of the solenoid coil is shown in Figure 6. The magnetic-field uniformity along the z-axis was 5.6 dB, which also complied with the standard.

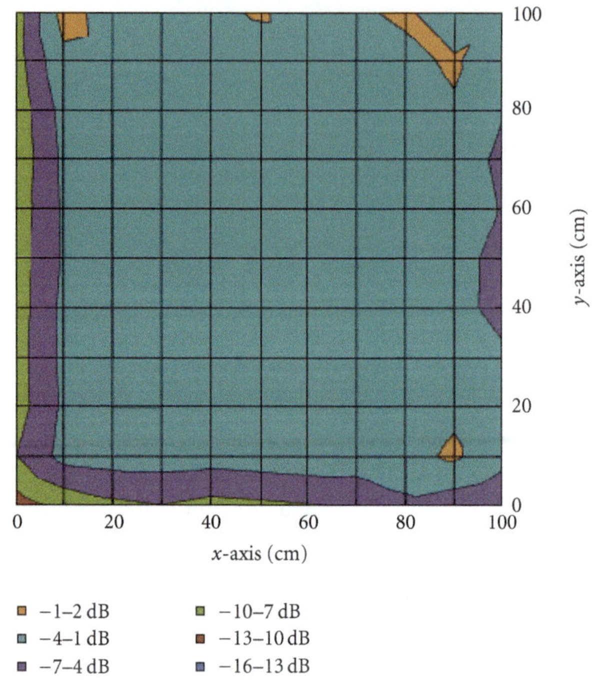

Figure 5: Magnetic-field distribution in xy-plane.

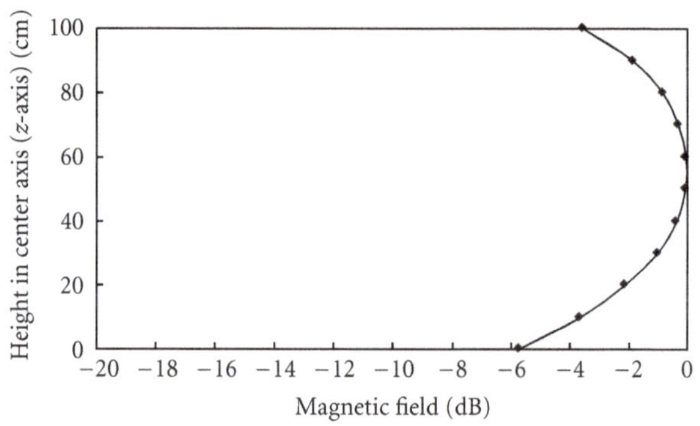

Figure 6: Magnetic-field distribution along z-axis (measured).

Therefore, the prototype coil was able to generate a uniform magnetic field in the required bandwidth of 3 MHz when wound with five turns like a solenoid. The uniformity

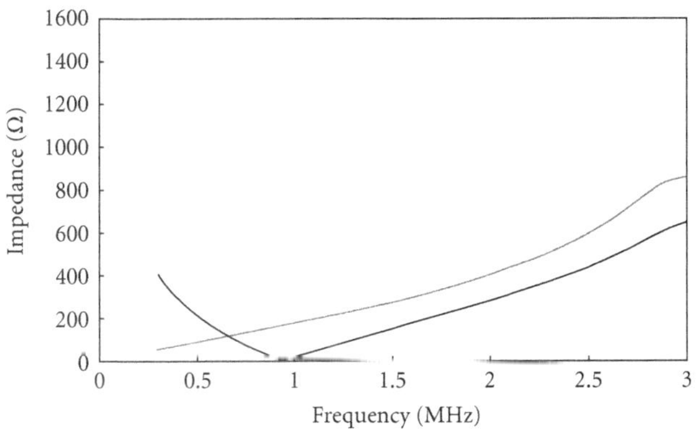

FIGURE 7: Absolute value of the impedance of the coil.

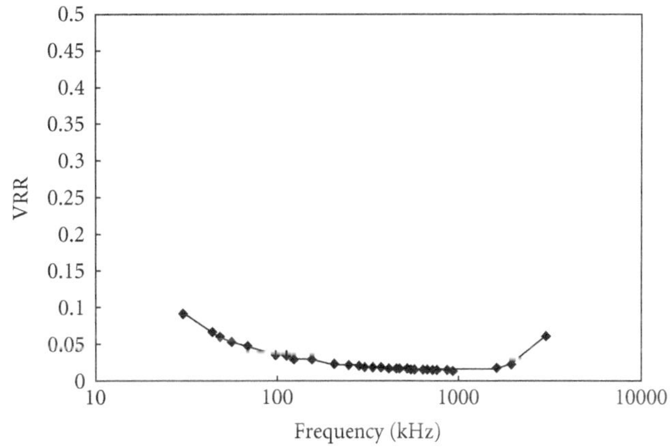

FIGURE 8: Voltage reduction ratio when the capacitor is inserted (measured).

of the generated magnetic field adhered to the standard defined by IEC61000-4-3.

3.3. Reduction in Coil Impedance. The voltage of the coil is given by expression (1), and it increases in proportion to frequency when a constant current flows into the coil. The reactance of the coil is proportional to the frequency:

$$V = j\omega LI. \quad (1)$$

The power-supply voltage necessary for the examination of the magnetic-field immunity increases at high frequencies, and the required test equipment becomes larger in scale. However, the power-supply facility should be as small as possible to ensure the easy examination. To achieve this, the necessary voltage can be decreased by connecting a capacitor in the series; this decreases the synthetic impedance to the coil. The resonance frequency when the inductance L and capacitance C are connected in the series is given by expression (2), and the capacitance is defined by expression (3):

$$f = \frac{1}{2\pi}\sqrt{LC}, \quad (2)$$

$$C = \frac{4\pi^2 f^2}{L}. \quad (3)$$

The examination standard of IEC dictates that the frequency be examined at 1% intervals. When the examination is done based on this standard, the number of test frequency points is $\log_{1.01} 3000 = 805$ in the frequency range from 1 kHz to 3 MHz (3000 times). A device that switches the capacitor automatically might be needed. This will be addressed in our future work.

As one example, Figure 7 shows the absolute value of the impedance of the coil and the absolute value of the impedance when a capacitor of $C = 1000$ pF and the coil are connected in series. The absolute value of the reactance becomes 0 Ω at 960 kHz when a capacitor of $C = 1000$ pF is inserted, and the impedance can be assumed to be minimal. The impedance of the coil can be kept low by sweeping the frequency of the current flowing in the coil while changing the capacitor connected in series to the coil. The voltage reduction ratio (VRR) is defined in expression (4) as follows:

$$\mathrm{VRR} = \frac{V_{\mathrm{in}}}{V_L}, \quad (4)$$

where V_{in} is the input power-supply voltage and V_L is the voltage in the coil. Figure 8 displays the characteristic VRR and frequency when the capacitor determined using expression (3) and the coil are connected in series. The voltage required is 9.9 V when the capacitance is inserted, and it is possible to decrease this voltage by 1.5%, although a voltage of 643 V is necessary at 940 kHz for a flow current of 4 A in the coil.

4. Magnetic-Field Immunity of the TET System

The magnetic-field immunity of the TET system was examined using a prototype solenoid coil with five turns. The TET system was set up for the examination in the center part of the coil, and a current of 3.5 A flowed into the coil. The induced voltage in the external and internal coils of the TET system was measured with an oscilloscope (Yokogawa, DL-1640, Tokyo, Japan). The generated magnetic field was 140 dBμA/m. This value was 12 dB stronger than the maximum magnetic-field strength of 128 dBμA/m 35 cm above the IH cooker with the water-filled pan, as shown in Figure 3. Figures 9 and 10 show the voltage induced in the external and internal coils of the TET system when the test frequency was assumed to be 30 kHz, which was the exciting frequency of the IH cooker. The voltages induced in the external and internal coils were 0.007 V and 0.079 V, respectively. It can be said that these voltage values are very smaller than the voltage rating of the artificial heart system (24 V); therefore, the TET system has sufficient magnetic immunity.

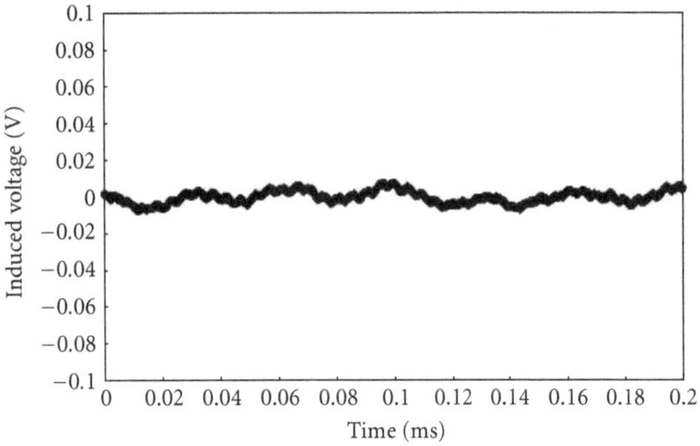

FIGURE 9: Induced voltage in external coil (measured).

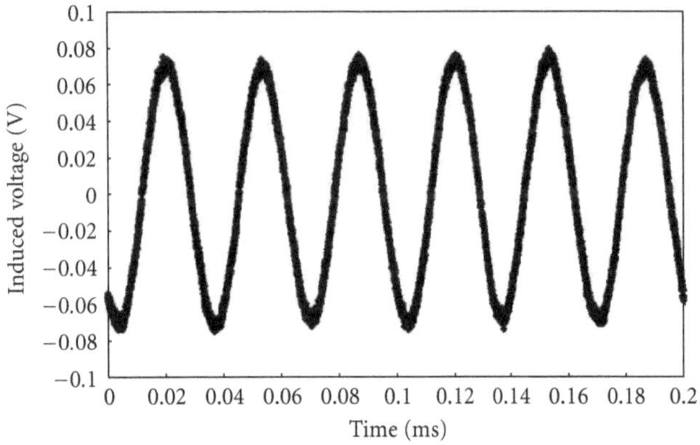

FIGURE 10: Induced voltage in internal coil (measured).

5. Conclusion

In this paper, we have reported on the design and development of a coil for the magnetic immunity test. This test is an important factor in guaranteeing the safety of TET systems for totally implantable artificial hearts or other medical devices. In addition, we detailed our investigation of the immunity of a transcutaneous transformer.

The results showed that an experimental coil with five turns like a solenoid was able to generate a uniform magnetic field in the necessary bandwidth. This uniformity of the generated magnetic field complied with the standard detailed in IEC61000-4-3. In addition, the power supply necessary for the examination could be reduced by inserting capacitors in series with the coil. Moreover, the magnetic-field immunity examination of the TET system was executed using the coil, and it was confirmed that the TET system had sufficient immunity to the magnetic field produced by the conventional usage of an IH cooker.

The high-frequency magnetic-field immunity examination proposed and discussed in this paper is important for the evaluation of equipment for home medical care, as there is a risk of exposure to a magnetic field while using such equipment. Therefore, the examination presented here is applicable not only for the TET system but also for evaluating the reliability of many other medical devices.

In future work, we will investigate the magnetic-field immunity of the TET system in other situations involving large magnetic fields and near power stations.

References

[1] Y. Yokoyama, O. Kawaguchi, T. Shinshi, U. Steinseifer, and S. Takatani, "A new pulse duplicator with a passive fill ventricle for analysis of cardiac dynamics," *Journal of Artificial Organs*, vol. 13, no. 4, pp. 189–196, 2010.

[2] K. Nawata, T. Nishimura, S. Kyo et al., "Outcomes of midterm circulatory support by left ventricular assist device implantation with descending aortic anastomosis," *Journal of Artificial Organs*, vol. 13, no. 4, pp. 197–201, 2010.

[3] T. Yamamoto, K. Koshiji, K. Tsukahara et al., "An externally-coupled transcutaneous energy transmission system for totally implantable artificial hearts-detection of abnormal coupling caused by misalignment and air gap in the ferrite core junction of the transcutaneous transformer," *Transaction of Japanese Society for Medical and Biological Engineering*, vol. 43, no. 2, pp. 261–267, 2005.

[4] T. Hongo, T. Yamamoto, H. Aoki et al., "Investigation on energy transmission efficiency of externally-coupled transcutaneous energy transmission system (ectets) miniaturized by using higher switching frequency for a totally-implantable artificial heart (tiah)," in *Proceedings of the American Society for Artificial Internal Organs 54th Annual Conference*, p. 49, 2008.

[5] T. Yamamoto, K. Koshiji, Y. Nawa et al., "Transcutaneous energy transmission system for a totally-implantable artificial heart in case using external battery," in *Proceedings of the World Congress on Medical Physics and Biomedical Engineering*, pp. 3026–3029, 2006.

[6] T. Nagatomo, H. Abe, R. Kohno et al., "Electromagnetic interference with a bipolar pacemaker by an induction heating (IH) rice cooker," *International Heart Journal*, vol. 50, no. 1, pp. 133–137, 2009.

[7] T. Yamamoto, K. Koshiji, A. Homma, E. Tatsumi, and Y. Taenaka, "Improvement in magnetic field immunity of externally-coupled transcutaneous energy transmission system for a totally implantable artificial heart," *Journal of Artificial Organs*, vol. 11, no. 4, pp. 238–240, 2008.

Novel Reduced Parts Online Uninterruptible Power Supply

Bahram Ashrafi and Mehdi Niroomand

Department of Electrical Engineering, University of Isfahan (UI), Isfahan 81746-73441, Iran

Correspondence should be addressed to Mehdi Niroomand, mehdi_niroomand@eng.ui.ac.ir

Academic Editor: Hadi Y. Kanaan

This paper presents design consideration and performance analysis of novel reduced parts online three-phase uninterruptible power supply (UPS) system. The proposed UPS system is based on reduced switch count dual bridge matrix converter. It employs only six power switches and results in reducing the cost of the system compared to conventional online UPS topologies, while achieving excellent performance. The performance of the proposed system is evaluated through simulation in terms of input/output waveforms quality and shows the viability of topology.

1. Introduction

Protecting sensitive loads like medical facilities, life supporting systems, data centers, telecommunication systems, and industrial processing systems are the main purpose of implementing uninterruptible power supplies (UPSs system) [1]. UPS system has to provide distortion free, high quality waveforms in order to ensure the reliable operation of the load, regardless of quality of the AC power source [2].

Different types of UPS systems are developed since the advent of UPS concept. Among these types, the online UPS is the most common type for low to medium power applications due to its superior performance such as high tolerance to the input voltage variation and precise regulation of the output voltage [3]. However, conventional online UPS topology employs big number of switches and as a consequence, suffers from high cost, which stops it from adopting in some cost sensitive applications like personal computers. So the idea of developing different topologies to reduce the cost of UPS systems attracts researcher's attention [4].

In general, the cost reduction is achieved by developing topologies in order to [5]

(i) reduce the number of power switches,

(ii) replace power switches with diodes, because diodes are cheaper than active power switches.

Employing the mentioned methods would further reduce the cost by eliminating the circuitry for driving power switches and simplifying the control circuit:

(i) lowering the voltage stress across power switches,

(ii) eliminating passive components like transformers, capacitors, and inductors.

Removing bulky passive elements and lowering the number of power switches would bring some advantages like greater compactness and higher reliability.

This paper present the design and performance analysis of a new online three-phase UPS system based on dual bridge matrix converter topology. The concept of reducing the number of switches is applied by using 9 switch dual bridge matrix converter. Further switch reduction is obtained by using capacitor midpoint connection (two series capacitors in dc-link) which leads to 6 switch topology.

The rest of the paper is organized as follows. In the first part, brief introduction to online UPS is presented and three different operation modes are discussed. In the next part the applied topology is introduced. It starts with dual bridge matrix converter and description of the switch reduction procedure from 24 to 9 and continues with capacitor midpoint connection concept, which further reduced the switches count to 6. Finally, simulation results are presented to demonstrate the feasibility of the proposed method.

2. Online Uninterruptible Power Supply (UPS)

Double conversion online UPS is depicted in Figure 1, based on IEC standard 62040-3.

As shown in Figure 1, this topology consists of rectifier, battery management set, battery pack, inverter, and static switches (bypass) [6]. As the term "Double Conversion" suggests, an online UPS converts power twice. At first, AC input voltage with all the distractions is converted into DC voltage. For stabilizing the DC voltage, a capacitor is used at the end of the rectification stage.

Second, DC voltage is converted back to regulated AC voltage to supply the load with high quality, uninterrupted power. The output AC voltage can even manage to have different frequency than input voltage, which is not possible with line-interactive UPS. When the AC input is available (normal mode of operation), described double conversion process would occur. In online UPS, inverter is rated at 100% of the load in order to supply the load, and rectifier power rating is set to meet the 100% of the power demanded by the load and power needed to charge the battery bank [2].

When AC input is outside of the desired range, the UPS is disconnected from AC source and the power is drawn from the battery, so the output is not affected. The transition time for switching from power source to the battery set is several milliseconds, and the capacitor in the dc-link supplies the load during transition time. So even there is a short interruption in power flow, the output voltage would remain continuous.

In brief, online UPS has 3 different operating modes: normal mode, stored energy mode, and bypass mode [1, 2].

2.1. Normal Mode of Operation. In this mode, input AC source supplies the load through double conversion process, which is AC-DC via front-end rectifier and DC-AC via back-end inverter. The dc-link voltage is also used to charge the batteries. In comparison with line-interactive UPS, online topology has higher power rating and results in higher cost. The efficiency of online UPS is usually lower than line-interactive one, so it consumes more power and produces more heat, which results in higher utility cost and additional cooling expense.

2.2. Stored-Energy Mode of Operation. In this mode the UPS is disconnected from AC input source, due to the distraction of the voltage source. The continuity of power flow is maintained by batteries and inverter. The duration of this mode depends on availability of AC line and batteries backup time.

2.3. Bypass Mode of Operation. In bypass mode, load is directly connected and fed by AC input source. It should be considered that the source frequency and load frequency should be equal in order to make the direct connection between source and load possible.

UPS operates in this mode when there is an internal malfunction or in case of maintenance.

The main advantages of online UPS are as follows [3, 4].

(i) High tolerance against the input voltage variations.

(ii) Regulation of the output voltage.

(iii) Controllable output frequency. It can perform frequency conversion (e.g., 50 Hz to 60 Hz).

The main disadvantages of this topology are low power factor, low efficiency, and high THD.

The rectification stage causes distortion to the input current. So an extra power factor correction stage is needed, which leads to higher cost:

(i) higher power losses compared to line-interactive UPS,

(ii) shorter life span compared to other topologies due to higher temperature.

Double conversion online UPS, with all mentioned drawbacks, is the most preferred topology in performance, power conditioning, and load protection. There is a large diversity in online UPS topologies to fulfill different tasks in different kind of applications, in range of few KVA to several MVA.

3. Matrix Converter

The matrix converter (Figure 2), first introduced in 1980, is an ac to ac force commutated converter structure. It is able to provide simultaneous voltage and frequency transformation [6].

Matrix converter employs four-quadrant switches, and by choosing proper modulation technique, it makes bidirectional power flow from source to load possible. So it inherently can regenerate energy back to source. Matrix converter does not employ any electrolytic capacitors or any other passive elements in dc-link, so it shows high reliability. These features make matrix converter suitable topology in applications like AC drives or power supplies [7, 8].

Recent trends to make higher quality, more efficient power supplies, caused renowned attention and extensive research work on matrix converter due to its superior characteristics like [5]:

(i) sinusoidal input and output waveforms,

(ii) adjustable power factor,

(iii) bidirectional power flow,

(iv) compact design (no bulky energy storage is needed),

(v) adjustable frequency.

In general, there are two major derivations in matrix converter topology. First one is conventional matrix converter in Figure 2, and the second one is dual bridge [9–11] approach which has "dc link" [5] and is depicted in Figure 3. Dual bridge matrix converter also called "indirect matrix converter" in the literature [11–13].

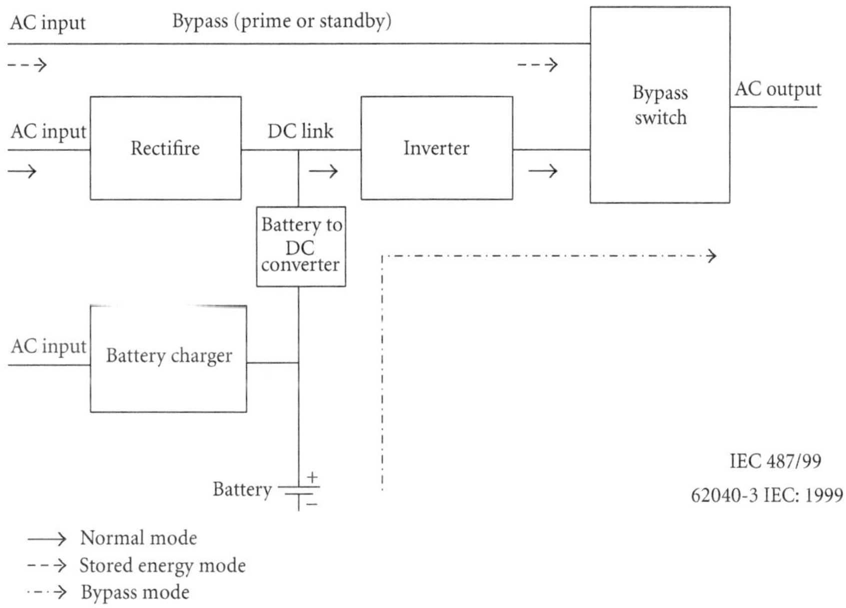

Figure 1: Double conversion online UPS based on IEC-62040 standard.

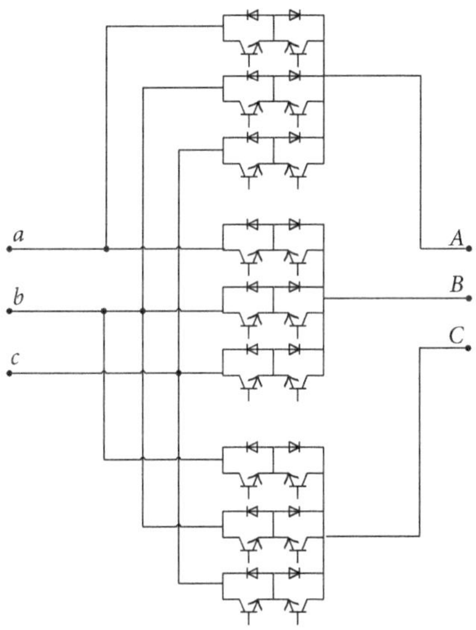

Figure 2: Conventional matrix converter.

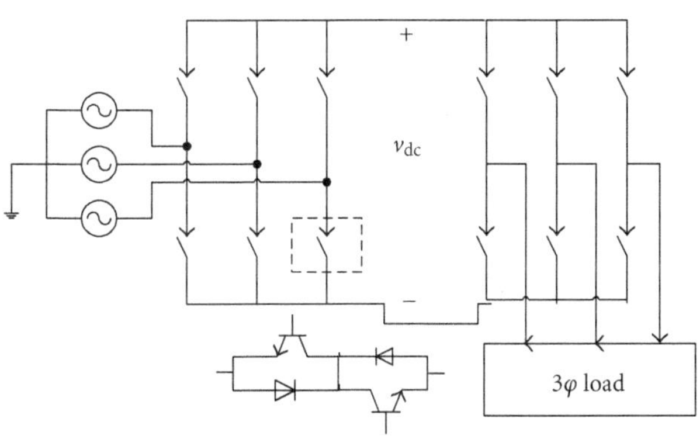

Figure 3: 24-switch topology.

3.1. Conventional Matrix Converter (CMC). CMCs have all the advantages mentioned above and its function, based on Figure 2, can be described as follows.

If the switching function of switches is defined as "0" when open and "1" when closed, the output voltage is defined by (1) as follows:

$$\begin{bmatrix} V_A \\ V_B \\ V_C \end{bmatrix} = \begin{bmatrix} S_{Aa} & S_{Ab} & S_{Ac} \\ S_{Ba} & S_{Bb} & S_{Bc} \\ S_{Ca} & S_{Cb} & S_{Cc} \end{bmatrix} \cdot \begin{bmatrix} V_a \\ V_b \\ V_c \end{bmatrix}. \quad (1)$$

But there are some drawbacks that stop CMCs from being utilized in industry such as the following:

(i) potential commutation problem,

(ii) complicated control algorithm with the need of high performance microprocessor,

(iii) realization of bidirectional switches,

(iv) elaborate clamp circuit.

As a result, in the past few years the concept of dual bridge matrix converter has developed.

3.2. Dual Bridge Matrix Converter (DBMC). Compared to CMCs, the recently introduced DBMC, Figure 3, has the following advantages [7].

(i) The difficulty of commutation is reduced. All line side switches turn on and off at zero current and all load side switches commutate similar to a conventional DC/AC inverter.

(ii) The number of switches can be reduced under certain constraints.

(iii) The clamp circuit can be greatly simplified, only one capacitor and one diode are used.

Considering all the advantages that DBMCs offer and the fact that this topology makes the switch reduction possible [7], this topology is chosen and in the next two sections the switch reduction from 24 to 6 is described.

3.3. Steps to Reduce the Switch Number. Reducing the number of switches in DBMC is extensively discussed in the literature such as [7, 12, 14]. In this part, a brief review is presented.

In general 12 bidirectional switches in Figure 3 require 24 unidirectional switches, but by proper assumptions the number of switches can be reduced [7].

In the following part, switch reduction procedure is described in four steps.

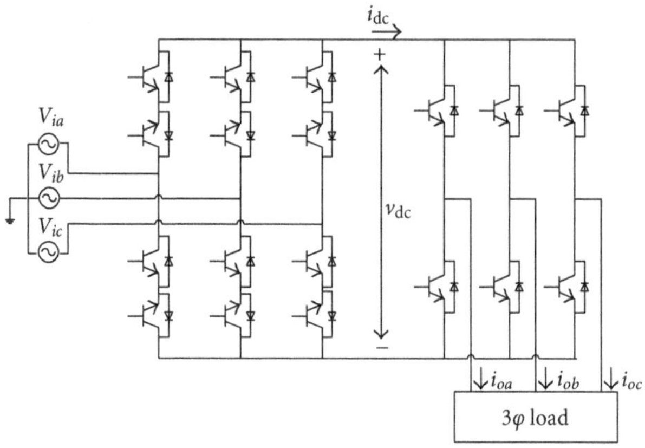

FIGURE 4: 18-switch topology.

Step 1. If, by employing appropriate control technique, the dc-link voltage is set to be positive, then replacing bidirectional switches with unidirectional switches at the load side is possible. This leads to 18-switch topology which is depicted in Figure 4 [7].

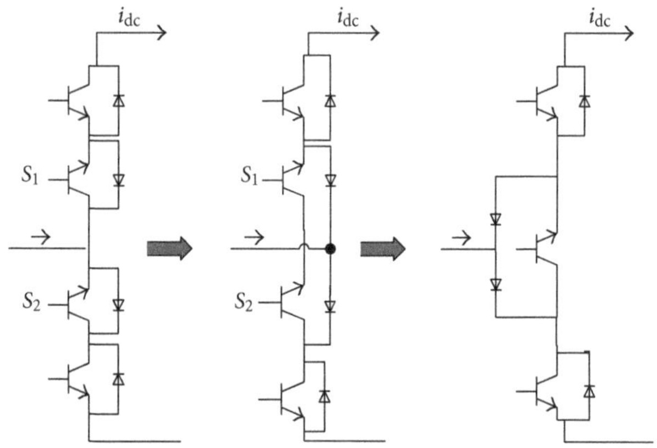

FIGURE 5: Switch reduction procedure as described in Step 2.

Step 2. According to Figure 5 [11, 14], S_1 and S_2 can share the same gate signal and can be replaced by one bidirectional switch and two diodes.

So 15-switch [10, 11] topology as shown in Figure 6 can be achieved. This topology is also called "sparse matrix converter" in [11–13, 15].

Step 3. By $I_{dc} > 0$ assumption, further switches reduction from 15 to 12 is possible. This leads to 12-switch topology in Figure 7. This topology is a standard 6-switch rectifier and 6-switch inverter connected back to back.

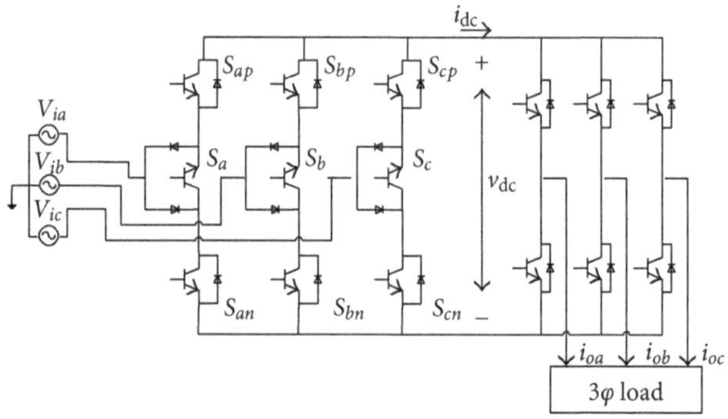

FIGURE 6: 15-switch topology.

Step 4. As far as switches S_{jp} and S_{jn} $(j \in a, b, c)$ in Figure 6 are not conducting current and only S_a, S_b, and S_c are conducting, the 9-switch topology (also named "ultra sparse matrix converter [11–13, 15]") is obtained and depicted in Figure 8.

By reducing the number of switches to 9, the presented topology has lost some of its desired features compared to standard and 24-switch topology, like bidirectional power flow between source and load. In 9-switch topology power only flows from source to load and power factor should always be higher than 0.866.

4. Proposed Topology

So far the 9-switch topology is obtained. In this section, we intend to further reduce the switch count by applying the concept of sharing leg between rectification and inversion stages [13–16]. Rectifier with capacitor dc-link is considered as one of the most promising structures for three-phase to three-phase power conversion [14].

In order to achieve this, a leg consists of two series capacitors is added and placed in parallel with rectification stage. The midpoint, tagged by "0" in Figure 9, is shared between an input leg and an output leg and results in elimination of

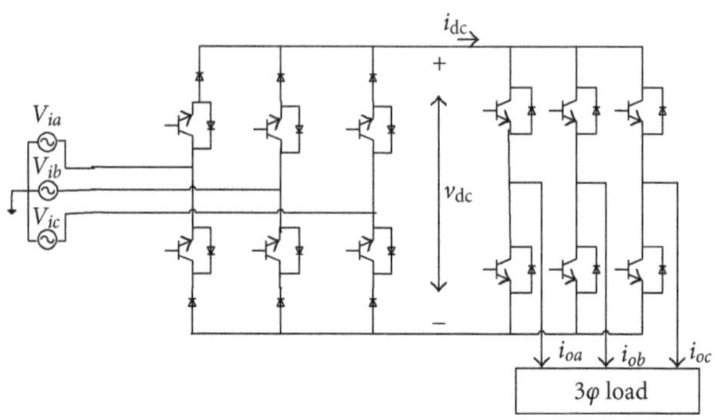

FIGURE 7: 12-switch topology.

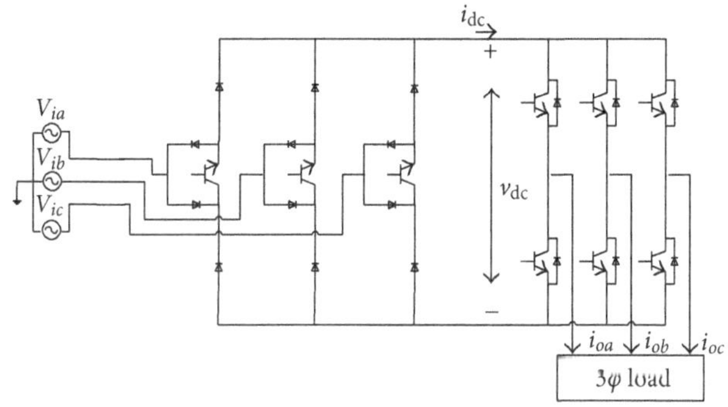

FIGURE 8: 9-switch topology according to [10, 13].

TABLE 1: System parameters.

Parameters	Rating
Power	1.2 kw
Input voltage (line-line)	420 v
Input filter	$L = 20$ uh, $C = 200$ uf
Power factor	0.94
Output voltage (line-line)	300 v
Output filter	$L = 15$ uh, $C = 10$ uf
THD	3.2%
Efficiency	89%
Load	$L = 10$ mh, $R = 40\,\Omega$
DC-Link voltage	650–700 v

one active switch and four diodes in rectification stage and two active switches in inversion stage.

By replacing the proposed 6-switches converter with rectifier block and inverter block in Figure 1, the novel 6-switches UPS is achieved which fulfill desired characteristics of double conversion UPS, such as compactness, output voltage regulation, and controllable frequency.

Three different operation mode of proposed UPS system are as follows.

(i) In normal mode, double conversion process take place through proposed 6-switches matrix converter.

(ii) In stored-energy mode, the batteries are in parallel with inversion stage and feed the load.

(iii) In by-pass mode, the by-pass switch in Figure 9 is closed and load is directly connected to the source.

Reliability is considered one of the most important characteristics of UPS system. It is indicated in the literature that 92% of all UPS failures are related to battery malfunction. This happens because of electrochemical nature of batteries. So providing the best operating condition for battery bank is of high importance when designing UPS system [1]. Since we can tightly regulate dc-link voltage in presented topology, the battery bank, which is consist of two set of batteries, is directly connected to the split capacitor dc bus.

By applying appropriate control technique, achieving sinusoidal output waveforms and controlled dc-link voltage is possible.

The control circuit in rectification stage consists of multiple control lops for each switches, one outer voltage loop and one inner current loop [1]. The first loop takes dc-link capacitor voltage as a feedback signal and compares it with reference voltage, which calculates error signal and by processing that signal, reference signal for second loop can be created. In second loop input current is sampled and compared with generated reference in first loop and PWM waveform with 30 kHz switching frequency is created and applied to active switches.

The 4 switches output inverter is controlled in a high frequency PWM pattern in order to produce high quality output voltage. Reference voltages for each of two legs are 120 degree shifted from each other. So produced line voltages are 120 degree shifted. As a result, the line voltage shared leg is also 120 degree shifted which completes symmetrical system of line voltages.

The control technique for output converter uses two control loops for each leg, one outer voltage loop and one inner current loop. As mentioned above, two reference voltages for each leg are 120 degree shifted from each other. Output voltage is sampled to calculate the error signal. For achieving stable voltage during steady state operation, PID module is used.

Calculated voltage error is also used as a reference signal for current loop. This loop uses the output line current as feedback signal. This loop is faster and is used to improve the dynamic response of the output converter.

In the next section, simulation result of the proposed UPS system is provided. Using the aforementioned control strategee will guaranty the performance of the system. In the rated power, during the steady state operation, as simulation results demonstrate, THD is lower than 3.2%, efficiency is about 89%, and sinusoidal output voltage and current for different load conditions are achieved.

5. Simulation Results

Simulation is carried out by using PSIM software. In the simulation, all switches are considered ideal. The voltage and current waveforms of the input, dc-link voltage as well as load current and voltage are depicted in this part to demonstrate the performance of proposed 6-switch topology (Figure 9) and applied control technique. The system parameters of simulated circuit are given in Table 1.

Figure 10 shows the simulation result for input current. In Figure 11, voltage and current phase "a" are shown. According to simulation result, power factor in input stage is 0.94.

Simulation result for dc-link voltage and current is shown in Figures 12 and 13.

DC-Link voltage is controlled using hysteresis controller. As shown in Figure 12, the voltage is fairly fitted between two set values.

Dc current is always positive which satisfies Step 3 in switch reduction procedure.

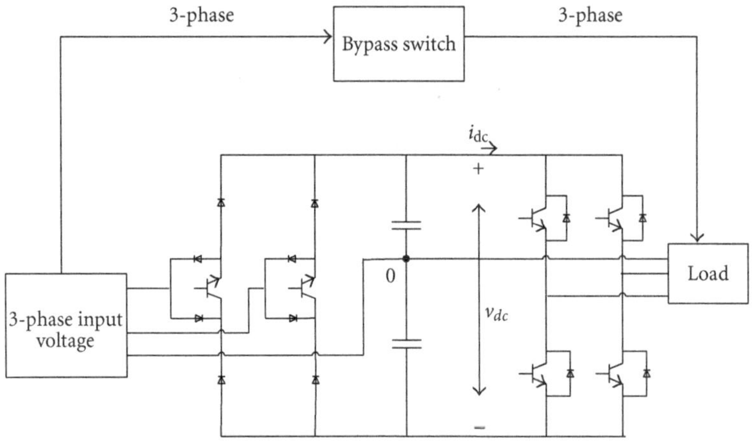

Figure 9: Proposed 6-switch topology.

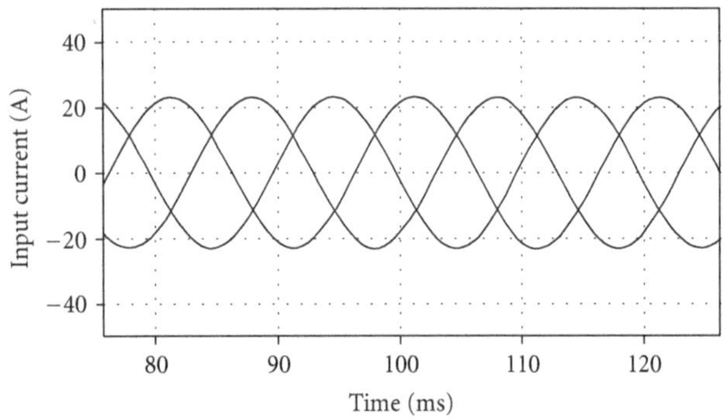

Figure 10: Input current of the 6-switch topology.

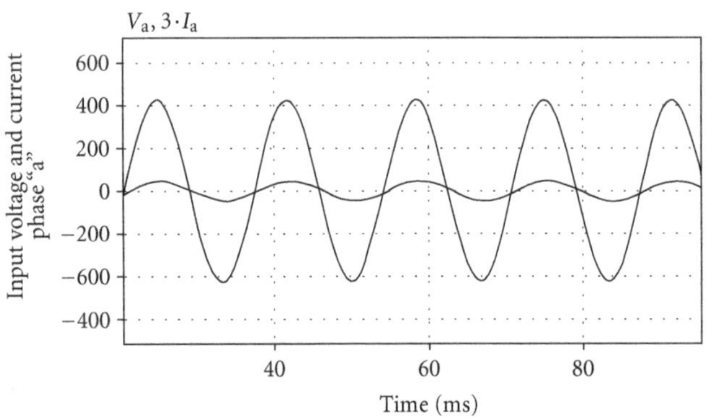

Figure 11: Input voltage and current of the 6-switch topology, Phase "a."

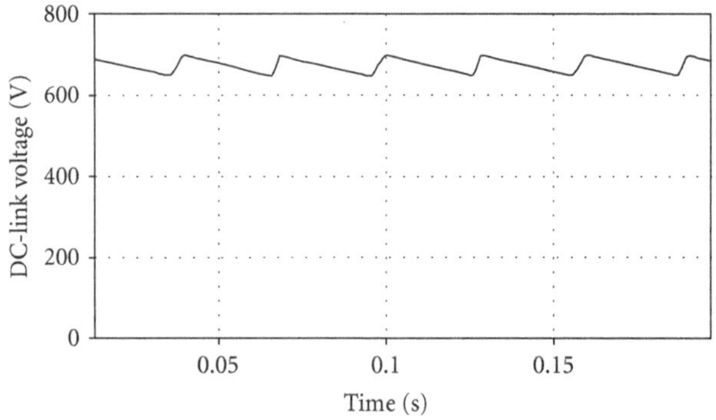

Figure 12: DC-link voltage of the 6-switch topology.

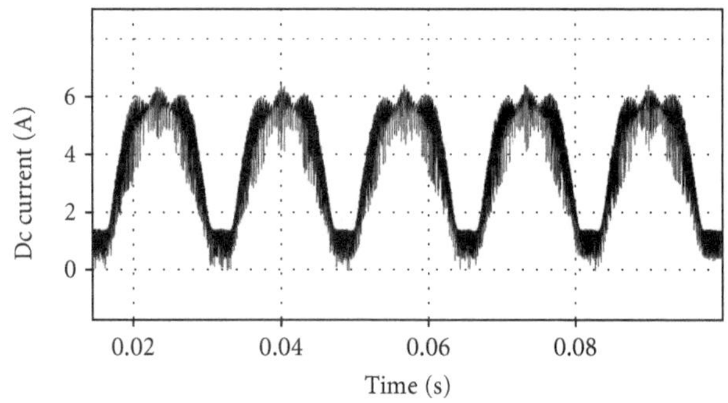

Figure 13: DC current of the 6-switch topology.

Figure 14 shows the output current for three-phase load connected to the back-end inverter. The applied current controller shows acceptable performance to supply load with three-phase sinusoidal current.

In Figure 15, the output voltage is shown. In order to demonstrate output voltage, a low pass filter is used and the cutoff frequency is set to 1000 Hz.

Demonstration of output voltage and current of phase "a" is presented in Figure 16. In this figure, the performance

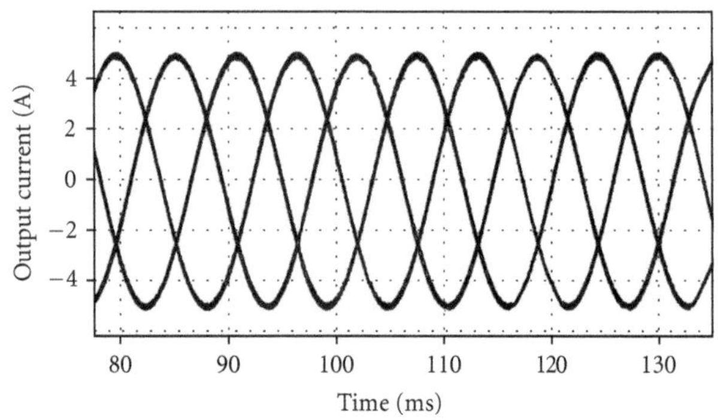

Figure 14: Output current of the 6-switches topology.

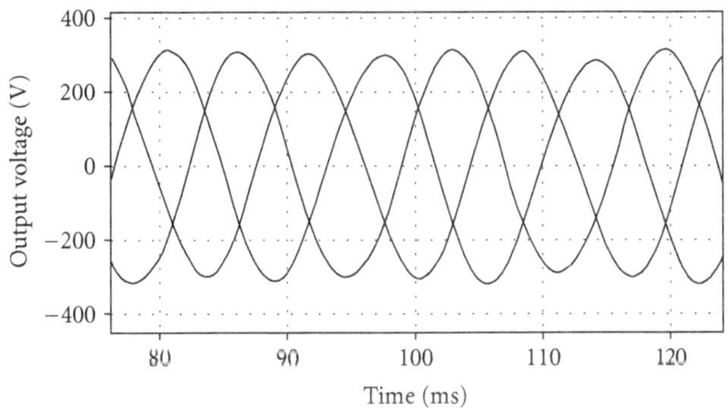

FIGURE 15: Output voltage of the 6-switch topology.

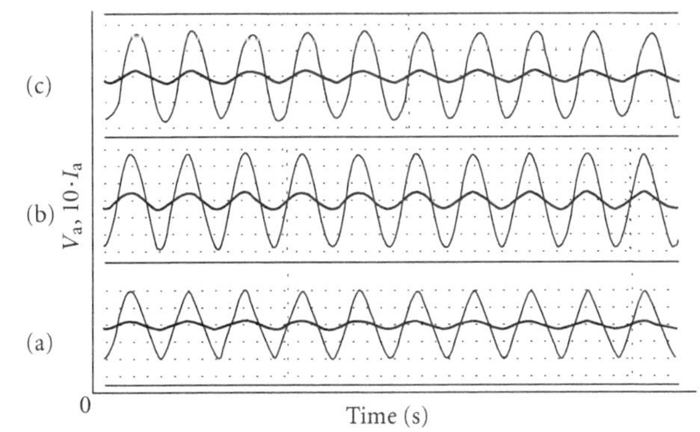

FIGURE 16: Output voltage and current at different load conditions (a) resistive load (b) 50% of rated Load (c) 120% rated load.

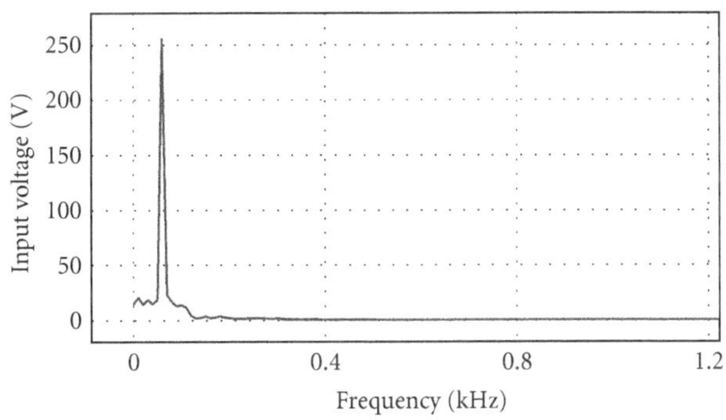

FIGURE 17: Output voltage of the 6-switch topology in frequency domain.

of the system for three different load conditions is shown, from resistive load to 120% of rated load. In this range, the proposed UPS system demonstrates a good performance in achieving sinusoidal waveforms and high power factor.

Figure 17 demonstrates output voltage in frequency domain. For calculating total harmonic distortion, the THD box in PSIM software is used. In steady state operation, THD is equal to 3.2%.

6. Conclusions

The performance of proposed 6-switch online UPS is investigated in this paper. Applied topology is based on the reduced switch count dual bridge matrix converter. Reducing the number of power switches leads to low price, highly reliable, and compact design. The proposed topology shows excellent performance, produced high quality, sinusoidal output voltage, and current waveforms. Simulation results have been presented to confirm the advantages of the proposed novel UPS system.

References

[1] A. Emadi, A. Nasiri, and S. B. Bekiarov, *Uninterruptible Power Supplies and Active Filters*, CRC Press, 2005.

[2] H. R. Karshenas and M. Niroomand, "Design and implementation of a single phase inverter with sine wave tracking method for emergency power supply with high performance reference," in *Proceedings of the 8th International Conference on Electrical Machines and Systems (ICEMS '05)*, pp. 1232–1237, September 2005.

[3] M. Niroomand and H. R. Karshenas, "Hybrid learning control strategy for three-phase uninterruptible power supply," *IET Power Electronics*, vol. 4, no. 7, pp. 799–807, 2011.

[4] M. Niroomand and H. R. Karshenas, "Design and implementation of a new hybrid learning control strategy for three-phase UPS," in *Proceedings of the 11th International Conference on Electrical Machines and Systems (ICEMS ;08)*, pp. 1732–1737, October 2008.

[5] A. Nasiri, S. B. Bekiarov, and A. Emadi, "Reduced parts three-phase series-parallel UPS system with active filter capabilities," in *Proceedings of the 38th IAS Annual Meeting, IEEE Industry Applications Conference*, vol. 2, pp. 963–969, October 2003.

[6] M. Niroomand and H. R. Karshenas, "Review and comparison of control methods for uninterruptible power supplies," in *Proceedings of the Power Electronic & Drive Systems & Technologies Conference (PEDSTC '10)*, pp. 18–23, IEEE, 2010.

[7] A. Nasiri, "Digital control of three-phase series-parallel uninterruptible power supply systems," *IEEE Transactions on Power Electronics*, vol. 22, no. 4, pp. 1116–1127, 2007.

[8] H. R. Karshenas and M. Niroomand, "Analysis and comparison of two control strategies for three-phase series-parallel UPS," in *Proceedings of the 11th International Conference on Electrical Machines and Systems (ICEMS '08)*, pp. 1726–1731, October 2008.

[9] J. Holtz and U. Boelkens, "Direct frequency converter with sinusoidal line currents for speed-variable AC motors," *IEEE Transactions on Industrial Electronics*, vol. 36, no. 4, pp. 475–479, 1989.

[10] L. Wei and T. A. Lipo, "A novel matrix converter topology with simple commutation," in *Proceedings of the 36th IAS Annual Meeting*, pp. 1749–1754, October 2001.

[11] J. W. Kolar, M. Baumann, F. Schafmeister, and H. Ertl, "Novel three-phase AC-DC-AC sparse matrix converter," in *Proceedings of the 17th Annual IEEE Applied Power Electronics Conference and Expositions*, pp. 777–791, Dallas, Tex, USA, March 2002.

[12] P. W. Wheeler, J. Rodríguez, J. C. Clare, L. Empringham, and A. Weinstein, "Matrix converters: a technology review," *IEEE Transactions on Industrial Electronics*, vol. 49, no. 2, pp. 276–288, 2002.

[13] J. W. Kolar, F. Schafmeister, S. D. Round, and H. Ertl, "Novel three-Phase AC-AC sparse matrix converters," *IEEE Transactions on Power Electronics*, vol. 22, no. 5, pp. 1649–1661, 2007.
[14] L. Wei, T. A. Lipo, and H. Chan, "Matrix converter topologies with reduced number of switches," in *Proceedings of the IEEE 33rd Annual Power Electronics Specialists Conference (PESC '02)*, pp. 57–63, June 2002.
[15] C. Klumpner, M. Lee, C. Pitic, P. Wheeler, and P. Zanchetta, "A new three-level indirect matrix converter with reduced number of switches," in *Proceedings of the IEEE Industry Applications Conference 42nd Annual Meeting*, pp. 186–193, September 2007.
[16] L. Wei, T. A. Lipo, and H. Chan, "Robust voltage commutation of the conventional matrix converter," in *Proceedings of the IEEE 34th Annual Power Electronics Specialists Conference*, pp. 717–722, June 2003.

Resistive Ferroresonance Limiter for Potential Transformers

Hamid Radmanesh,[1,2] G. B. Gharehpetian,[2] and Hamid Fathi[2]

[1] *Department of Electrical Engineering, Islamic Azad University, Takestan Branch, Takestan 1995755681, Iran*
[2] *Electrical Engineering Department, Amirkabir University of Technology, Tehran, Iran*

Correspondence should be addressed to Hamid Radmanesh, hamid.nsa@gmail.com

Academic Editor: Hadi Y. Kanaan

The ferroresonance or nonlinear resonance is a complex phenomenon, which may cause overvoltage in the electrical power system and endangers the system reliability and operation. The ability to predict the ferroresonance in the transformer depends on the accuracy of the transformer model used. In this paper, the effect of the new suggested ferroresonance limiter on the control of the chaotic ferroresonance and duration of chaotic transients in a potential transformer including nonlinear core losses is studied. To study the proposed ferroresonance limiter, a single phase 100 VA, 275 kV potential transformer is simulated. The magnetization characteristic of the potential transformer is modeled by a single-value two-term polynomial. The core losses are modeled by third order power series in terms of voltage and include core nonlinearities. The simulation results show that the ferroresonance limiter has a considerable effect on the ferroresonance overvoltage.

1. Introduction

The ferroresonance is typically initiated by saturable magnetizing inductance of a transformer and a capacitive distribution cable or transmission line connected to the transformer. In most practical situations, ferroresonance results in dominated currents, but in some operating "mode", may cause significant high value distorted winding voltage waveform, which is typically referred to as ferroresonance [1]. Although occurrences of the "resonance" involves a capacitance and an inductance, there is no definite resonant frequency. In this phenomenon, more than one response is possible for the same set of parameters, and drifts or transients may cause the response to jump from one steady-state response to another one. Its occurrence is more likely to happen in the absence of adequate damping [1]. Researches on ferroresonance in transformers have been conducted for more than 80 years. The word ferroresonance first appeared in the literature in 1920. Nonlinear dynamical tools for studying ferroresonance have been used in [2]. Practical interests have shown that the use of series capacitors for voltage regulation could cause ferroresonance in distribution systems [3]. Ferroresonant behavior of a 275 kV potential transformer, fed from a sinusoidal supply via circuit breaker grading capacitance, has been studied in [4, 5]. The potential transformer ferroresonance from an energy transfer point of view has been presented in [6]. A systematical method for suppressing ferroresonance at neutral-grounded substations has been studied in [7]. A sensitivity analysis on power transformer ferroresonance of a 400 kV double circuit transmission line has been reviewed in [8]. A novel analytical solution to the fundamental ferroresonance has been given in [9, 10]. In that paper, the problem with the traditional excitation characteristic (TEC) of nonlinear inductors has been investigated. The TEC contains harmonic voltages and/or currents. The stability domain calculations of the period-1 ferroresonance have been investigated in [11]. The application of the wavelet transform and MLP neural network for the ferroresonance identification has been used in [12]. The impact of the transformer core hysteresis on the stability domain of ferroresonance modes has been studied in [13, 14]. A 2D finite-element electromagnetic analysis of an autotransformer experiencing ferroresonance is given in [15, 16]. In [17], a new modeling of transformers enabling to simulate slow transients more accurate than the existing models has been presented. In the current paper,

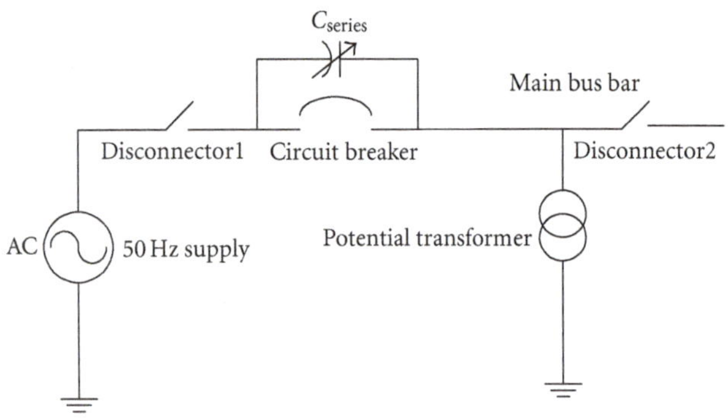

FIGURE 1: Power system single-line diagram.

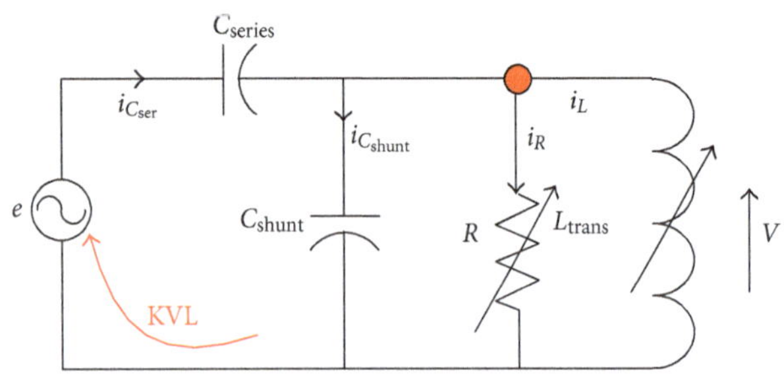

FIGURE 2: Basic reduced equivalent ferroresonance circuit.

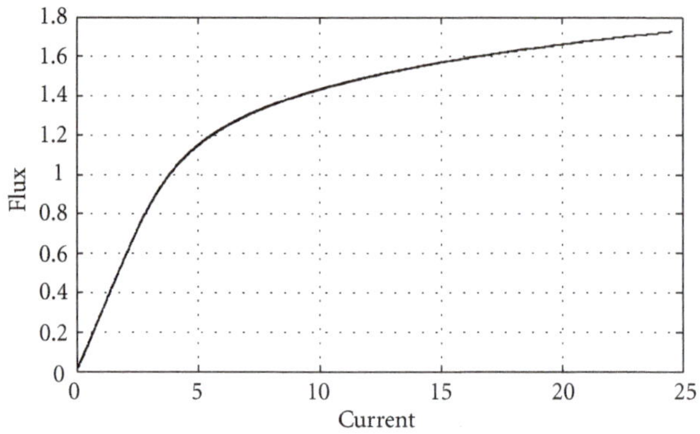

FIGURE 3: Flux-current characteristic of transformer core.

new ferroresonance limiter is suggested as a compact circuit including one resistor, power electronic switch, and control circuit for limiting and stabilizing the unstable and high amplitude ferroresonance oscillations. The resistance is connected to the grounding point of the potential transformer (PT) and during ferroresonance occurrence, the power electronic switch connects the resistor to the transformer. In this paper the simulation results of the case study confirm that system states lead to chaos, and bifurcation occurs in the proposed model. The presence of the proposed ferroresonance limiter tends to clamp the ferroresonance overvoltage. The ferroresonance limiter successfully reduces the chaotic region for higher exponents.

2. System Modeling

The ferroresonance or nonlinear resonance can occur on circuits with power transformers connected to overhead lines and also potential transformers connected to isolated sections of bus bars. Energy is coupled via intercircuit capacitance of parallel lines or open circuit breaker grading capacitance [18].

By careful system design or PT placement it may be possible to increase the phase-to-earth capacitance of the PT ferroresonance circuit, for example, using a longer bus bar, short cables, CVTs, and bushing capacitance. Specifically, the connection of PTs to isolated section of bus, bar, that is, to a low capacitance, should be avoided. Figure 1 shows circuit diagram of the system components at a 275 kV substation. PT is isolated from sections of bus bars via disconnector. In Figure 2, C_{series} models the circuit breaker grading capacitance and C_{shunt} is the total bus bar capacitance to earth. The ferroresonance condition occurs upon closure of disconnector 1 and circuit breaker, leading to a system fault caused by the failure of the voltage transformer primary winding. This figure shows the basic ferroresonance equivalent circuit used to analyze the ferroresonance phenomena. The resistor, R, represents the transformer nonlinear core losses.

In Figure 2, E is the rms supply phase voltage. The nonlinear resistor R is an important factor in the initiation of the ferroresonance. In the peak current range for steady-state operation, the flux-current linkage can be approximated by a linear characteristic such as $i_L = a\lambda$, where the coefficient of the linear term (a) corresponds closely to the reciprocal of the inductance ($a \cong 1/L$). However, for very high currents the iron core might be driven into saturation and the flux-current characteristic becomes highly nonlinear, here the $\lambda - i$ characteristic of the potential transformer is modeled as in [20] by the following polynomial:

$$i = a\lambda + b\lambda^7, \quad (1)$$

where, $a = 3.14$, and $b = 0.41$. Figure 3 shows the $\lambda - i$ iron core characteristic for $q = 7$.

The basic ferroresonance circuit of the PT can be presented by the differential equation. A small change in the value of the system voltage, capacitance, or losses may lead to dramatic changes in PT behavior. A more suitable representation for studying the ferroresonance and other nonlinear systems is provided by nonlinear dynamic methods based on phase plan diagram, time domain simulation, and bifurcation diagram.

3. System Dynamic and Equation

Mathematical analysis of the equivalent circuit by applying *KVL* and *KCL* laws the equations of the system can be expressed by the following equations:

$$e = \sqrt{2}E\sin(\omega t),$$

$$i_L = a\lambda + b\lambda^7,$$

$$v_L = \frac{d\lambda}{dt},$$
$$i_{C_{\text{series}}} = C_{\text{ser}}\frac{d}{dt}(e - v_L),$$
$$i_{C_{\text{shunt}}} = C_{\text{sh}}\frac{dv_L}{dt} = C_{\text{sh}}\frac{d^2\lambda}{dt^2},$$
$$i_R = h_0 + h_1 v_L + h_2 v_L^2 + h_3 v_L^3, \quad (2)$$

$$\frac{(C_{\text{ser}} + C_{\text{sh}})}{C_{\text{ser}}}\frac{d^2\lambda}{dt^2} = -\frac{1}{C_{\text{ser}}}\left(h_0 + h_1 v_L + h_2 v_L^2 + h_3 v_L^3\right) - \frac{1}{C_{\text{ser}}}(a\lambda + b\lambda^7) + \sqrt{2}E\omega\cos\omega t, \quad (3)$$

where, ω is the supply frequency, and E is the rms supply phase voltage. The time behavior of the basic ferroresonance circuit is described by (3). The results for two different parameter sets showing the two possible types of ferroresonance are discussed. Table 1 listed base values of parameters and the parameters are given in Table 2. Also, the coefficient values of the nonlinear core model are presented in Table 3.

For the initial conditions, we have

$$\lambda(0) = 0.0, \qquad v_l = \frac{d\lambda}{dt}(0) = \sqrt{2}. \quad (4)$$

4. System Analysis Considering Ferroresonance Limiter

The system, which has been simulated, is shown in Figure 4. In this figure, the ferroresonance limiter is connected to the transformer via power electronic switches (sw_1 and sw_2). In this case, the system time domain simulations have been carried out by using fourth order Runge-Kutta method and validated against MATLAB SIMULINK.

The primary purpose of inserting ferroresonance limiter between the neutral point of the transformer and the earth is to limit the ferroresonance overvoltage and overcurrent. Low impedance earthing is conventionally defined as impedance that limits the prospective ferroresonance current to the full load current of the transformer. The value of the impedance required is easily calculated to a reasonable approximation by dividing the rated phase voltage by the rated phase current of the transformer. The ferroresonance limiter impedance is conventionally achieved using resistors rather than inductors, to limit the tendency for the fault arc to persist due to the inductive started energy. In Figure 4, R_{FLR} is the ferroresonance limiter resistance. The typical values for system parameters, presented in Table 2, are used for the simulations, while the ferroresonance limiter is added to the system and its value is equal to:

$$R_{\text{FLR}} = 50\,\text{k}\Omega. \quad (5)$$

The power electronic switches need proper switching pulse, in order to connect and disconnect the ferroresonance limiter resistance to the grounding point of the transformer.

Table 1: Simulation parameters base values.

Base value of input voltage	(275/square3) kV
Base value of volt-amperes	100 VA
Base angular frequency	$(2\pi 50t)$ rad/sec

Table 2: Parameters of simulations without using ferroresonance limiter effect.

C_{shunt} (nf)	C_{series} (nf)	ω (rad/sec)	E (kV)	R_{FLR} (kΩ)
3	0.5	314	275	50

Table 3: Parameters of nonlinear core model [19].

h_0	h_1	h_2	h_3
$-3.5213E-03$	$5.7869E-07$	$-1.4167E-12$	$1.21105E-18$

Figure 5 shows the switching process diagram with one decision box. V_{ref} is adjusted between 0.9–1.2 p.u of the source voltage and this reference voltage is compared with the measurement voltage on the transformer coil. When reference and measurement voltages are equal, then the decision box generates the pulse for sw_1 and in other cases, this circuit can operate the second switch. To protect the power electronic switches against overvoltages, the step down transformer is used here. The differential equation for the ferroresonance circuit, shown in Figure 4, can be written as follows:

$$\begin{aligned} C_{\text{ser}}C_{\text{sh}}R_n\frac{d^2v_L}{dt^2} &= C_{\text{ser}}\sqrt{2}E\omega\cos\omega t \\ &\quad - (C_{\text{ser}} + C_{\text{sh}} + C_{\text{ser}}R_n h_1 \\ &\quad + 2C_{\text{ser}}R_n h_2 v_L + 3C_{\text{ser}}R_n h_3 v_L^2)\frac{dv_L}{dt} \\ &\quad - (C_{\text{ser}}R_n a + C_{\text{ser}}R_n q b\lambda^6 + h_1)v_L \\ &\quad - \left(h_0 + h_2 v_L^2 + h_3 v_L^3 + a\lambda + b\lambda^7\right). \end{aligned} \quad (6)$$

5. Simulation Results

5.1. Ferroresonance in Potential Transformer. The phase space and waveform of the ferroresonance overvoltage are shown in Figures 6 and 7, respectively. The phase plan diagram clearly shows the chaotic trajectory characteristic of a periodic waveform and it can be seen in Figure 7 that the amplitude of the overvoltage reaches to 6 p.u, which is very dangerous for the system equipments; especially, it can cause the PT failure.

5.2. Effect of Ferroresonance Limiter. This section discusses the effect of ferroresonance limiter on the ferroresonance overvoltage by using bifurcation and phase plan diagram. The previous case with the same parameters but with

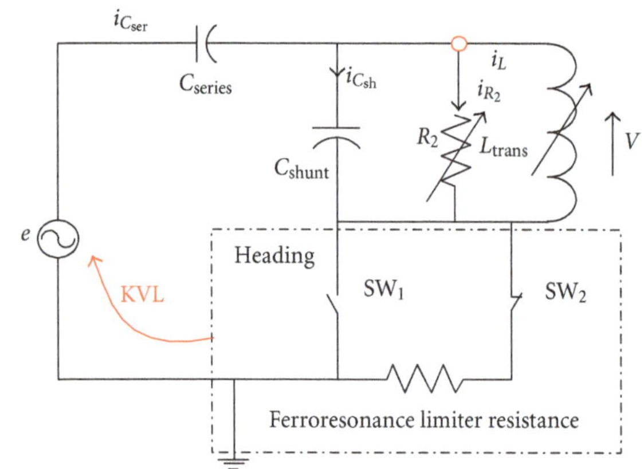

FIGURE 4: Equivalent ferroresonance circuit and ferroresonance limiter.

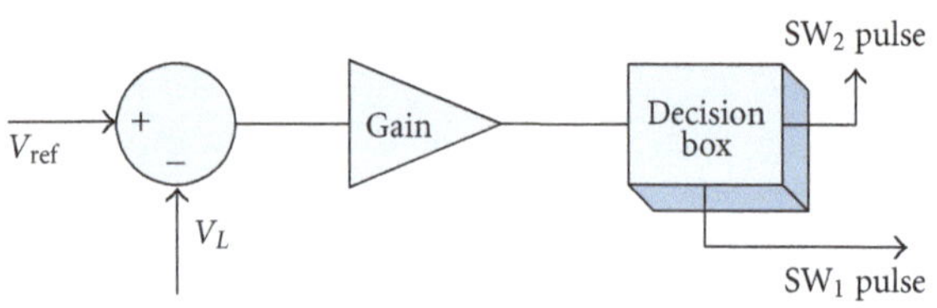

FIGURE 5: Controlling diagram of switching pulses.

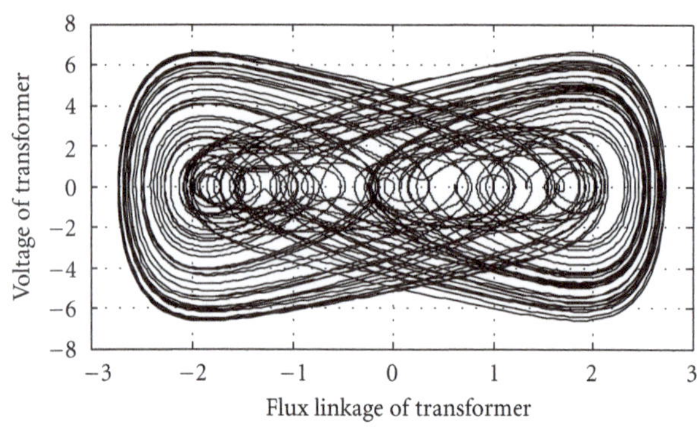

FIGURE 6: Phase plan diagram for chaotic motion (without using ferroresonance limiter).

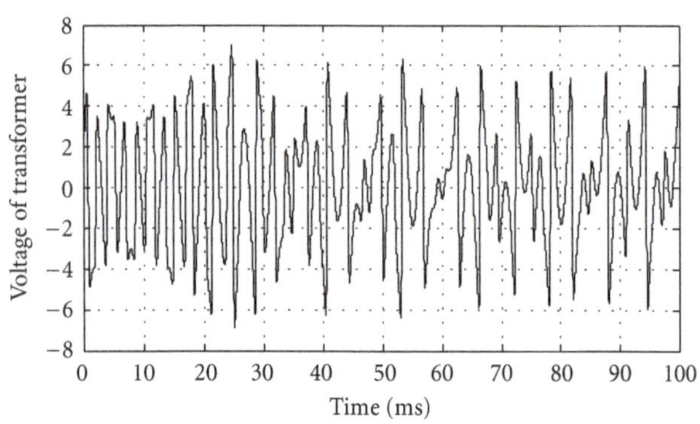

FIGURE 7: Time domain simulation for chaotic motion (without using ferroresonance limiter).

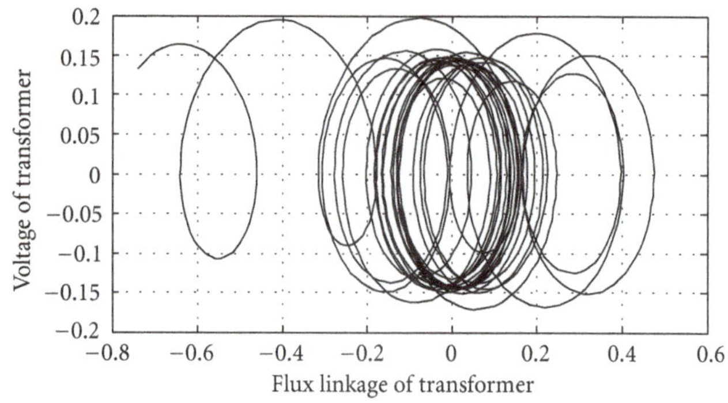

FIGURE 8: Phase plan diagram for quasiperiodic motion, after using ferroresonance limiter.

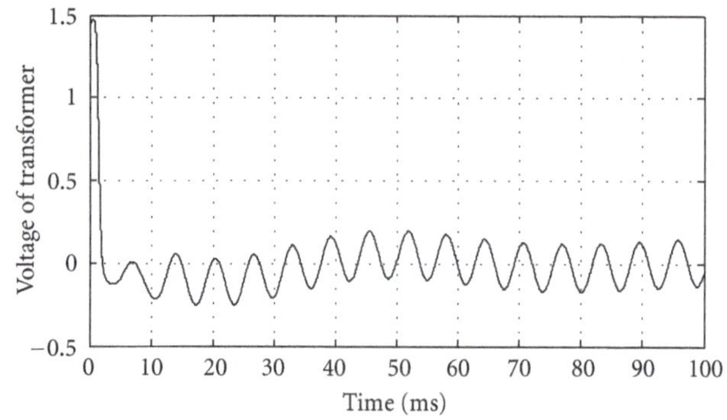

FIGURE 9: Time domain simulation for quasiperiodic motion, after using ferroresonance limiter.

the ferroresonance limiter is modeled in this section. The simulation results indicate that the chaotic ferroresonance is changed to the periodic oscillation, and the amplitude of the overvoltage is decreased to 1.5 p.u. By considering the ferroresonance limiter effect, the chaotic signal shown in Figure 6 is changed to signal shown in Figure 8. Also, time domain simulation for quasiperiodic motion, after using ferroresonance limiter has been shown in Figure 9. According to this plot, system behavior is changed to the torus oscillation, and the amplitude of the overvoltage, shown in Figure 10, is highly controlled and reaches 0.2 p.u.

5.3. Bifurcation Diagram Analysis. Another tool that is used for studying nonlinear equations is the bifurcation diagram [13]. Figure 10(a) clearly shows the ferroresonance overvoltage in PT when the nominal voltage of the power system increases to 5 p.u.

According to this plot, the chaotic trajectory and the amplitude of the ferroresonance overvoltage reach 2.5 p.u. This overvoltage can cause PT failure. Figure 10(b) shows that there is no chaotic region in the system behavior after using the ferroresonance limiter. Also, the fundamental resonance in the system and the amplitude of the overvoltage decreases less than 2 p.u and the system can work in the safe operation region.

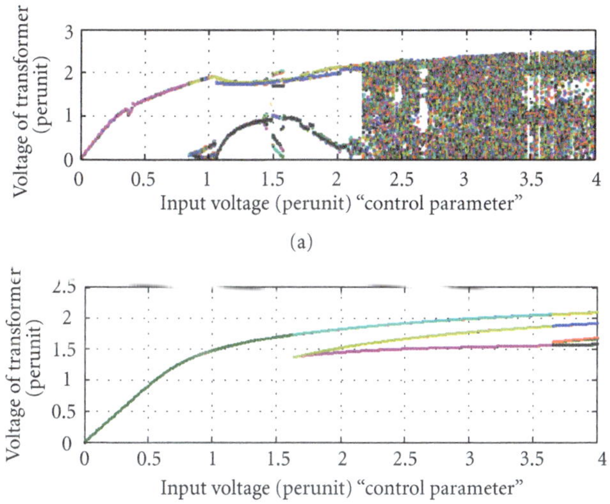

FIGURE 10: (a) Bifurcation diagram for voltage of transformer versus voltage of system, without using ferroresonance limiter. (b) Bifurcation diagram for voltage of transformer versus voltage of system, after using ferroresonance limiter.

6. Conclusion

In this paper, one of the nonlinear phenomena in the power transformer, known as ferroresonance, has been studied. The ferroresonance overvoltages are temporary overvoltages and they can cause insulation failures. This paper has suggested a new ferroresonance limiter for PTs. The proposed ferroresonance limiter results in reduction of the amplitude of the overvoltage and successfully eliminates chaotic behaviours. Based on simulations, it has been shown that after using ferroresonance limiter, the periodic oscillation was obvious and the chaotic oscillation has been changed to the fundamental and periodic resonances.

Nomenclature

h_0, h_1, h_2, h_3: Coefficients for core losses nonlinear function (per unit)
n: Index for the neutral connection
a: Coefficient for linear part of magnetizing curve (per unit)
b: Coefficient for nonlinear part of magnetizing curve (per unit)
q: Index of nonlinearity of the magnetizing curve
C_{series}: Linear capacitor of circuit breaker (F)
C_{shunt}: Linear capacitor of the power system (F)
R_m: Core losses resistance (ohm)
i: Instantaneous value of branch current (A)
v: Instantaneous value of the voltage across a branch element (V)
E: Instantaneous value of driving source (V)
φ: Flux linkage in the nonlinear inductance (Weber)
ω: Angular frequency of the driving force (rad/sec)
PT: Potential transformer
C.B: Circuit breaker
N.R: Neutral earth resistance
DS1: Disconector1
DS2: Disconnector2
MOSA: Metal oxide
FLR: Ferroresonance limiter resistance.

References

[1] H. Radmanesh and F. S. Hamid, "Analyzing ferroresonance phenomena in power transformers including zinc oxide arrester and neutral resistance," *Applied Computational Intelligence and Soft Computing*, vol. 2012, Article ID 525494, 5 pages, 2012.

[2] B. A. Mork and D. L. Stuehm, "Application of nonlinear dynamics and chaos to ferroresonance in distribution sysytems," *IEEE Transactions on Power Delivery*, vol. 9, no. 2, pp. 1009–1017, 1994.

[3] J. W. Butler and C. Concordia, "Analysis of series capacitor application problems," *AIEE Transactions*, vol. 56, pp. 975–988, 1937.

[4] Z. Emin, B. A. T. Al Zahawi, and Y. K. Tong, "Voltage transformer ferroresonance in 275 kV substation," in *Proceedings of the 11th International Symposium on High Voltage Engineering*, pp. 283–286, August 1999.

[5] H. Radmanesh and M. Rostami, "Effect of circuit breaker shunt resistance on chaotic ferroresonance in voltage transformer," *Advances in Electrical and Computer Engineering*, vol. 10, no. 3, pp. 71–77, 2010.

[6] R. G. Andrei and B. R. Halley, "Voltage transformer ferroresonance from an energy transfer standpoint," *IEEE Transactions on Power Delivery*, vol. 4, no. 3, pp. 1773–1778, 1989.

[7] Y. Li, W. Shi, R. Qin, and J. Yang, "A systematical method for suppressing ferroresonance at neutral-grounded substations," *IEEE Transactions on Power Delivery*, vol. 18, no. 3, pp. 1009–1014, 2003.

[8] C. Charalambous, Z. D. Wang, M. Osborne, and P. Jarman, "Sensitivity studies on power transformer ferroresonance of a 400kV double circuit," *IET Journal of Generation, Transmission and Distribution*, vol. 2, no. 2, pp. 159–166, 2008.

[9] Y. Li, W. Shi, and F. Li, "Novel analytical solution to fundamental ferroresonance—part I: power frequency excitation characteristic," *IEEE Transactions on Power Delivery*, vol. 21, no. 2, pp. 788–793, 2006.

[10] H. Radmanesh, G. B. Gharehpetian, and H. Fathi, "Ferroresonance of power transformers considering nonlinear core losses and metal oxide surge arrester effects," *Electric Power Components and Systems*, vol. 40, no. 5, pp. 463–479, 2012.

[11] D. A. N. Jacobson, P. W. Lehn, and R. W. Menzies, "Stability domain calculations of period-1 ferroresonance in a nonlinear resonant circuit," *IEEE Transactions on Power Delivery*, vol. 17, no. 3, pp. 865–871, 2002.

[12] G. Mokryani and M. R. Haghifam, "Application of wavelet transform and MLP neural network for Ferroresonance identification," in *Proceedings of the IEEE Conference of Conversion and Delivery of Electrical Energy in the 21st Century*, pp. 1–6, July 2008.

[13] A. Rezaei-Zare, R. Iravani, and M. Sanaye-Pasand, "Impacts of transformer core hysteresis formation on stability domain of

ferroresonance modes," *IEEE Transactions on Power Delivery*, vol. 24, no. 1, pp. 177–186, 2009.

[14] H. Radmanesh and G. B. Gharehpetian, "Ferroresonance suppression in power transformers using chaos theory," *International Journal of Electrical Power and Energy Systems*, vol. 45, no. 1, pp. 1–9, 2013.

[15] C. A. Charalambous, Z. D. Wang, P. Jarman, and M. Osborne, "2-D finite-element electromagnetic analysis of an autotransformer experiencing ferroresonance," *IEEE Transactions on Power Delivery*, vol. 24, no. 3, pp. 1275–1283, 2009.

[16] H. Radmanesh, "Controlling chaotic ferroresonance oscillations in autotransformers including linear and nonlinear core losses effect," *International Review of Electrical Engineering*, vol. 5, no. 6, pp. 2644–2652, 2010.

[17] P. G. Khorasani and A. Deihimi, "A new modeling of matlab transformer for accurate simulation of ferroresonance," in *Proceedings of the 2nd International Conference on Power Engineering, Energy and Electrical Drives (POWERENG '09)*, pp. 529–534, March 2009.

[18] H. Radmanesh and H. Fathi, "Studying voltage transformer ferroresonance," *Research Journal of Applied Sciences, Engineering and Technology*, vol. 4, no. 19, pp. 3680–3686, 2012.

[19] K. Al-Anbarri, R. Ramanujam, R. Saravanaselvan, and K. Kuppusamy, "Effect of iron core loss nonlinearity on chaotic ferroresonance in power transformers," *Electric Power Systems Research*, vol. 65, no. 1, pp. 1–12, 2003.

[20] E. P. Dick and W. Watson, "Transformer models for transient studies based on field measurements," *IEEE Transactions on Power Apparatus and Systems*, vol. 100, no. 1, pp. 409–419, 1981.

17

Comparison of Duty Cycle Generator Algorithms for SPICE Simulation of SMPS

Alexander Abramovitz

Department of Electrical Engineering and Computer Science, University of California, Irvine, CA 92697-2625, USA

Correspondence should be addressed to Alexander Abramovitz, alabr@hotmail.com

Academic Editor: Henry S. H. Chung

The paper presents and discusses an algorithm for average modeling of the PWM modulator in switch-mode power systems by general purpose electronic circuit simulators such as PSPICE. A comparison with previous theoretical models is conducted. To test the accuracy of the average PWM models comparison to cycle-by-cycle simulation was conducted. The proposed algorithm shows better accuracy than earlier counterparts.

1. Introduction

Today circuit simulation and computer-aided design are universally accepted engineering tools and have become industry standard method of product development. Two approaches are possible for simulation of switched mode systems: cycle-by-cycle simulation and average behavior simulation. Cycle-by-cycle simulation is a quite straightforward approach. Cycle-by-cycle simulation can be performed programming the complete power electronic circuit to the simulator. Cycle-by-cycle simulation allows studding the power stage at the switching frequency scale and observing the instantaneous voltages and currents at any point in the circuit. First disadvantage of cycle-by-cycle simulation is that simulating the detailed switching process is time consuming. This is particularly true for nontrivial practical cases. The second and by far more important limitation is that the cycle-by-cycle model of a switching circuit does not lend itself to frequency response analysis. This is because the switching stage has no stable operating point and, hence, does not allow the PSPICE simulator to perform linearization and calculate the small signal gains required for frequency response analysis. Therefore, a different approach is needed to attain frequency domain simulation of the control loop.

State Space Averaging is a classical theoretical analysis method of switch-mode power electronics systems [1–5]. Average modeling of the power stage can also be helpful in simulation as they can be readily implemented using PSPICE behavioral sources [6–8]. Average models are continuous and, hence, can be automatically linearized by the PSPICE simulator and prepared for the frequency domain analysis. The ability to obtain the frequency response of the feedback loop allows the practicing engineer to evaluate the system's stability and to design the compensator network to meet the design objectives.

The distinct characteristic of SMPS is that a switched-mode stage is employed as power processor, whereas the control circuits are mostly analog where Pulse Width Modulator (PWM) is used as an interface. A typical structure of a PWM switch-mode power system (SMPS) is illustrated in Figure 1 [8]. Here, as an example, an average current mode (ACM) system is shown. There are two major challenges in simulation of a switch-mode system. The first is modeling the switcher, whereas the second is modeling the PWM modulator. To model the average behavior of switch mode power stages Switched Inductor Model (SIM) was proposed [6–8], whereas the PWM duty cycle generation process can be modeled by software Duty Cycle Generator (DCG) approach [8]. The PWM modeling problem is that in practice the switching ripple propagates into the control loop and affects the switch on and off times. However, the average SIM model has no ripple components; therefore, in order to obtain accurate

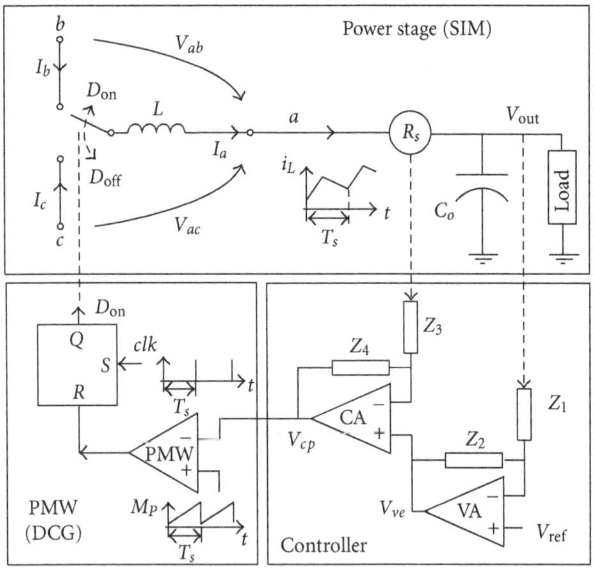

FIGURE 1: A typical ACM SMPS. Modeling the SMPS requires the SIM to model the power stage and the DCG to model the PWM [8].

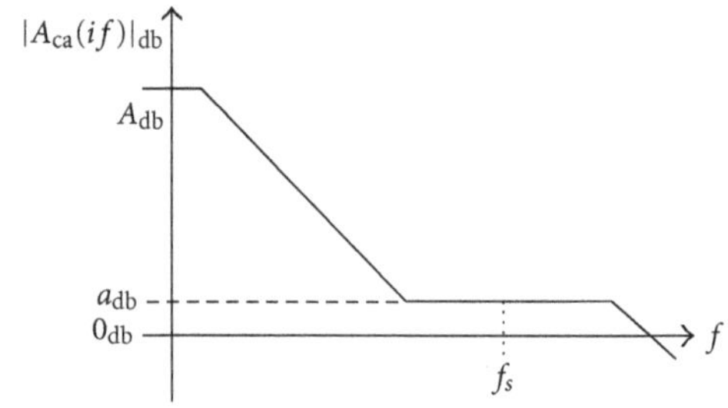

FIGURE 2: The inner loop amplifier, CA, frequency response.

simulation results, software DCG should be programmed to predict the switching ripple effects using only the average signals of the SIM model. Another task of the DCG is to anticipate the mode changes of the power stage and calculate the correct off duty cycle in case of CCM-DCM transition.

This paper proposes a more precise, PSPICE compatible, average DCG algorithm for modeling the PWM comparator. The operation of the proposed average PWM model is demonstrated by time domain and frequency domain simulations. The paper also conducts a comparison with previously reported results. To validate the model's accuracy, the proposed average algorithm and its earlier counterparts are compared to cycle-by-cycle simulation. Particularly, in the discontinuous current mode the proposed algorithm shows better accuracy than earlier counterparts.

2. Software Duty Cycle Generators

2.1. PWM Relationships. In recent years ACM control has become the method of choice for many advanced SMPS. The principle of ACM is to implement a multiloop control system in which the inner loop (see Figure 1), controls the average current and makes it tightly follow the outer loop command. The inner/current loop amplifier, CA, provides higher gain in the low-frequency region and extends the inner loop bandwidth. These features are very desirable and provide good tracking performance.

Usually, the average current loop amplifier (CA) is designed to have a high/low-frequency gain and a flat response in the vicinity of the switching frequency as shown in Figure 2 [9]. Consequently, the current programming signal at the output of CA, v_{cp}, has an average component, $\langle v_{cp}\rangle$, with a superimposed attenuated inductor current ripple, as shown in Figure 3. The ripple is scaled by the gain of the current sensing network, R_s, and the CA amplifier gain at the switching frequency, $a = |A_{ca}(f_s)|$.

Referring to Figures 1 and 3, the switch on time is initiated by a clock pulse and terminated by the PWM comparator at the moment the current programming signal, v_{cp}, intersects the external ramp. The comparator is usually followed by a latch, which remains reset till the end of the switching cycle and prevents output chatter. A real world comparator uses instantaneous ramp and CA output voltages, whereas in the average system represented by SIM these variables do not exist. The essence of the modeling problem is to accurately determine the on duty cycle, D_{on}, relaying on the knowledge of the average CA output voltage, $\langle v_{cp}\rangle$, the ramp amplitude, V_p, and other average system parameters.

Let the external, charging, and discharging slopes of PWM comparator input signals (see Figure 3), be defined following the notation of [8]:

$$\begin{aligned} S_e &= \frac{V_p}{T_s}, \\ S_{on} &= \frac{aR_s}{L} V_{ab}, \\ S_{off} &= \frac{aR_s}{L} V_{ac}, \end{aligned} \tag{1}$$

where (see Figure 1) V_{ab} and V_{ac} are the SIM terminal voltages; L is the SIM inductor value; T_s is the switching cycle; V_p is the external ramp peak voltage; a, R_s are as defined above.

The intersection instance of current error amplifier and the ramp voltages (see Figure 3) uniquely determines the on duty cycle, D_{on}, as well as all the average quantities for the given switching cycle. Investigating the modulator waveforms of Figure 3(a), and applying basic geometrical considerations to calculate the area under the v_{cp} curve, as suggested in [10], reveals that the average current programming signal, $\langle v_{cp}\rangle$, D_{on} and D_{off} duty cycles and other circuit parameters are related as follows:

$$\begin{aligned} \langle v_{cp}\rangle = T_s\Bigg[& S_e D_{on} + \frac{S_{on} D_{on}^2}{2} + \frac{S_{off} D_{off}^2}{2} \\ & + S_{off} D_{off}(1 - D_{on} - D_{off})\Bigg]. \end{aligned} \tag{2}$$

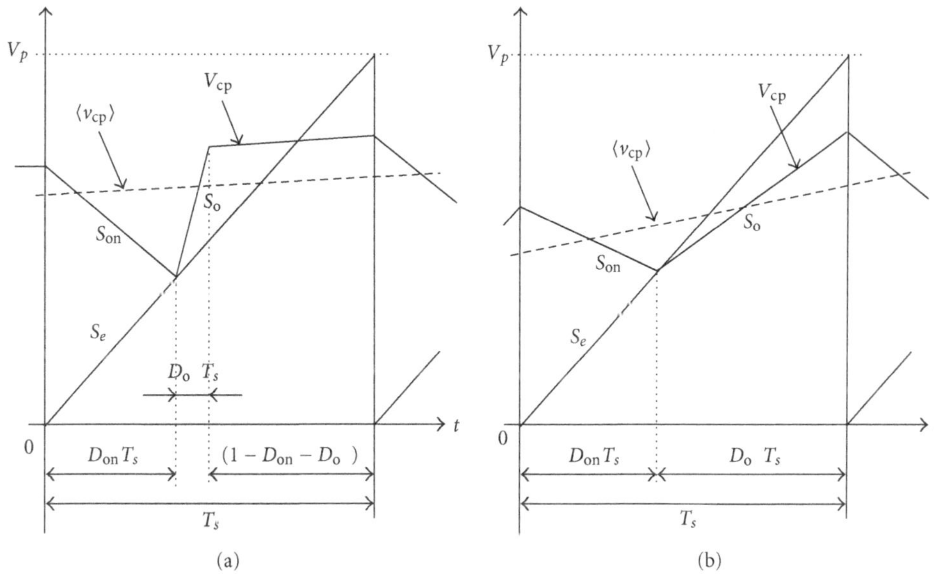

FIGURE 3: PWM comparator input waveforms in DCM (a) and CCM (b) modes.

The required on duty cycle, D_{on}, is the solution of the quadratic equation (2), which takes into consideration the CA average current programming signal $\langle v_{cp} \rangle$ as well as other operating conditions. The switcher terminal voltages are imbedded in the slopes S_{on} and S_{off}, as defined in (1), whereas D_{off} in DCM depends on D_{on} as well as on the average current and system's parameters.

2.2. Comparison with Other Theoretical Results. The proposed software DCG algorithm (2) can generate the required average duty cycle, D_{on}, in the steady state, in transient conditions, CCM or DCM modes.

Inspection of (2) reveals that under CCM conditions, the term $(1 - D_{on} - D_{off})$ vanishes and (2) is reduced to the expression, previously reported in [10]:

$$\langle v_{cp} \rangle = T_s \left[S_e D_{on} + \frac{S_{on} D_{on}^2}{2} + \frac{S_{off} D_{off}^2}{2} \right]. \quad (3)$$

DCG based on (3) can be used in CCM under transient conditions for which S_{on} and S_{off} may assume an arbitrary value. Obviously, with the term $(1 - D_{on} - D_{off})$ missing, (3) is less accurate than (2) in DCM.

Considering that the power stage is operating with slow varying signals in the vicinity of the CCM steady-state equilibrium so that $S_e D_{on} \approx S_{off} D_{off}$, (2) takes the following form:

$$\langle v_{cp} \rangle = T_s \left[S_e D_{on} + \frac{1}{2} S_{on} D_{off} \right], \quad (4)$$

which is identical to that suggested in [11].

For a CA amplifier with a significant attenuation of the switching ripple, $S_e \gg S_{on}, S_{off}$, (2) is further simplified yielding

$$\langle v_{cp} \rangle = T_s S_e D_{on}, \quad (5)$$

which is identical to the classical Voltage Mode PWM function [12].

2.3. On Duty Cycle Programming. Algorithms (2)–(5) are arranged in a descending order of accuracy and programming complexity. Explicit solution for D_{on} duty cycle could be obtained from (4) and (5). The case of (5) could be implemented in PSPICE by a simple "E" source yielding D_{on} duty cycle, *coded in voltage* at the output. The "E" source should have a gain constant $(1/V_P)$ and controlled by a single input variable, that is, by the average current programming signal $\langle v_{cp} \rangle$:

$$D_{on} = \left(\frac{1}{V_p} \right) \langle v_{cp} \rangle. \quad (6)$$

The modulator model (6) totally disregards the ripple component. As it is clearly shown at Figure 3, (6) aims at the intersection of the ramp and the average CA voltage, $\langle v_{cp} \rangle$, somewhat to the right of the correct value of D_{on}, thus, providing only an approximate results for both the CCM and the DCM modes. Therefore, (6) is generally recognized as a simplistic model that can be used under low current ripple conditions arising in Voltage Mode converters, whose loop gain provides heavy attenuation of the switching ripple.

Implementation of (4) requires E-value source performing a division:

$$D_{on} = \frac{\langle v_{cp} \rangle}{V_p + k V_{ab}}. \quad (7)$$

Here, the constant $k = aR_s/2Lf_s$, V_p, and the switching frequency, $f_s = 1/T_s$, should be defined as parameters in PSPICE program. This result is analogous to that suggested by [11]. DCG programmed according to (7) is relatively simple; however, it accounts for the current ripple. Implementation of (7) requires an E-value source governed by two control variables $\langle v_{cp} \rangle$ and V_{ab}. Note that application of (7) requires a preliminary calculation or simulation to correctly establish the system constant k [7, 8].

A different programming approach is required in order to apply the nonlinear DCG algorithms (2). Using (1) and rearranging terms, the proposed DCG algorithm (2) could be realized as follows:

$$D_{on} = \frac{\langle v_{cp} \rangle - kV_{ac}D_{off}(2 - 2D_{on} - D_{off})}{V_p + kV_{ab}D_{on}}. \quad (8)$$

Due to its relative complexity, the proposed DCG implementation (8) is hardly useful as an analytical tool; however, PSPICE can obtain a numerical solution of recursive equations of the $D_{on} = f(D_{on}, \ldots)$ type. Equation (8) holds several advantages for simulation purposes. First is that accurate D_{on} duty cycle could be generated of the given average current programming signal, $\langle v_{cp} \rangle$, and the average SIM terminal voltages V_{ab} and V_{ac}.

As suggested by [6], in order to take account of the DCM-CCM mode transitions frequently encountered in switch-mode systems, the Don generator should also be accompanied by the Doff software generator. A quick reference to Doff generator programming is given in the Appendix.

3. Simulation Results

3.1. Comparison of the DCG Algorithm's Performance through Simulation. In order to compare the performance of the previously described software duty cycle generator algorithms, a PSPICE simulation program was created. The program simulated the time domain response of an ACM dc-dc boost converter in DCM and CCM regimes. Simulation diagram of the circuit is shown in Figure 4(a). The circuit parameters were input dc voltage 12 Vdc, output dc voltage 48 Vdc, boost inductor 200 uHy, switching frequency 100 kHz, peak ramp voltage 5 Vpk. Other circuit parameters are given in the diagram. The circuit was commanded to operate in the DCM and in CCM modes by appropriately stepped current reference signal. The simulation was run cycle-by-cycle to obtain the exact time domain behavior of a real circuit. The obtained results were used as a reference for comparison. The average values of the current amplifier output signal and the inductor current were obtained by heavily filtering. The filtering was done using ORCAD lowpass filter ABM blocks with passband attenuation of 1 db up to 75 kHz, whereas attenuation of 50 db was attained at 100 kHz which effectively removed the switching ripple components. The so-obtained averaged variables were used to command the average models of the DCG generators according to (6), (7) and (8). Doff duty cycle was obtained using (A.5) as described in the Appendix.

The simulated waveforms of the key variables in both the DCM and the CCM modes are shown in Figures 4(b)–4(e). The normalization of the saw-tooth ramp and the current programming signals was done relatively to the peak ramp voltage, V_p. The average duty cycles, generated from the averaged variables, were compared to the waveforms of the cycle-by-cycle-simulated PWM comparator. The closer the average D_{on} signal is to the intersection point of the ramp and CA voltage, the accurate the D_{on} algorithm is. As expected, for both CCM and DCM modes DCG generator (8) proved to be the most accurate, (7) had good accuracy and (6) was only fairly accurate. Particularly, in DCM, see Figure 4(c), the accuracy of the proposed algorithm (8) is noticeably better than that of (7), and much better than that of (6). In CCM (see Figure 4(e)) algorithm (6) achieves only fair accuracy, whereas both algorithms (7) and (8) are of comparable performance providing excellent results where only a negligible advantage in favor of (8) can be seen.

3.2. Demonstration of Autonomous Operation of the Proposed DCG. Next, a complete model of the inner loop of an ACM Boost PFC converter based on benchmark design [9] was derived using the methodology described in the previous sections. The SIM equivalent circuit was used to model the boost power stage and DCG duty cycle generators to model the PWM. The DCGs (7) and (8) were compared. The simulation diagram of the circuit is shown in Figure 5(a). The simulation program operates only with the average variables and allows performing Transient as well as AC study of the current loop. The circuit should be prepared for the frequency response analysis by properly placing an AC test source, V_{test}, within the current loop as shown and introducing offset value to the Vline source to preserve the operating point. Introducing initial conditions helps guiding the simulator to settle on a correct operating point. Comparisons of the simulated frequency response plots due to the DCG (7) and DCG (8) of the inner loop are shown in Figure 5(b) and are hardly distinguishable from each other. The steady-state time domain waveforms of the average APFC model with DCG (8) are shown in Figure 5(c).

4. Conclusion

The paper presented a PSPICE software algorithm for modeling the average behavior of PWM modulator in switch-mode systems. The proposed duty cycle algorithm describes the PWM function in both CCM and DCM operating modes and correctly predicts the duty cycle under any operating conditions. Comparison with previously published results supports the theoretical validity of the proposed approach. The proposed average PWM model (8) was implemented in PSPICE software and compared with cycle-by-cycle simulation yielding excellent results. The proposed algorithm (8) was also compared to the earlier counterparts and was indeed found to be more accurate. Particularly, in DCM, the accuracy of the proposed algorithm (8) is noticeably better than that of (7), whereas in CCM (7) and (8) are of comparable performance with only a negligible advantage in favor of (8). In the frequency domain, which uses log scale for the amplitude, the responses were comparable. However, the higher accuracy of the proposed algorithm (8) comes at the price of added programming complexity, which is its main disadvantage. The proposed algorithm (8) can be

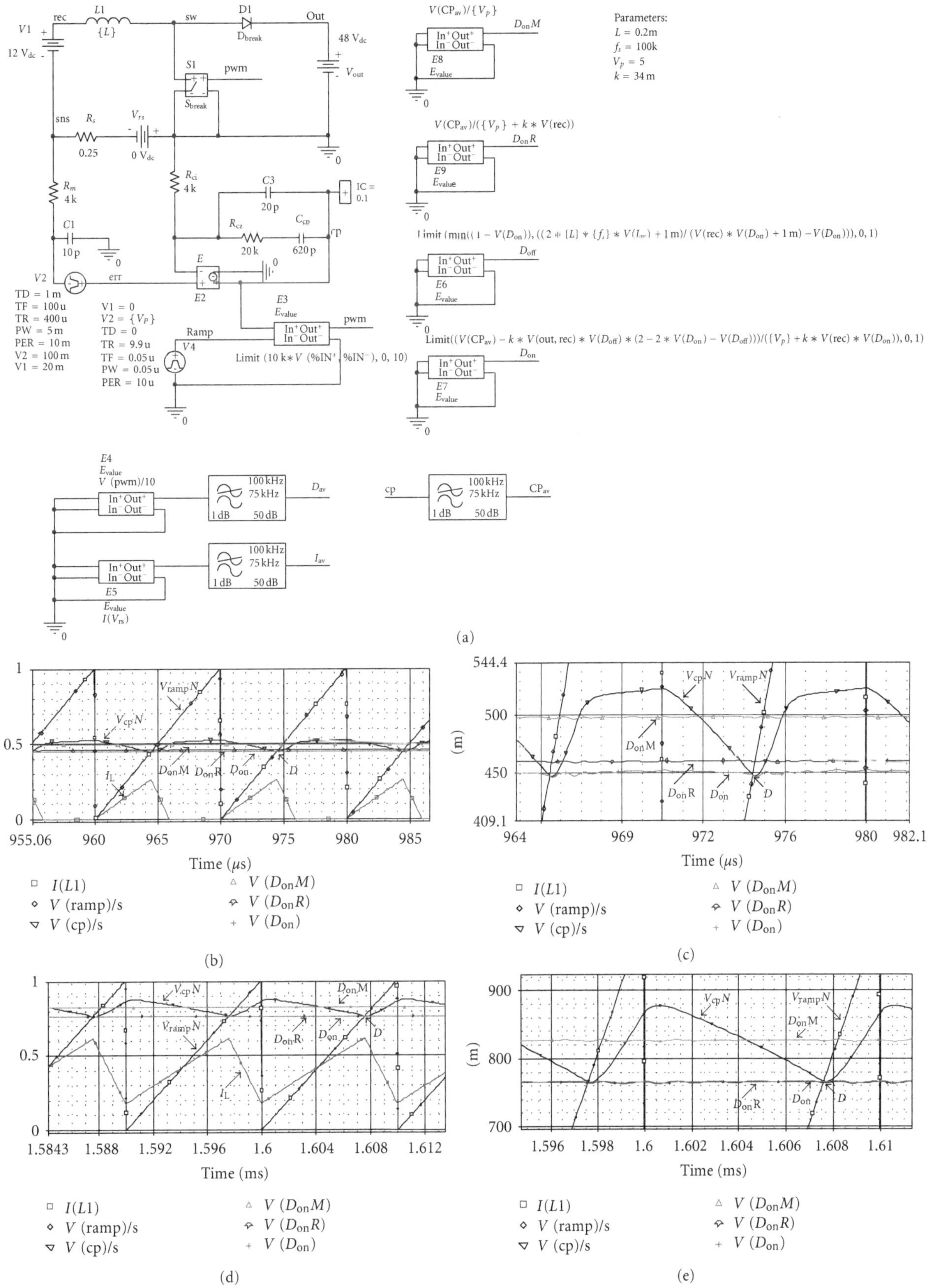

FIGURE 4: Simulation diagram of the ACM Boost converter (a): simulated waveforms of the instantaneous inductor current IL, instantaneous normalized ramp voltage VrampN, instantaneous normalized current programming voltage VcpN, cycle-by-cycle duty cycle D, average duty cycle DonM according to (6), average duty cycle DonR according to (7), average duty cycle Don according to (8). The DCM mode waveforms (b); exploded view of the DCM mode waveforms (c); the CCM mode waveforms (d); exploded view of the CCM mode waveforms (e).

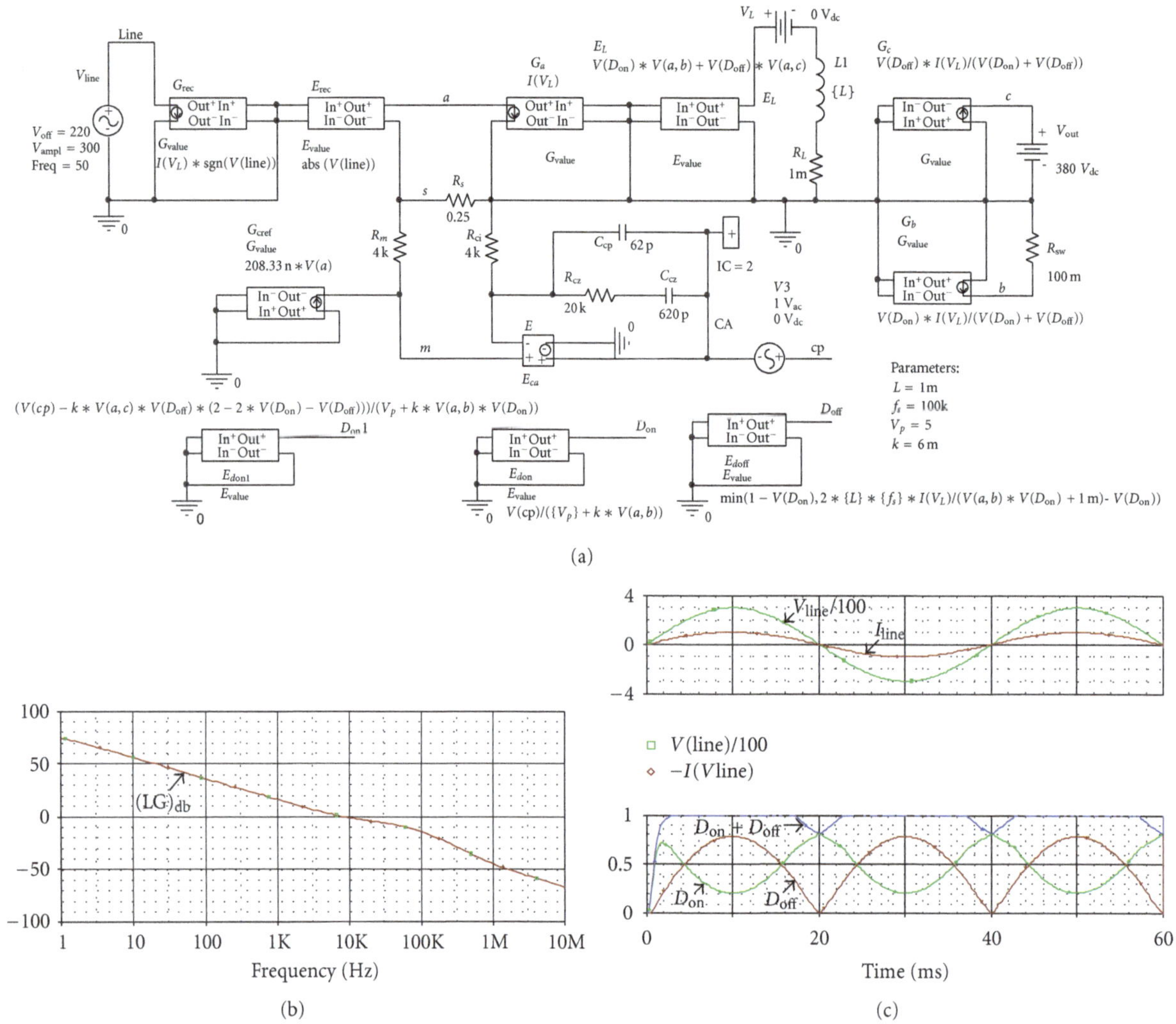

FIGURE 5: PSPICE simulation diagram of the Average Current mode APFC current loop (a); Simulation of the current loop frequency responses (b); time domain waveforms of the scaled line voltage Vline/100 and line current I(Vline) (top), and the duty cycles Don, Doff; note that the sum $D_{\text{off}} + D_{\text{off}} < 1$ as DCM commences (bottom) (c).

successfully applied in simulation; however, it is inconvenient to be used as a theoretical tool. Also, the recursive nature of (8) aggravates convergence problems. While having somewhat lesser accuracy in DCM, the software duty cycle generation algorithm (7) has the advantage of simplicity and lesser convergence problems. Moreover, the expression (7) is identical to the PWM modulator function proposed by [11], which is currently considered as a widely accepted method for theoretical analysis of current mode converters. Therefore, the results of this study provide an additional verification and support the validity and soundness of [11].

Appendix

A. A Quick Reference to SIM and D_{off} DCG

Energy transfer in hard switched power converters is accomplished by periodic charging and discharging of an inductor. The topology of the switched inductor is shown in Figure 6(a). The switched inductor model (SIM) is an average behavioral model of the switched inductor block. SIM was developed by [6] and is shown in Figure 6(b).

SIM derives the average inductor current applying an average voltage, E_L, across the power stage inductor, L [6]. The average voltage, E_L, is a function of the terminal voltages, V_{ab} and V_{ac}, and the duty cycles D_{on} and D_{off}:

$$E_L = V_{\text{ab}} D_{\text{on}} + V_{\text{ac}} D_{\text{off}}. \tag{A.1}$$

The physical feature of current steering between the terminals b and c is modeled by the dependent current sources of the SIM model [8] (Figure 6(b)) which are defined as follows:

$$\begin{aligned} G_a &= I_L, \\ G_b &= \frac{I_L D_{\text{on}}}{D_{\text{on}} + D_{\text{off}}}, \\ G_c &= \frac{I_L D_{\text{off}}}{D_{\text{on}} + D_{\text{off}}}. \end{aligned} \tag{A.2}$$

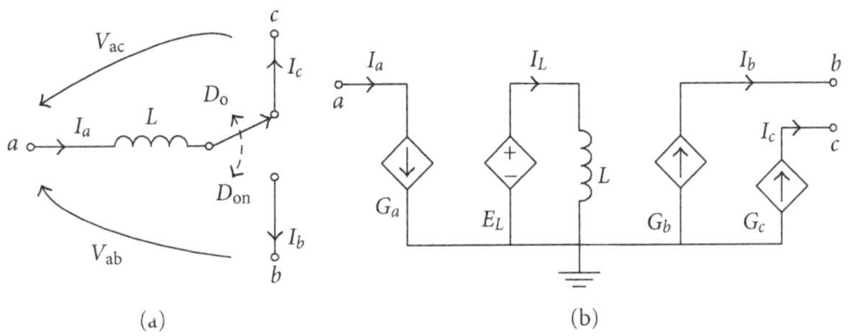

FIGURE 6: The topology of the switched inductor in PWM converters (a); and the Switched Inductor Model (SIM) equivalent circuit (b) [6].

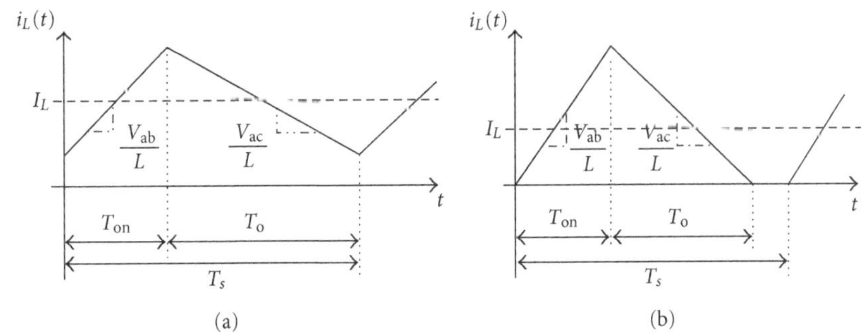

FIGURE 7: Definition of the inductor discharge duty cycle Doff: CCM mode (a) and the DCM (b).

To implement (A.2) the simulation program is also required to produce the D_{off} duty cycle. In case the power stage operates in CCM (see Figure 7(a)) the Off Duty Cycle depends solely on D_{off}:

$$D_{\text{off}} = 1 - D_{\text{on}}. \quad \text{(A.3)}$$

In the case of DCM, however (see Figure 7(b)), D_{off} depends also on other system parameters and operating conditions:

$$D_{\text{off}} = \frac{2Lf_sI_L}{V_{\text{ab}}D_{\text{on}}} - D_{\text{on}} < 1 - D_{\text{on}}. \quad \text{(A.4)}$$

The DCM D_{off} by (A.4) is always smaller than CCM D_{off} by (A.3). Hence, the software Off-Duty Cycle Generator is programmed to calculate both expressions (A.3) and (A.4) and decide on the proper, D_{off}, using the PSPICE minimum function [7]:

$$D_{\text{off}} = \min\left\{(1 - D_{\text{on}}), \left(\frac{2Lf_sI_L}{V_{\text{ab}}D_{\text{on}}} - D_{\text{on}}\right)\right\}. \quad \text{(A.5)}$$

As a result, PSPICE is able to identify CCM-DCM mode changes and calculate the correct D_{off} under any operating conditions.

References

[1] R. D. Middlebrook and S. Cuk, "A general unified approach to modeling switched-converter power stages," in *Proceedings of the Annual IEEE Power Electronics Specialists Conference (PESC '76)*, pp. 18–34, 1976.

[2] R. Holloway and G. Eirea, "Model current-mode control with ease and accuracy," *Power Electronics Technology*, vol. 34, no. 11, pp. 22–25, 2008.

[3] R. E. Griffin, "Unified power converter models for continuous and discontinuous conduction mode," in *Proceedings of the 20th Annual IEEE Power Electronics Specialists Conference (PESC '89)*, pp. 853–860, June 1989.

[4] G. C. Verghese, C. A. Bruzos, and K. N. Mahabir, "Averaged and sampled-data models for current mode control: a reexamination," in *Proceedings of the 20th Annual IEEE Power Electronics Specialists Conference (PESC '89)*, pp. 484–491, June 1989.

[5] F. D. Tan and R. D. Middlebrook, "Unified model for current-programmed converters," *IEEE Transactions on Power Electronics*, vol. 10, no. 4, pp. 397–408, 1995.

[6] S. Ben-Yaakov, "Spice simulation of PWM DC-DC convertor systems: voltage feedback, continuous inductor conduction mode," *Electronics Letters*, vol. 25, no. 16, pp. 1061–1063, 1989.

[7] Y. Amran, F. Huliehel, and S. Ben-Yaakov, "A unified SPICE compatible average model of PWM converters," *IEEE Transactions on Power Electronics*, vol. 6, no. 4, pp. 585–594, 1991.

[8] S. Ben-Yaakov, D. Vardy, and Z. Gaaton, "A unified model of current feedback in switch mode converters," in *Proceedings of the International Conference on Circuits and Systems*, pp. 1891–1894, 1992.

[9] P. C. Todd, "UC3854 controlled power factor correction circuit design," in *Product and Application Handbook*, U-134 Application note, Unitrode Integrated Circuits, 1993.

[10] S. Ben-Yaakov and Z. Gaaton, "Generic SPICE compatible model of current feedback in switch mode convertors," *IEEE Electronic Letters*, vol. 28, no. 14, pp. 1356–1358, 1992.

[11] R. B. Ridley, "A new continuous-time model for current-mode control," in *Proceedings of the Power Conversion and Intelligent Motion (PCIM '89)*, pp. 455–464, 1989.

[12] R. D. Middlebrook, "Predicting modulator phase lag in PWM converter feedback loops," in *Proceedings of the Advances in Switchedmode Power Conversion*, paper H-4, pp. 245–250, 1981.

18

A 25 kW, 25 kHz Induction Heating Power Supply for MOVPE System Using L-LC Resonant Inverter

Mangesh Borage and Sunil Tiwari

Power Supplies and Industrial Accelerator Division, Raja Ramanna Centre for Advanced Technology, Indore 452013, India

Correspondence should be addressed to Mangesh Borage; mbb@rrcat.gov.in

Academic Editor: Jose Pomilio

A topology named L-LC resonant inverter (RI) for induction heating (IH) applications takes most of the merits of the conventional series and parallel resonant schemes, while eliminating their limitations. In this paper, fundamental frequency AC analysis of L-LC RI is revisited, and a new operating point is suggested featuring enhanced current gain and near in-phase operation as compared to the conventional operating point. An approximate analysis of the circuit with square-wave voltage source is also described highlighting the effect of auxiliary inductor on the source current waveform. The analysis also leads to an optimum choice of the auxiliary inductance. The requirements of the metal organic vapour phase epitaxy (MOVPE) system in which a graphite susceptor is required to be heated to 1200°C demanding a 25 kW, 25 kHz IH power supply, the configuration of developed IH system, and experimental results are presented.

1. Introduction

The induction heating (IH) [1] is commonly used for heat treatment of metals (hardening, tempering, and annealing), heating prior to deformation (forging, swaging, upsetting, bending, and piercing), brazing and soldering, shrink fitting, coating, melting, crystal growing, cap sealing, sintering, carbon vapor deposition, epitaxial deposition, and plasma generation. IH is a noncontact method. The heat is generated only in the part, not in the surrounding area except by radiation. The location of the heating can be defined to a specified area on the metal component, thereby achieving accurate and consistent results. As heating occurs in the object itself, IH is considered more efficient than alternative methods.

An IH system comprises a basic induction power source which provides the required power output at the required power frequency, complete with matching components, an induction coil assembly, a method of material handling, and some method of cooling. Generally, full-bridge or half-bridge resonant inverters (RIs) are most commonly used as the power supplies for IH. The equivalent model of an IH coil with work piece can be represented in simplified form by an equivalent inductance (L_{eq}) and resistance (R_{eq}) as shown in Figure 1. If the IH coil is directly fed from a supply, the ratio of apparent to real power will be large. Therefore, the IH coil is properly compensated by capacitors and additional inductors in suitable configuration so that minimum reactive power is drawn from the source. Additionally to match the load voltage-current requirements to the available source, a matching network is required. The matching is normally achieved with the help of an isolation transformer of suitable turns ratio.

Based on the connection of compensating capacitor with the IH coil, the two most commonly used RI topologies are as follows.

(1) Series resonant inverter (SRI): the compensating capacitor is placed in series with the IH coil, and it is fed by a voltage source [2–6].

(2) Parallel resonant inverter (PRI): the compensating capacitor is placed in parallel with the IH coil, and it is fed by the current source [7–11].

FIGURE 1: Equivalent circuit representation of an IH coil.

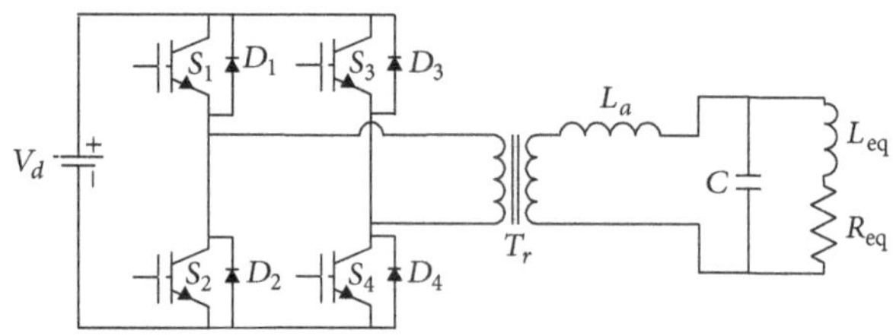

FIGURE 2: Circuit diagram of L-LC RI for IH application.

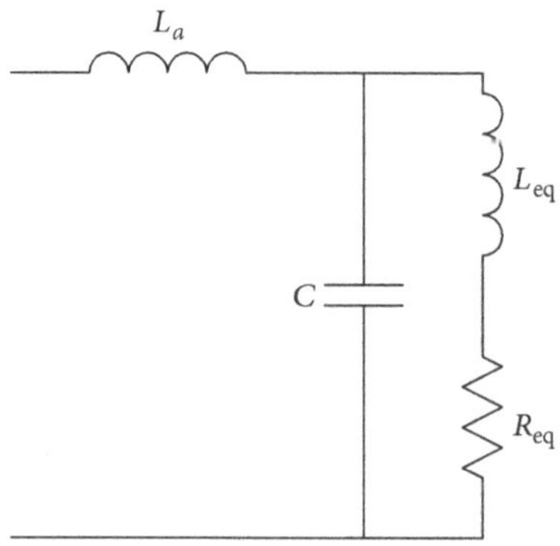

FIGURE 3: Equivalent circuit diagram of L-LC RI for analysis.

The analysis of these circuits has been performed in great details, and the comparative assessment is also reported in the literature [12, 13].

A topology named L-LC RI has been proposed for IH applications [14–21], which takes most of the merits of SRI and PRI while eliminating their limitations. It operates with input dc voltage source, thereby eliminating bulky input current smoothening inductor. It offers high current gain, which in turn reduces the current rating of the secondary winding of matching transformer and the feeder to coil.

Metal organic vapour phase epitaxy (MOVPE) [22] is a highly controlled method for the deposition of semiconductor epitaxial layers and heterostructures, which are required for the development of several optoelectronic and electronic devices. The process of MOVPE involves a vapour phase reaction between metalorganic compound and a hydride gas, which are transported to a heated (around 1200°C) graphite susceptor, resulting in a growth of the desired material. IH is one of the preferred methods for noncontact heating of the susceptor.

The paper aims to investigate the characteristics of L-LC RI for an application demanding a 25 kW, 25 kHz IH power supply to heat graphite susceptor to 1200°C in a MOVPE system for the growth of nitride semiconductors being developed at our Institute. Section 2 describes AC analysis of L-LC CN and investigates various characteristics when the converter is operated at the resonant frequency. The suggested operating point is different than the previously proposed operating point which provides enhanced current gain with smaller auxiliary inductance and results in near in-phase operation. Section 3 describes the operation of high-Q L-LC compensating network (CN) with square-wave voltage source leading to an optimum choice of the auxiliary inductor. The requirements, description, and design of the practical system are discussed in Section 4. Experimental results are presented in Section 5.

2. Analysis of L-LC CN

Figure 2 shows the L-LC RI. Input dc source V_d could be an unregulated source (obtained with single-phase or three-phase diode rectifier and filter) or could be a regulated voltage source (obtained with single-phase or three-phase diode/SCR rectifier and filter or another front-end switch-mode regulator). In the former case, regulation of the power to the work piece should be done in the RI stage using frequency variation [23, 24], fixed-frequency pulse width modulation (PWM) [25–27], or pulse density modulation (PDM) [28, 29]. In the latter case, the output control can be done by varying V_d, providing easy control of the output power over a wide range. However, two cascaded converters tend to reduce the overall efficiency.

The following analysis based on the fundamental frequency approximation examines the important characteristics of L-LC topology. In an equivalent circuit of L-LC CN, shown in Figure 3, input voltage source is assumed to be a sinusoidal voltage source whose rms value is equal to the rms value of the fundamental component of the square-wave excitation. Following definitions are made for the analysis.

Angular resonant frequency:

$$\omega_o = 2\pi f_o = \frac{1}{\sqrt{L_{eq}C}}. \tag{1}$$

Normalized operating frequency:

$$\omega_n = \frac{\omega}{\omega_o}, \tag{2}$$

where $\omega = 2\pi f$ is the angular operating frequency and f is the operating or switching frequency.

Characteristic Impedance:

$$Z_n = \sqrt{\frac{L_{eq}}{C}}. \tag{3}$$

Circuit Q:

$$Q = \frac{\omega_o L_{eq}}{R_{eq}} = \frac{Z_n}{R_{eq}}. \tag{4}$$

Inductance ratio:

$$\gamma = \frac{L_a}{L_{eq}}. \tag{5}$$

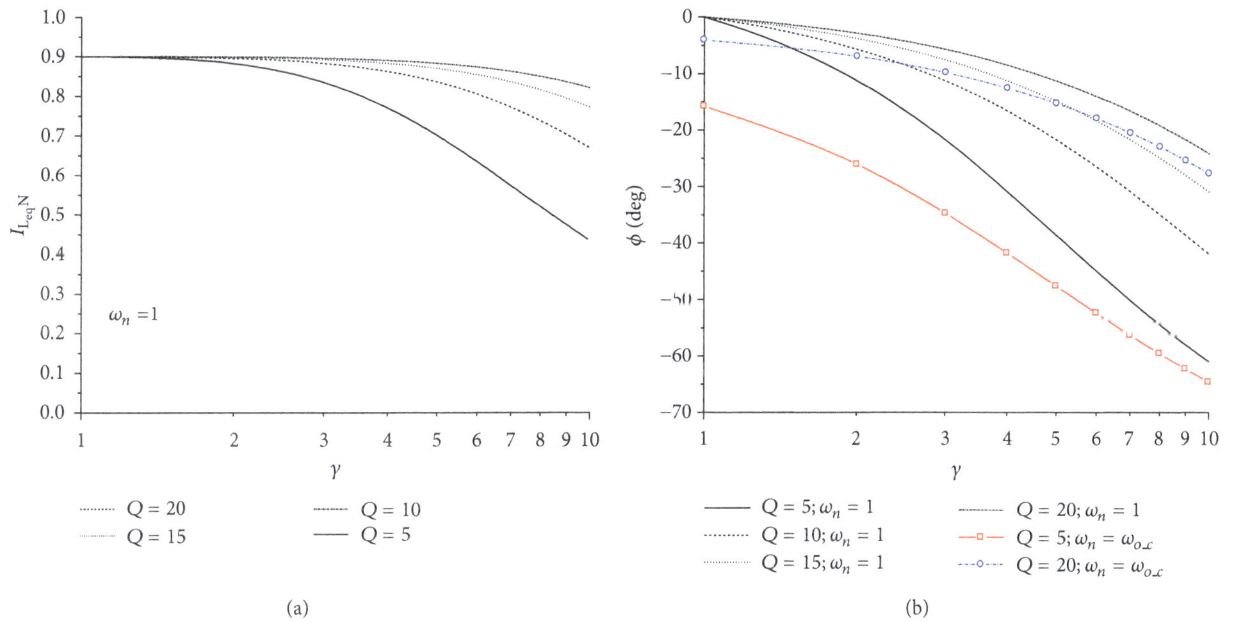

Figure 4: The plots of $I_{L_{eq}N}$ (a) and ϕ (b) as a function of γ for different values of Q at two operating points.

The expression for normalized current in coil and normalized source current (or the current in inductor L_a) can be, respectively, derived as

$$I_{L_{eq}N} = \frac{I_{L_{eq}}}{V_d/Z_n} = \frac{2\sqrt{2}}{\pi}\frac{1}{(1/Q)\left(1-\omega_n^2\gamma\right)+j\left[\omega_n(1+\gamma)-\omega_n^3\gamma\right]},$$
$$I_{L_aN} = \frac{I_{L_a}}{V_d/Z_n} = \frac{2\sqrt{2}}{\pi}\frac{\left(1-\omega_n^2\right)+j\omega_n(1/Q)}{(1/Q)\left(1-\omega_n^2\gamma\right)+j\left[\omega_n(1+\gamma)-\omega_n^3\gamma\right]}. \tag{6}$$

Next, the salient characteristics of the converter operating at the proposed operating point, $\omega_n = 1$, are investigated, leading to the following observations:

(1)

$$\left|I_{L_{eq}N}\right|_{\omega_n=1} = \frac{2\sqrt{2}}{\pi}\frac{Q}{\sqrt{Q^2+(1-\gamma)^2}}. \tag{7}$$

For $Q \gg |(1-\gamma)|$,

$$\left|I_{L_{eq}N}\right|_{\omega_n=1} \approx \frac{2\sqrt{2}}{\pi}, \tag{8}$$

(2)

$$\left|I_{L_aN}\right|_{\omega_n=1} = \frac{2\sqrt{2}}{\pi}\frac{1}{\sqrt{Q^2+(1-\gamma)^2}}, \tag{9}$$

$$\phi|_{\omega_n=1} = \angle I_{L_aN}\big|_{\omega_n=1} = -\tan^{-1}\left(\frac{\gamma-1}{Q}\right). \tag{10}$$

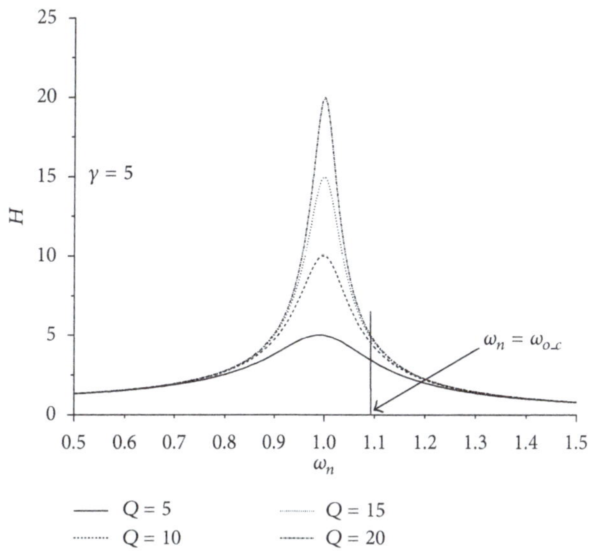

Figure 5: The plots of H as a function of ω_n for $\gamma = 5$.

The source current at $\omega_n = 1$ is also seen to be inductive for all values of $\gamma > 1$, and the phase angle is function of Q and γ. If $\gamma = 1$, the source current is always in-phase with voltage.

Figure 4(a) shows the plot of $I_{L_{eq}N}$ as a function of γ for different values of Q. When the converter is operated at $\omega_n = 1$, $I_{L_{eq}N}$ is seen to be relatively insensitive to γ, particularly at high values of Q—a typical operating condition in an IH application.

Figure 4(b) shows the plot of ϕ as a function of γ. The source current lags the applied square-wave voltage $\omega_n = 1$. While this lagging current is beneficial for zero-voltage-switching (ZVS) of the semiconductor switches, it can be anticipated that a higher value of ϕ results in higher source

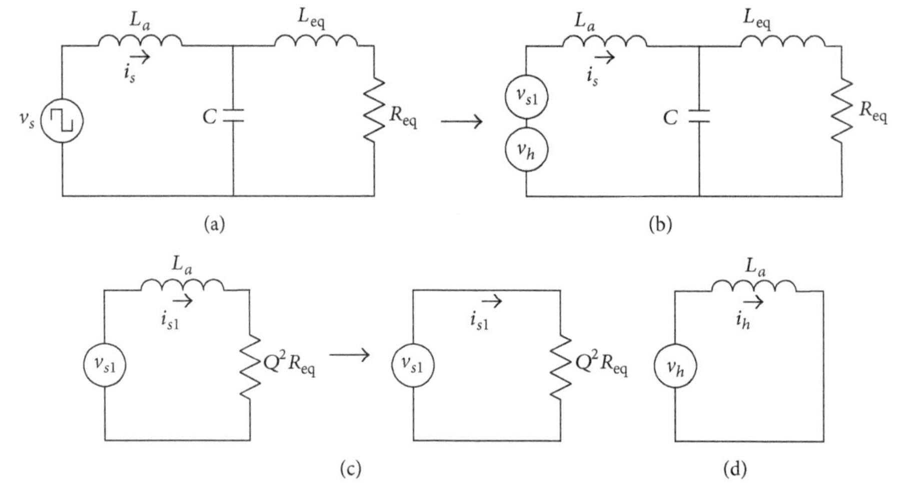

FIGURE 6: (a) Equivalent circuit for L-LC RI for analysis with v_s. (b) Equivalent circuit redrawn with v_s decomposed into v_{s1} and v_h. (c) Approximate equivalent circuit for fundamental frequency. (d) Approximate equivalent circuit for harmonics frequencies.

current causing more conduction loss in the switches. It is evident from the plots of Figure 4(b) that ϕ is small for operation at $\omega_n = 1$, particularly for operation with high Q. Therefore, the source current and conduction loss in the switches will be smaller at this operating point.

The current gain of the CN, H, is given by

$$H = \frac{I_{L_{eq}N}}{I_{L_aN}} = \frac{1}{(1-\omega_n^2) + j\omega_n(1/Q)}, \tag{11}$$

which can further be simplified at $\omega_n = 1$ as

$$|H|_{\omega_n=1} = Q. \tag{12}$$

Figure 5 shows the plot of H as a function of ω_n for $\gamma = 5$ and different values of Q. Maximum current gain (equal to Q) is observed for operation at the proposed operating point $\omega_n = 1$.

The characteristics and design of L-LC RI for IH applications operating at $\omega_n = \omega_{o\mathcal{L}} = \sqrt{(1+\gamma)/\gamma}$ have been reported extensively in the literature [15–21]. At this operating point,

$$|H|_{\omega_n=\omega_{o\mathcal{L}}} = \frac{Q\gamma}{\sqrt{Q^2 + \gamma(1+\gamma)}}. \tag{13}$$

For $Q \gg \gamma$, $|H|_{\omega_n=\omega_{o\mathcal{L}}} \approx \gamma$, which is less than theoretical maximum value given by (12) at the proposed operating point. Thus, as also shown in Figure 5, the proposed operating point of L-LC RI results in the enhanced current gain. Further, the expression for ϕ at the conventional operating point can be written as

$$\phi|_{\omega_n=\omega_{o\mathcal{L}}} = \tan^{-1}\left(-\frac{\gamma}{Q}\sqrt{\frac{1+\gamma}{\gamma}}\right). \tag{14}$$

The plots of (14) as a function of γ are shown in Figure 4(b) for Q = 5 and Q = 20. It can be seen that phase angle is smaller at the proposed operating point than the conventional operating point, thereby resulting in less conduction loss in the switches.

3. Behaviour of L-LC CN with Square-Wave Voltage Source

Analysis based on the fundamental frequency approximation presented in Section 2 suggests that the converter, when operated at $\omega_n = 1$ offers highest current gain and small phase angle between source voltage and current resulting in low reactive power loading and low conduction loss in the semiconductor devices of the square-wave inverter. The analysis presented in Section 2 assumes sine wave voltage source at the input. In practice, the input voltage is a square-wave generated by operating switches S_1–S_4 in Figure 2 at 50 per cent duty cycle. In this section, the behaviour of L-LC resonant inverter for operation at $\omega_n = 1$ with square-wave voltage source examined. Equivalent circuit for L-LC RI for analysis with square-wave voltage source is shown in Figure 6(a). The source voltage v_s can be defined as

$$v_s(t) = V_d \cdot \text{sign}(\sin(\omega_o t)), \tag{15}$$

v_s can be decomposed into its fundamental component v_{s1} and harmonics v_h as

$$v_s(t) = v_{s1}(t) + v_h(t) = \frac{4V_d}{\pi}\sin(\omega_o t) + v_h(t). \tag{16}$$

From (15) and (16)

$$\begin{aligned} v_h(t) &= V_d \cdot \text{sign}(\sin(\omega_o t)) - \frac{4V_d}{\pi}\sin(\omega_o t) \\ &= V_d - \frac{4V_d}{\pi}\sin(\omega_o t) \quad \text{for } 0 \le t \le \frac{T_s}{2} \\ &= -V_d - \frac{4V_d}{\pi}\sin(\omega_o t) \quad \text{for } \frac{T_s}{2} \le t \le T_s, \end{aligned} \tag{17}$$

wherein T_s is the switching period. Thus, equivalent circuit of Figure 6(a) can be redrawn as Figure 6(b), wherein v_s decomposed into v_{s1} and v_h.

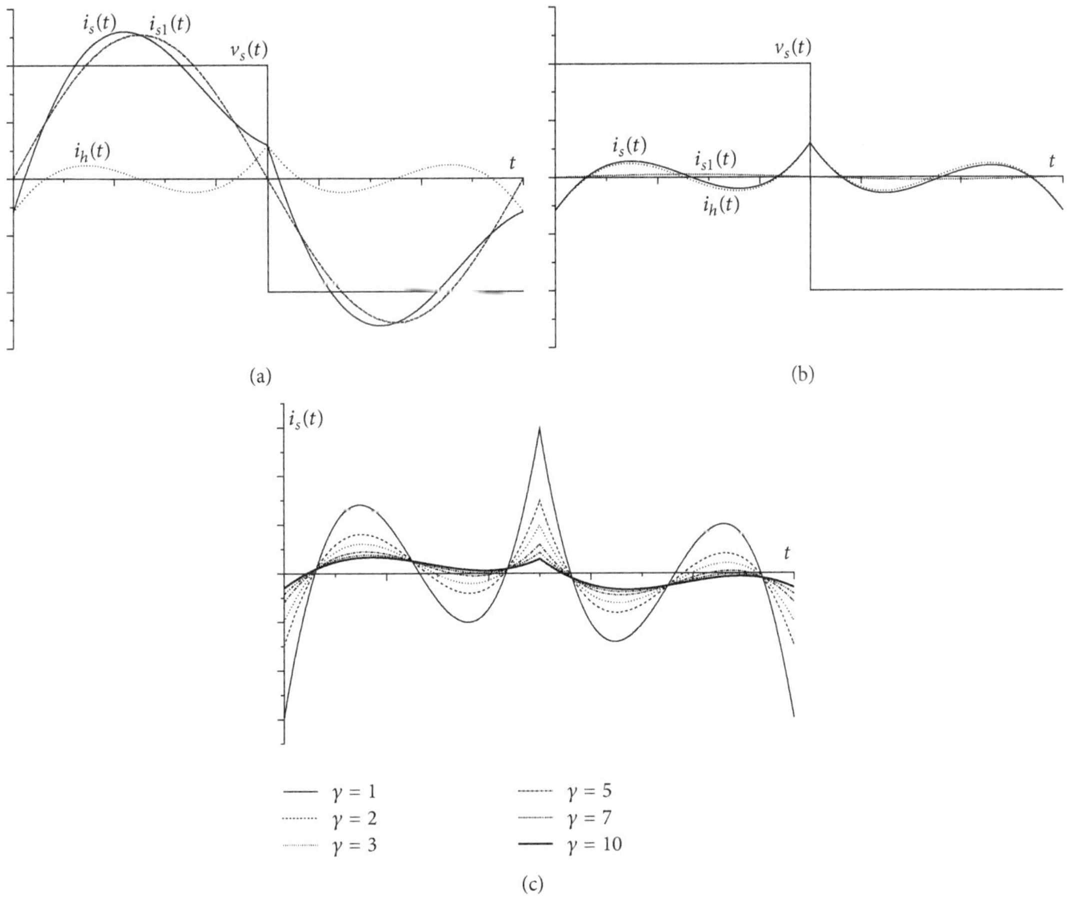

FIGURE 7: Typical calculated waveforms of $v_s(t)$, $i_{s1}(t)$, $i_h(t)$, and $i_s(t)$ for low-Q ($Q = 1$; $\gamma = 1$) (a) and high-Q ($Q = 50$; $\gamma = 1$) (b) operation. The effect of γ on the waveform of $i_s(t)$ is illustrated by the calculated waveforms in (c) for $Q = 50$.

Equivalent impedance (Z_{eq}) of the L_{eq}, R_{eq}, and C resonant circuit at $\omega_n = 1$ is given by

$$\left|Z_{eq}\right|_{\omega_n=1} = R_{eq}Q\sqrt{1+Q^2} \approx R_{eq}Q^2 \quad \text{for } Q \gg 1. \tag{18}$$

Further, for high-Q resonant circuit in IH applications, Z_{eq} can be approximated to be zero for all operating points other than $\omega_n = 1$. Under this assumption, equivalent circuit for Figure 6(b) can be decoupled for operation at $\omega_n = 1$ and at harmonic frequencies, as shown in Figures 6(c) and 6(d), respectively. Assuming $R_{eq}Q^2 \gg \omega_o L_a$, equivalent circuit at $\omega_n = 1$ can further be simplified as shown in Figure 6(c). With this simplification, the circuit can be analyzed for fundamental and harmonic frequencies separately to find out individual source current and then be added to find out resultant source current. An expression for i_s can be derived as

$$i_s(t) = i_{s1}(t) + i_h(t),$$

$$i_{s1}(t) = \frac{4V_d}{\pi Q^2 R_{eq}} \sin(\omega_o t),$$

$$\begin{aligned} i_h(t) &= \frac{V_d}{L_a}\left(t - \frac{T_s}{4}\right) - \frac{4V_d}{\pi\omega_o L_a}\sin\left(\omega_o t - \frac{\pi}{2}\right) \quad \text{for } 0 \le t \le \frac{T_s}{2} \\ &= -\frac{V_d}{L_a}\left(t - \frac{3T_s}{4}\right) - \frac{4V_d}{\pi\omega_o L_a}\sin\left(\omega_o t - \frac{\pi}{2}\right) \quad \text{for } \frac{T_s}{2} \le t \le T_s. \end{aligned} \tag{19}$$

Figure 7(a) shows the calculated waveforms for the operation of the circuit under low-Q condition. It can be easily noticed that $i_{s1}(t)$, being inversely proportional to Q, is significantly larger than $i_h(t)$ for low-Q values. Therefore, resulting $i_s(t)$ is also nearly sinusoidal. Under this condition, the predictions of the ac analysis (Section 2) are reasonably accurate. However, under high-Q condition, as shown in Figure 7(b), amplitude of $i_{s1}(t)$ is much smaller than $i_h(t)$. Therefore, $i_s(t)$ is nearly the same as $i_h(t)$ and is highly nonsinusoidal. Under this condition, the predictions of Section 2 tend to be erroneous. Since the analysis of Section 2 does not account for harmonics, the actual peak and rms values of i_s calculated by (3) are significantly more than those predicted by (9). This will result in degradation of the current gain from its value predicted by (12). However, the amplitude and rms value of $i_h(t)$ can be controlled by choosing a proper value of L_a. Figure 7(c) shows the calculated waveform of $i_s(t)$ for different values of γ when $Q = 50$. Since the rms value of $i_h(t)$ can be reduced by increasing the value of L_a, the degradation in the current gain can be corrected to some extent by choosing higher value of γ. Figure 8 shows the plots of current gain as a function of Q for different values of γ. A plot of the current gain predicted by (12) is also shown for direct comparison. The figure also shows the evolution of the current gain at the conventional operating point predicted by (13) for $\gamma = 1$ and $\gamma = 5$. It is noticed that the actual current gain at the proposed operating point is higher than that at the conventional operating point.

The physical size of an inductor depends on its value, peak current, and rms current. Since, peak and rms value of $i_s(t)$ decreases with increasing L_a, it is intuitive to expect that the physical size of L_a will first reduce as L_a increased, reach a minimum and subsequently increase with further increase in L_a. To obtain a value of L_a resulting in minimum size, the following term is defined as its normalized size index:

$$K_{L_a} = \frac{L_a I_{s,\mathrm{pk}} I_{s,\mathrm{rms}}}{L_{\mathrm{eq}} I_{L_{\mathrm{eq}},\mathrm{pk}} I_{L_{\mathrm{eq}},\mathrm{rms}}}. \quad (20)$$

Figure 9 shows the plots of (20) for different values of Q. It is seen that for high-Q applications, L_a should typically be 5 times L_{eq}.

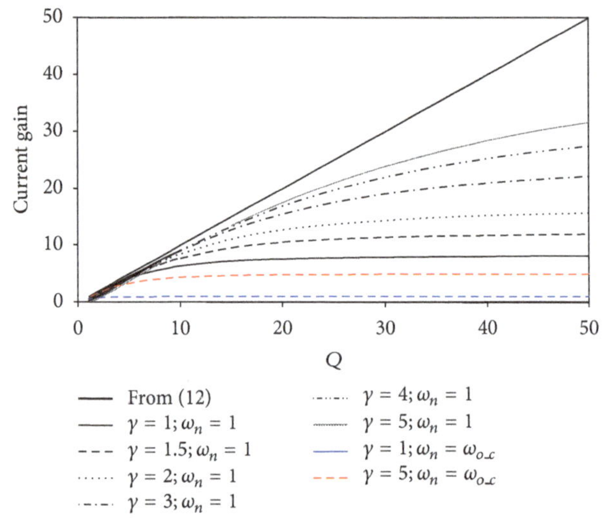

FIGURE 8: The plots of current gain as a function of Q for different values of γ at the at the proposed operating point $\omega_n = 1$ and conventional operating point $\omega_n = \omega_{o\_c}$.

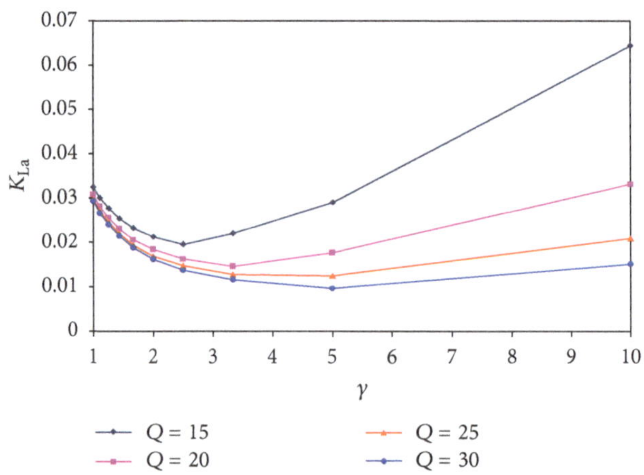

FIGURE 9: The plots of (20) showing the variation of K_{L_a} as a function of γ for different values of Q.

4. System Description

MOVPE is a highly controlled method for the deposition of semiconductor epitaxial layers and heterostructures, which are required for the development of several optoelectronic and electronic devices. The process of MOVPE involves a vapour phase reaction between metalorganic compound and a hydride gas, which are transported to a heated (around 1200°C) graphite susceptor, resulting in a growth of the desired material. A power supply is required to heat graphite susceptor (50 mm diameter and 20 mm length) to 1200°C in a quartz reactor of 80 mm diameter.

4.1. Work Coil and Resonant Capacitor. While recommended coil design practices suggest that the coil should be as close to the susceptor and the coil length should be larger than that of the susceptor [1], physical dimensions of the susceptor and the reactor force the inner diameter of the coil to be 100 mm. Further, since a stainless steel showerhead with nozzles for flowing gases into the reactor is attached to one end of the reactor, the maximum length of the coil is also limited to 50 mm. The coil is made up of hollow copper conductor. The number of turns, conductor diameter, and its wall thickness are optimized considering various parameters like coil resistance, power loss, electrical efficiency, required cooling water flow rate, and pressure drop. The designed coil has 4 turns of hollow copper conductor with 3/8 inch OD and SWG 19 wall thickness. The coil resistance is estimated to be 11.3 mΩ, and coil inductance is 2.70 μH. The operating frequency of 25 kHz has been fixed to ensure uniform heating of the susceptor (the skin depth of graphite is nearly 17 mm at 25 kHz). Equivalent resistance of work piece is estimated to be 16.6 mΩ. The coil is designed to carry a maximum of 1000 A rms.

A bank of 6 conduction-cooled capacitors of 3 μF each has been used as the resonant capacitor C. These capacitors are mounted on a cold plate, which in turn is water cooled. However, only 5 capacitors are used to get the resonant frequency of 25 kHz. The capacitor bank is kept very close to the coil to reduce the circulating path.

4.2. Power Circuit. A schematic circuit diagram of the power circuit is shown in Figure 10, which can broadly be divided into three sections: front-end-three phase diode rectifier with filter, a dc-dc converter, and the L-LC RI. In addition to

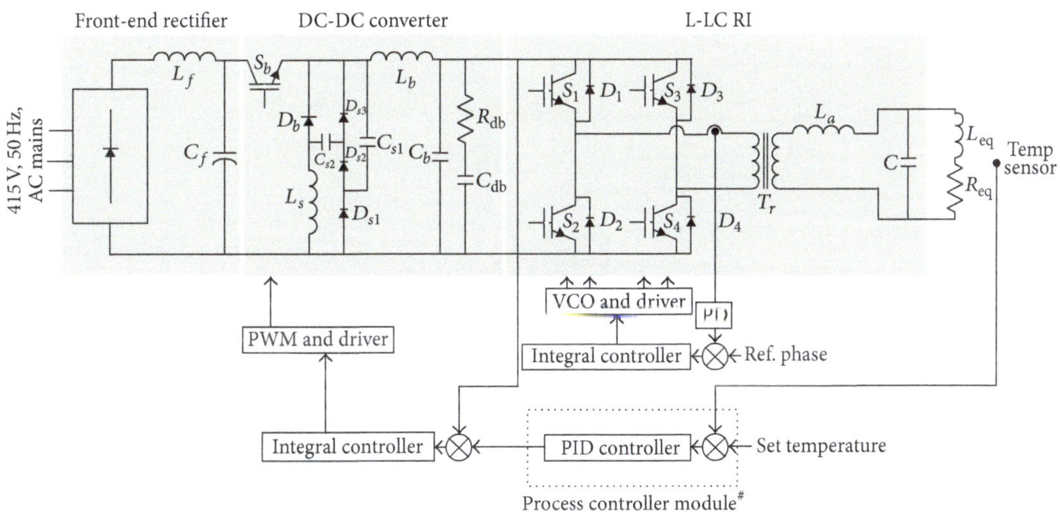

FIGURE 10: The power circuit and the control block diagram of a 25 kHz, 25 kW induction heating power supply using L-LC RI for MOVPE system. #The process controller, shown for the completeness, is not used in the laboratory testing of the system.

these, the actual power circuit also consists of breakers, EMI filter, and inrush-current-limiting circuit in the rectifier stage. However, these are not shown in Figure 10 for clarity.

The input to the power supply is 415 V, 50 Hz, and 3-phase ac. The values of filter inductor (L_f) and capacitor (C_f) in the input rectifier section are 2.2 mH and 3 mF, respectively, giving the cut-off frequency of 70 Hz.

It is seen in Section 2 that the L-LC RI exhibits desired behaviour only when it is operated at $\omega_n = 1$. Therefore, variation of switching frequency for output power control is not permitted. Therefore, fixed-frequency control methods such as pulse width modulation (PWM) or the quantized control methods such as pulse density modulation can be used. However, the soft-switching operation of the switches in RI stage may not be guaranteed over the entire operating range, and the quantized control methods, such as PDM, result in discrete output levels and also have limited range of output power control. If, on the other hand, input voltage to the RI stage is controlled, output power control over a wider range and operational flexibility can be ascertained. Demerits of this scheme, namely, requirement of an intermediate dc-dc converter stage and reduction of overall conversion efficiency, are sacrificed for aforementioned advantages.

A dc-dc buck converter is chosen as the intermediate dc-dc converter stage. IGBT S_b and diode D_b constitute the main switching cell. A passive loss-less snubber circuit (composed of snubber inductor L_s, capacitors C_{s1}, C_{s2}, and diodes D_{s1}, D_{s2}, and D_{s3}) is used to limit switching losses in this stage. It is worthwhile to note that apart from the careful choice of the component values, described in [30], an equal attention to several practical aspects [31], such as forward recovery of snubber diodes and stray inductances, is also important for effective snubber action. Inductor L_b, capacitors C_b, C_{bd}, and resistor R_{bd} constitute a high-frequency damped low-pass filter.

Two half-bridge IGBT modules SKM100GB123D are used to realize the H-bridge in the inverter section feeding square-wave voltage to the high-frequency isolation transformer. The transformer T_r having 1:1 turns ratio has been developed using 7 pairs of EE80 core with 7 turns of the primary winding and 7 turns of the secondary winding. While the primary is wound using 0.2 × 50 mm copper foil, oxygen-free high conductivity hollow copper conductor is used for water-cooled secondary winding. The results for Figure 9 suggest that the additional resonant inductor L_a should approximately be 5 times L_{eq} for high-Q operation. Therefore, in the present system, $L_a = 13.5\,\mu$H is chosen. The transformer leakage inductance is estimated to be 5 μH. The additional 8.5 μH inductance is realized by running the secondary turns around auxiliary gapped core (2 pairs of EE80), following the configuration suggested in [32].

A capacitor has been placed in series with the transformer primary (not shown in Figure 10) to prevent the transformer saturation in the event of unsymmetrical excitation. Table 1

FIGURE 11: The photograph of the developed induction heating power supply. The graphite susceptor heated to 1200°C is shown in the inset.

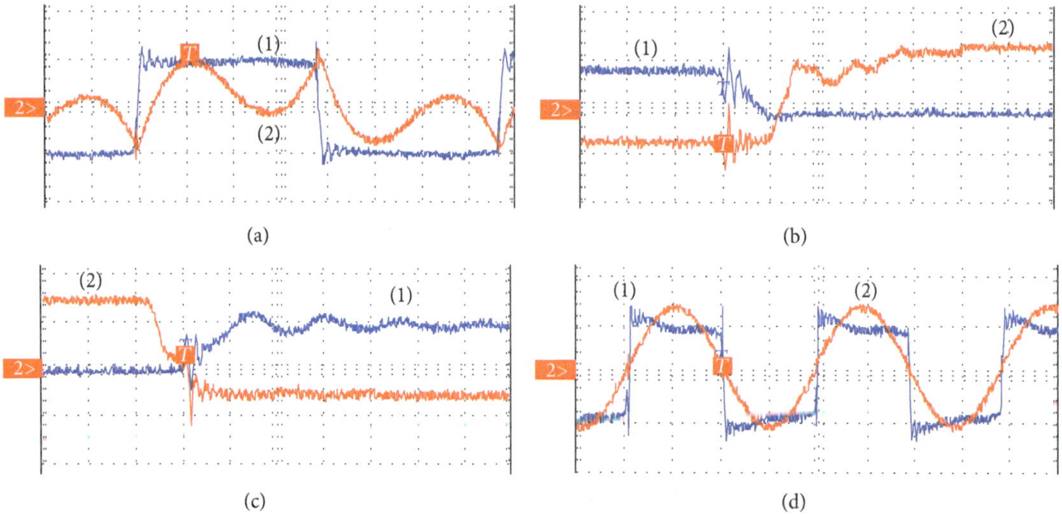

Figure 12: Experimental waveforms. (a) v_s (trace 1, 200 V/div) and i_s (trace 2, 25 A/div); x-scale: 5 μs/div. (b) Waveforms at turn-on: IGBT collector-to-emitter voltage (trace 1, 200 V/div) and gate-to-emitter voltage (trace 2, 10 V/div) x-scale: 500 ns/div. (c) Waveforms at turn-off: IGBT collector-to-emitter voltage (trace 1, 200 V/div) and gate-to-emitter voltage (trace 2, 10 V/div) x-scale: 500 ns/div. (d) v_s (trace 1, 200 V/div) and the voltage across the work coil (trace 2, 200 V/div) x-scale: 10 μs/div.

Table 1: Component values and device part numbers used in the system.

Component	Value/Part number
L_f	2.2 mH
C_f	3 mF
L_b	300 μH
C_b	30 μF
C_{bd}	150 μF
R_{bd}	2.2 Ω
L_s	5 μH
C_{s1}	0.03 μF
C_{s2}	0.33 μF
L_a	13.5 μH
C	15 μF
D_{r1}-D_{r6}	VUO 82
S_b-D_b	SKM100GAR123D
S_1-S_4 and D_1-D_4	SKM100GB123D, two modules
D_{s1}, D_{s2}, D_{s3}	DSEI 31 × 100

summarizes the values of the components and part numbers of the semiconductor devices used in the system.

4.3. Control Circuit. The input dc voltage to the square-wave inverter is controlled to regulate the susceptor temperature. The block diagram of overall control system is also shown in Figure 10. A thermocouple-type temperature sensor and process PID controller module are planned to be used for programming and controlling susceptor temperature. The output of PID module acts as the reference for the inner control loop to control the input dc voltage to the square-wave inverter by controlling the duty ratio of the buck converter. An additional phase control loop is employed to maintain the tune-in condition of the RI stage against the slow drift of resonant frequency over varied operating conditions and time. This loop senses the phase of the inverter output current using a phase detector, PD. The output of the controller drives a voltage controlled oscillator (VCO) that adjusts the switching frequency of the inverter in such a way that the inverter output current lags the voltage slightly.

5. Results

The photograph of the developed induction heating power supply, being tested in laboratory to heat a graphite susceptor, is shown in Figure 11. The graphite susceptor heated to 1200°C in air is shown in the inset.

The inverter output voltage, v_s, (trace 1, 200 V/div) and current waveforms, i_s, (trace 2, 25 A/div) are shown in Figure 12(a). The nature of current waveform matches with the predicted waveform given in Figure 7. The waveforms of collector-to-emitter (trace 1, 200 V/div) and gate-to-emitter (trace 2, 10 V/div) voltage of an IGBT in H-bridge inverter during turn-on and turn-off transitions are shown in Figures 12(b) and 12(c), respectively, demonstrating the soft-switching. Figure 12(d) shows the waveforms of v_s (trace 1, 200 V/div) and the voltage across the work coil (trace 2, 200 V/div) showing that only fundamental component of input square-wave voltage has been passed to the work coil.

6. Conclusion

The paper reports the various issues in the development of an IH power supply for MOVPE application using L-LC RI. The fundamental frequency ac analysis of L-LC RI is revisited, and it is shown that the converter exhibits enhanced current gain and near-unity power factor operation when operated at the resonant frequency. Further analysis of the circuit with square-wave voltage source highlights the effect of the auxiliary inductor on the source current waveform leading to the optimum choice for the value of auxiliary inductor. The requirements of MOVPE system demanding a 25 kW, 25 kHz IH power supply to heat a graphite susceptor to 1200°C, the configuration of developed IH system, and experimental results are presented, thereby demonstrating the suitability of L-LC RI operating at its resonant frequency for this application.

References

[1] C. A. Tudbury, *Basics of Induction Heating. vol. 1*, Library of Congress Catalog Number 60-8958, 1960.

[2] I. Millán, J. M. Burdío, J. Acero, O. Lucía, and S. Llorente, "Series resonant inverter with selective harmonic operation applied to all-metal domestic induction heating," *IET Power Electronics*, vol. 4, no. 5, pp. 587–592, 2011.

[3] Ó. Lucía, J. M. Burdio, I. Millan, J. Acero, and D. Puyal, "Load-adaptive control algorithm of half-bridge series resonant inverter for domestic induction heating," *IEEE Transactions on Industrial Electronics*, vol. 56, no. 8, pp. 3106–3116, 2009.

[4] P. K. Jain, J. R. Espinoza, and S. B. Dewan, "Self-started voltage-source series-resonant converter for high-power induction heating and melting applications," *IEEE Transactions on Industry Applications*, vol. 34, no. 3, pp. 518–525, 1998.

[5] H. P. Ngoc, H. Fujita, K. Ozaki, and N. Uchida, "Phase angle control of high-frequency resonant currents in a multiple inverter system for zone-control induction heating," *IEEE Transactions on Power Electronics*, vol. 26, no. 11, pp. 3357–3366, 2011.

[6] F. Forest, E. Labouré, F. Costa, and J. Y. Gaspard, "Principle of a multi-load/single converter system for low power induction heating," *IEEE Transactions on Power Electronics*, vol. 15, no. 2, pp. 223–230, 2000.

[7] R. Bonert and J. D. Lavers, "Simple starting scheme for a parallel resonance inverter for induction heating," *IEEE Transactions on Power Electronics*, vol. 9, no. 3, pp. 281–287, 1994.

[8] E. J. Dede, V. Esteve, J. Garcia, A. E. Navarro, E. Maset, and E. Sanchis, "Analysis of losses and thermal design of high-power, high-frequency resonant current fed inverters for induction heating," in *Proceedings of the 19th International Conference on Industrial Electronics, Control and Instrumentation*, pp. 1046–1051, November 1993.

[9] E. J. Dede, J. Jordan, J. V. Gonzalez, J. Linares, V. Esteve, and E. Maset, "Conception and design of a parallel resonant converter for induction heating," in *Proceedings of the 6th Annual Applied Power Electronics Conference and Exposition (APEC '91)*, pp. 38–44, March 1991.

[10] E. J. Dede, J. Jordan, J. A. Linares et al., "On the design of medium and high power current fed inverters for induction heating," in *Proceedings of the IEEE Industry Applications Society Annual Meeting*, pp. 1047–1053, October 1991.

[11] V. Chudnovsky, B. Axelrod, and A. L. Shenkman, "An approximate analysis of a starting process of a current source parallel inverter with a high-Q induction heating load," *IEEE Transactions on Power Electronics*, vol. 12, no. 2, pp. 294–301, 1997.

[12] L. Hobson and D. W. Tebb, "Transistorized power supplies for induction heating," *International Journal of Electronics*, vol. 59, no. 5, pp. 543–552, 1985.

[13] F. P. Dawson and P. Jain, "A comparison of load commutated inverter systems for induction heating and melting applications," *IEEE Transactions on Power Electronics*, vol. 6, no. 3, pp. 430–441, 1991.

[14] G. L. Fischer and H.-C. Doht, "Inverter system for inductive tube welding utilizing resonance transformation," in *Proceedings of the 29th Institute for Advanced Study (IAS) Annual Meeting*, pp. 833–840, October 1994.

[15] J. Espi, E. Dede, A. Ferreres, and R. Garcia, "Steady-state frequency analysis of the L-LC resonant inverter for induction heating," in *Proceedings of the IEEE International Power Electronics Congress Technical Proceedings (CIEP '96)*, pp. 22–28, 1999.

[16] J. M. Espi, E. J. Dede, E. Navarro, E. Sanchis, and A. Ferreres, "Features and design of the voltage-fed L-LC resonant inverter for induction heating," in *Proceedings of the 30th Annual IEEE Power Electronics Specialists Conference (PESC '99)*, pp. 1126–1131, July 1999.

[17] J. M. Espí and E. J. Dede, "Design considerations for three element L-LC resonant inverters for induction heating," *International Journal of Electronics*, vol. 86, no. 10, pp. 1205–1216, 1999.

[18] S. Chudjuarjeen, A. Sangswang, and C. Koompai, "An improved LLC resonant inverter for induction-heating applications with asymmetrical control," *IEEE Transactions on Industrial Electronics*, vol. 58, no. 7, pp. 2915–2925, 2011.

[19] E. J. Dede, J. Gonzalez, J. A. Linares, J. Jordan, D. Ramirez, and P. Rueda, "25-kW/50-kHz generator for induction heating," *IEEE Transactions on Industrial Electronics*, vol. 38, no. 3, pp. 203–209, 1991.

[20] S. Dieckerhoff, M. J. Ryan, and R. W. De Doncker, "Design of an IGBT-based LCL-resonant inverter for high-frequency induction heating," in *Proceedings of the IEEE Industry Applications Conference and 34th IAS Annual Meeting*, pp. 2039–2045, October 1999.

[21] J. M. E. Huerta, E. J. D. G. Santamaría, R. G. Gil, and J. Castelló-Moreno, "Design of the L-LC resonant inverter for induction heating based on its equivalent SRI," *IEEE Transactions on Industrial Electronics*, vol. 54, no. 6, pp. 3178–3187, 2007.

[22] K. Nishida, S. Haneda, K. Hara, H. Munekata, and H. Kukimoto, "MOVPE of GaN using a specially designed two-flow horizontal reactor," *Journal of Crystal Growth*, vol. 170, no. 1-4, pp. 312–315, 1997.

[23] Y.-S. Kwon, S.-B. Yoo, and D.-S. Hyun, "Half-bridge series resonant inverter for induction heating applications with load-adaptive PFM control strategy," in *Proceedings of the 14th annual Applied Power Electronics Conference and Exposition (APEC '99)*, pp. 575–581, March 1999.

[24] H. W. Koertzen, J. D. Van Wyk, and J. A. Ferreira, "Comparison of swept frequency and phase shift control for forced commutated series resonant induction heating converters," in *Proceedings of the Conference Record of the IEEE Industry Applications and 30th IAS Annual Meeting*, pp. 1964–1969, October 1995.

[25] J. M. Burdío, L. A. Barragán, F. Monterde, D. Navarro, and J. Acero, "Asymmetrical voltage-cancellation control for full-bridge series resonant inverters," *IEEE Transactions on Power Electronics*, vol. 19, no. 2, pp. 461–469, 2004.

[26] L. Grajales and F. C. Lee, "Control system design and small-signal analysis of a phase-shift-controlled series-resonant inverter for induction heating," in *Proceedings of the IEEE Power Electronics Specialist Conference*, pp. 450–456, 1995.

[27] L. Grajales, J. A. Sabate, K. R. Wang, W. A. Tabisz, and F. C. Lee, "Design of a 10 kW, 500 kHz phase-shift controlled series-resonant inverter for induction heating," in *Proceedings of the 28th Annual Meeting of the IEEE Industry Applications Conference*, pp. 843–849, October 1993.

[28] N. A. Ahmed, "High-frequency soft-switching AC conversion circuit with dual-mode PWM/PDM control strategy for high-power IH applications," *IEEE Transactions on Industrial Electronics*, vol. 58, no. 4, pp. 1440–1448, 2011.

[29] V. Esteve, E. Sanchis-Kilders, J. Jordán et al., "Improving the efficiency of IGBT series-resonant inverters using pulse density modulation," *IEEE Transactions on Industrial Electronics*, vol. 58, no. 3, pp. 979–987, 2011.

[30] C.-J. Tseng and C.-L. Chen, "Passive lossless snubbers for dc/dc converters," in *Proceedings of the 13th Annual Applied Power Electronics Conference and Exposition (APEC '98)*, pp. 1049–1054, February 1998.

[31] M. Borage and S. Tiwari, "On the development of 30 kVA, 430 Hz sine wave inverter for DC accelerator application," in *Proceedings of the Indian Particle Accelerator Conference (InPAC '11)*, IUAC, New Delhi, India, February 2011.

[32] A. Kats, G. Ivensky, and S. Ben-Yaakov, "Application of integrated magnetics in resonant converters," in *Proceedings of the IEEE 12th Applied Power Electronics Conference*, pp. 925–930, February 1997.

19

Harmonic Mitigation Techniques Applied to Power Distribution Networks

Hussein A. Kazem

Faculty of Engineering, Sohar University, P.O. Box 44, 311 Sohar, Oman

Correspondence should be addressed to Hussein A. Kazem; h.kazem@soharuni.edu.om

Academic Editor: Hadi Y. Kanaan

A growing number of harmonic mitigation techniques are now available including active and passive methods, and the selection of the best-suited technique for a particular case can be a complicated decision-making process. The performance of some of these techniques is largely dependent on system conditions, while others require extensive system analysis to prevent resonance problems and capacitor failure. A classification of the various available harmonic mitigation techniques is presented in this paper aimed at presenting a review of harmonic mitigation methods to researchers, designers, and engineers dealing with power distribution systems.

1. Introduction

The nonlinear characteristics of many industrial and commercial loads such as power converters, fluorescent lamps, computers, light dimmers, and variable speed motor drives (VSDs) used in conjunction with industrial pumps, fans, and compressors and also in air-conditioning equipment have made the harmonic distortion a common occurrence in electrical power networks. Harmonic currents injected by some of these loads are usually too small to cause a significant distortion in distribution networks. However, when operating in large numbers, the cumulative effect has the capability of causing serious harmonic distortion levels. These do not usually upset the end-user electronic equipment as much as they overload neutral conductors and transformers and, in general, cause additional losses and reduced power factor [1–5]. Large industrial converters and variable speed drives on the other hand are capable of generating significant levels of distortion at the point of common coupling (PCC), where other users are connected to the network [6, 7].

Because of the strict requirement of power quality at the input AC mains, various harmonic standards and engineering recommendations such as IEC 1000-3-2, IEEE 519 (USA), AS 2279, D.A.CH.CZ, EN 61000-3-2/EN 61000-3-12, and ER G5/4 (UK) are employed to limit the level of distortion at the PCC. To comply with these harmonic standards, installations utilizing power electronic and nonlinear loads often use one of the growing numbers of harmonic mitigation techniques [8]. Because of the number and variety of available methods, the selection of the best-suited technique for a particular application is not always an easy or straightforward process. Many options are available, including active and passive methods. Some of the most technically advanced solutions offer guaranteed results and have little or no adverse effect on the isolated power system, while the performance of other simple methods may be largely dependent on system conditions. This paper presents a comprehensive survey on harmonic mitigation techniques in which a large number of technical publications have been reviewed and used to classify harmonic mitigation techniques into three categories: passive techniques, active techniques, and hybrid harmonic reduction techniques using a combination of active and passive methods. A brief description of the electrical characteristics of each method is presented with the aim of providing the designer and site engineer with a more informed choice regarding their available options when dealing with the effects and consequences of the presence of these harmonics in the distribution network.

2. Passive Harmonic Mitigation Techniques

Many passive techniques are available to reduce the level of harmonic pollution in an electrical network, including the

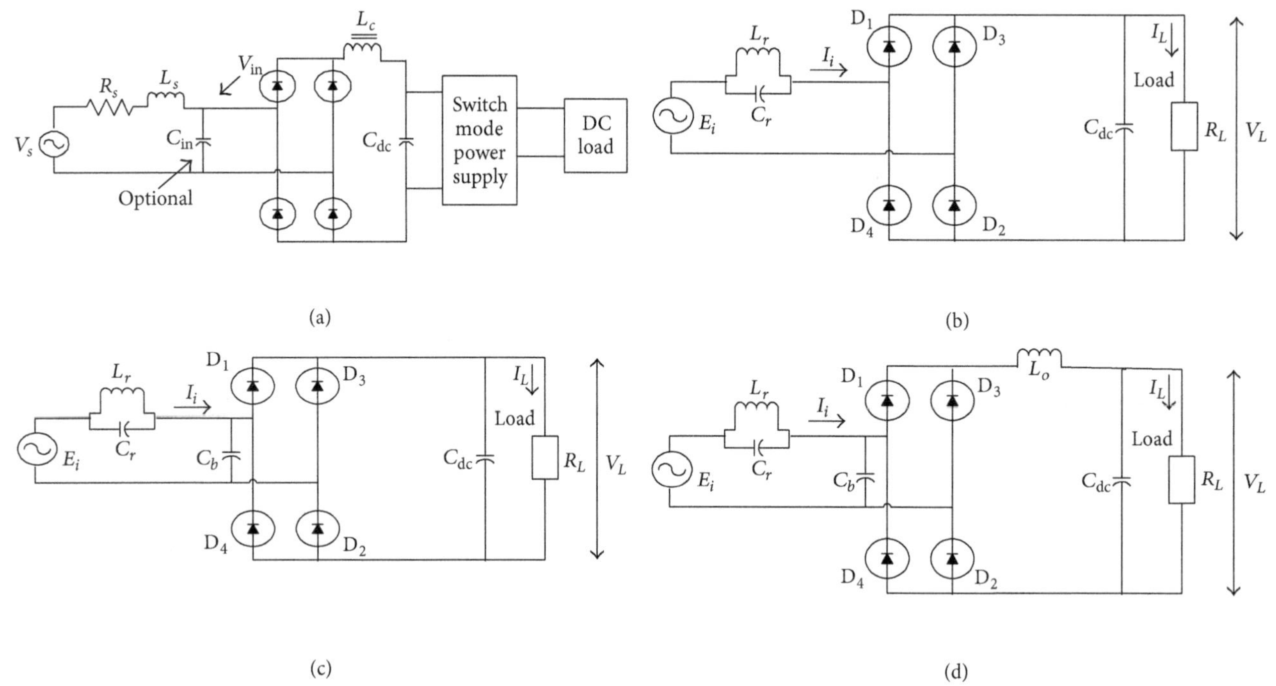

FIGURE 1: (a) The Series inductor filters for current shaping, (b) The Ziogas inductor capacitor filter, (c) The Yanchao improvement on Ziogas filter, and (d) The Hussein improvement on Yanchao filter.

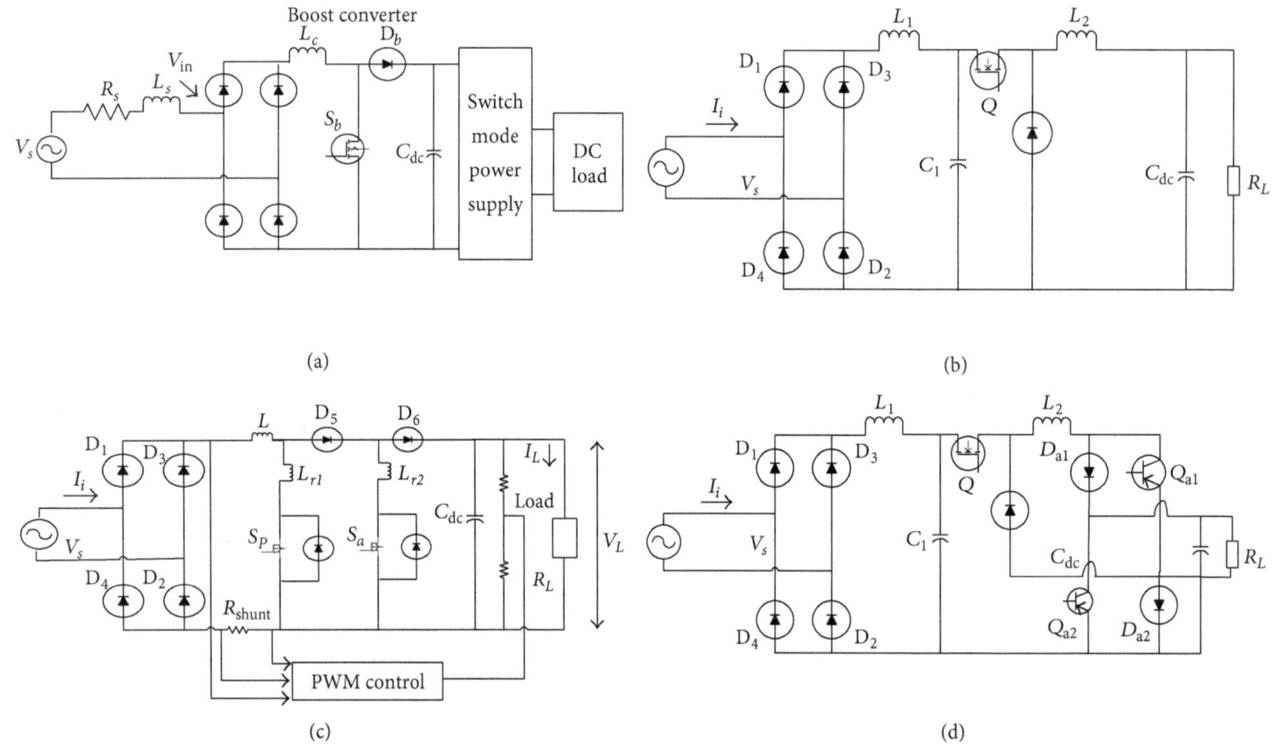

FIGURE 2: (a) Boost converter current shaping circuit, (b) buck converter current shaping circuit, (c) improve boost converter current shaping circuit, and (d) improve buck converter current shaping circuit.

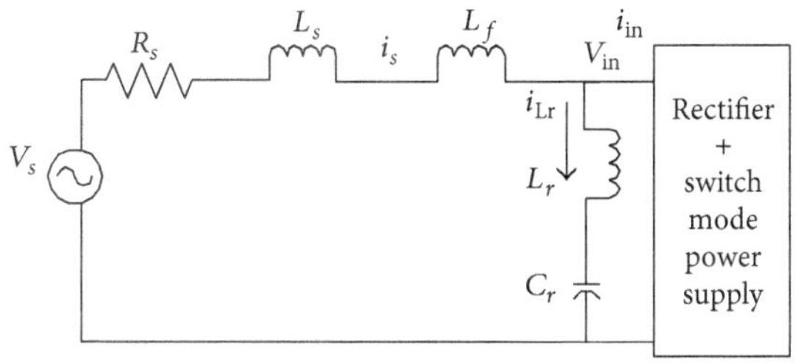

FIGURE 3: A parallel-connected resonant filter.

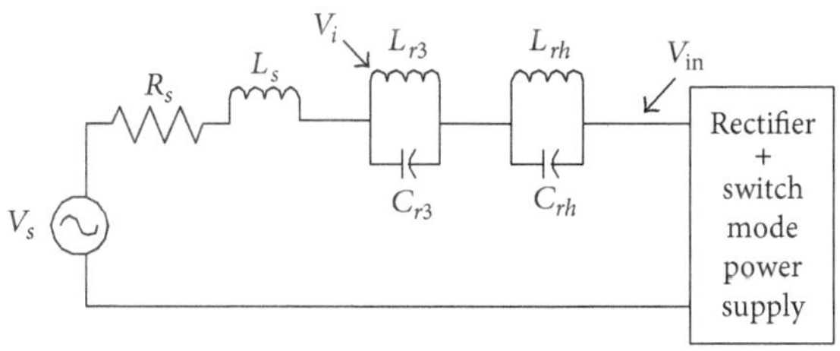

FIGURE 4: Double-tuned series-connected resonant filter.

connection of series line reactors, tuned harmonic filters, and the use of higher pulse number converter circuits such as 12-pulse, 18-pulse, and 24-pulse rectifiers. In these methods, the undesirable harmonic currents may be prevented from flowing into the system by either installing a high series impedance to block their flow or diverting the flow of harmonic currents by means of a low-impedance parallel path [9].

Harmonic mitigation techniques used for supply power factor correction and harmonics mitigation in two ways to qualify the products performance. One is to put a limit on the PF for loads above a specified minimum power. Utility companies often place limits on acceptable power factors for loads (e.g., <0.8 leading and >0.75 lagging). A second way to measure or specify a product is to define absolute maximum limits for current harmonic distortion. This is usually expressed as limits for odd harmonics (e.g., 1st, 3rd, 5th, 7th, etc.). This approach does not need any qualifying minimum percentage load and is more relevant to the electric utility.

Harmonic regulations or guidelines are currently applied to keep current and voltage harmonic levels in check. As an example, the current distortion limits in Japan illustrated in Tables 1 and 2 represent the maximum and minimum values of total harmonic distortion (THD) in voltage and the most dominant fifth harmonic voltage in a typical power system [10].

Certain techniques, such as the use of tuned filters, require extensive system analysis to prevent resonance problems and capacitor failures, while others, such as the use of 12-pulse or 24-pulse converters, can be applied with virtually no system analysis.

2.1. Effect of Source Reactance. Typical AC current waveforms in single-phase and three-phase rectifiers are far from a sinusoid. The power factor is also very poor because of the high harmonic contents of the line current waveform. In rectifier with a small source reactance, the input current is highly discontinuous, and, as a consequence, the power is drawn from the utility source at a very poor power factor.

TABLE 1: Voltage THD and fifth harmonic voltage in a high-voltage power transmission system.

	Over 154 kV		154–22 kV	
	THD	5th harmonic	THD	5th harmonic
Max.	2.8%	2.8%	3.3%	3.2%
Min.	1.1%	1.0%	1.4%	1.3%

TABLE 2: Voltage THD and fifth harmonic voltage in a 6.6 kV power distribution system.

	6.6 kV			
	Residential		Commercial	
	THD	5th harmonic	THD	5th harmonic
Max.	3.5%	3.4%	4.6%	4.3%
Min.	3.0%	2.9%	2.1%	1.2%

The magnitude of harmonic currents in some nonlinear loads depends greatly on the total effective input reactance, comprised of the source reactance plus any added line reactance. For example, given a 6-pulse diode rectifier feeding a DC bus capacitor and operating with discontinuous DC current, the level of the resultant input current harmonic spectrum is largely dependent on the value of AC source reactance and an added series line reactance; the lower the reactance, the higher the harmonic content [1–3].

Other nonlinear loads, such as a 6-pulse diode rectifier feeding a highly inductive DC load and operating with continuous DC current, act as harmonic current sources. In such cases, the amount of voltage distortion at the PCC is dependant on the total supply impedance, including the effects of any power factor correction capacitors, with higher impedances producing higher distortion levels [7, 11].

2.2. Series Line Reactors. The use of series AC line reactors is a common and economical means of increasing the source impedance relative to an individual load, for example, the input rectifier used as part of a motor drive system. The harmonic mitigation performance of series reactors is a function of the load; however, their effective impedance reduces proportionality as the current through them is decreased [12].

2.3. Tuned Harmonic Filters. Passive harmonic filters (PHF) involve the series or parallel connection of a tuned LC and high-pass filter circuit to form a low-impedance path for a specific harmonic frequency. The filter is connected in parallel or series with the nonlinear load to divert the tuned frequency harmonic current away from the power supply. Unlike series line reactors, harmonic filters do not attenuate all harmonic frequencies but eliminate a single harmonic frequency from the supply current waveform. Eliminating harmonics at their source has been shown to be the most effective method to reduce harmonic losses in the isolated power system. However, the increased first cost entailed presents a barrier to this approach. If the parallel-connected filter is connected further upstream in the power network, higher day-to-day costs will accumulate due to I^2R losses in the conductors and other plant items that carry the harmonic currents. Conversely, for series-connected filter at the load, there are increased losses in the filter itself. These losses are simply the result of the higher series impedance, which blocks the flow of harmonics but increases the line loss as a result of the flow of the remaining components of the load current [12, 13]. The quality factor of the filter inductor Q affects the actual value of the low-impedance path for each filter. Usually, a value of Q ranges between 20 and 100 [14]. Many types of harmonic filters are commonly employed, including the following:

2.3.1. Series Induction Filters. Harmonic currents produced by switched-mode power supplies and other DC-to-DC converter circuits can be significantly lowered by the connection of a series inductor that can be added on either the AC or DC power circuit [15–17], as shown in Figure 1. So many improvements on these filters have been made.

Ziogas passive filter for single-phase rectifiers has some reduction in Total Harmonics Distortion THD and improvement in PF in comparison with conventional rectifier. Also, Yanchoa waveshaping filter used to reduce THD and increase power factor. Connecting author filter at the output terminal of the rectifier will improve power factor and reduce input current THD of the supply.

2.3.2. DC-DC Converter Current Shaping. Like the series induction filter, this circuit (Figure 2) can greatly reduce current distortion produced by switched-mode power supplies and other DC converter circuits by modulating the duty cycle of switch S_b to control the shape of input supply current to track a desired sine waveshape [5, 18–20]. So many improvements on these filters have been made.

2.3.3. Parallel-Connected Resonant Filter. Passive LC filters tuned to eliminate a particular harmonic are often used to reduce the level of low-frequency harmonic components like the 5th and 7th produced by three-phase rectifier and inverter circuits. The filter is usually connected across the line as shown in Figure 3. If more than one harmonic is to be eliminated, then a shunt filter must be installed for each harmonic. Care must be taken to ensure that the peak impedances of such an arrangement are tuned to frequencies between the required harmonic frequencies to avoid causing high levels of voltage distortion at the supply's PCC because of the presence of an LC resonance circuit [7, 12].

2.3.4. Series-Connected Resonant Filter. This work on a similar in principle to the parallel version, but with the tuned *LC* circuits connected in series with the supply. The series filter can be tuned to a single harmonic frequency, or it may be multituned to a number of harmonic frequencies. The multituned arrangement connects multiple tuned filters in series as shown in Figure 4 showing a third harmonic tuned *LC* circuit, L_{r3} and C_{r3}, and a high-frequency tuned *LC* circuit, L_{rh} and C_{rh}, to eliminate high-order harmonics [5, 7, 12].

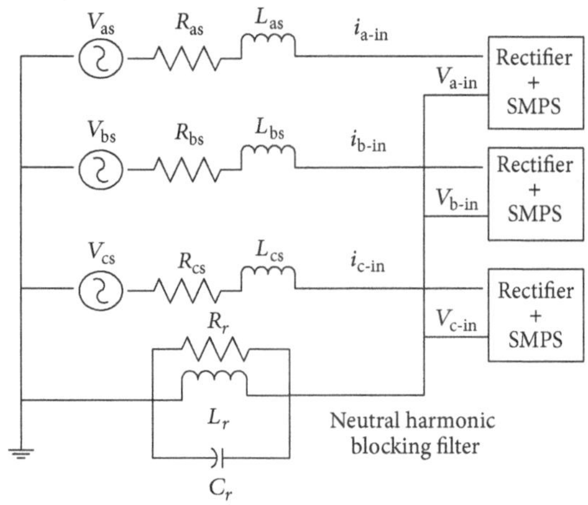

FIGURE 5: A neutral current blocking filters.

2.3.5. Neutral Current Filter. This filter is connected in the neutral conductor between the site transformer and the three-phase load to block all triple frequency harmonics, as shown in Figure 5. Because these triple zero-sequence harmonics are in phase with each other, they all flow through the neutral conductor, and it is more economical to block them in the neutral instead of individual phases [5, 12].

2.3.6. Zigzag Grounding Filter. By integrating phase shifting into a single or multiphase transformer with an extremely low zero-sequence impedance, substantial reduction of triple, 5th, and 7th harmonics can be achieved. This method provides an alternative to protect the transformer neutral conductor from triple harmonics by canceling these harmonics near the load. In this method, an autotransformer connected in parallel with the supply can provide a zero-sequence current path to trap and cancel triple harmonics as shown in Figure 6 [16].

2.4. Higher Pulse Converters. Three phases, 6-pulse static power converters, such as those found in VSD, generate low-frequency current harmonics. Predominantly, these are the 5th, 7th, 11th, and 13th with other higher orders harmonics also present but at lower levels. With a 6-pulse converter circuit, harmonics of the order $6k \pm 1$, where $k = 1, 2, 3, 4$, and so forth, will be present in the supply current waveform. In high-power applications, AC-DC converters based on the concept of multipulse, namely, 12, 18, or 24 pulses, are used to reduce the harmonics in AC supply currents. They are referred to as a multipulse converters. They use either a diode bridge or thyristor bridge and a special arrangement of phase-shifting magnetic circuit such as transformers and inductors to produce the required supply current waveforms [9, 21–27].

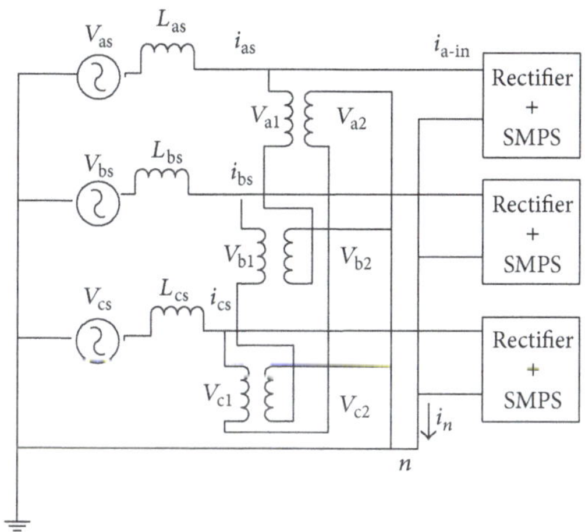

FIGURE 6: Zigzag autotransformer connected to three-phase nonlinear loads.

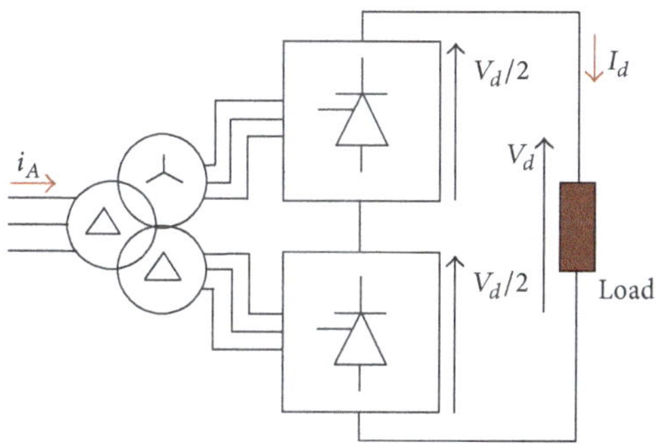

FIGURE 7: Series 12-pulse rectifier connection.

2.4.1. 12-Pulse Rectification. In large converter installations, where harmonics generated by a three-phase converter can reach unacceptable levels, it is possible to connect two 6-pulse converters in series with star/delta phase-shifting transformers to generate a 12-pulse waveform and reduce the harmonics on the supply and load sides, as shown in Figure 7. This could be beneficial despite the considerable extra cost of the transformers. Twelve-pulse rectifier is frequently specified by consulting engineers for heating, ventilating, and air-conditioning applications because of their theoretical ability to reduce harmonic current distortion.

Instead of connecting the two converter bridges in series, they could also be connected in parallel to give 12-pulse operation. A parallel 12-pulse arrangement is shown in Figure 8. Parallel connections require special care to ensure adequate balance between the currents drawn by each bridge. Secondary leakage reactance must be carefully matched, and extra reactors are needed on the DC side to absorb the instantaneous differences between the two DC voltage waveforms, [9, 22, 28].

When using a 12-pulse system, the 5th and 7th harmonics disappear from line current waveforms leaving the 11th as the first to appear. Only harmonics of the order $12k \pm 1$, where $k = 1, 2, 3, 4$, and so forth, will be present in the supply current waveform, thus resulting in a high power factor, low THD at input AC mains, and ripple-free DC output of high quality.

2.4.2. 18-Pulse Rectification. Eighteen-pulse converter circuits, shown in Figure 9, use a transformer with three sets of secondary windings that are phase-shifted by 20 degrees with respect to each other. Only harmonics of the order $18k \pm 1$, where $k = 1, 2, 3, 4$, and so forth, will be present in the supply current waveform [9, 29].

2.4.3. 24-Pulse Rectification. Connecting two 12-pulse circuits with a 15° phase shift produces a 24-pulse system. Figure 10 shows one such system in which the two 12-pulse circuits are connected in parallel to produce the required 24-pulse system. The 11th and 13th harmonics now disappear from the supply current waveform leaving the 23rd as the first to appear. Only harmonics of the order $24k \pm 1$, where $k = 1, 2, 3, 4$, and so forth, will be present in a 24-pulse system [9, 30].

3. Active Harmonic Mitigation Techniques

When using active harmonic reduction techniques, the improving in the power quality came from injecting equal-but-opposite current or voltage distortion into the network, thereby canceling the original distortion. Active harmonic filters (AHFs) utilize fast-switching insulated gate bipolar transistors (IGBTs) to produce an output current of the required shape such that when injected into the AC lines, it cancels the original load-generated harmonics. The heart of the AHF is the controller part. The control strategies applied to the AHF play a very important role on the improvement of the performance and stability of the filter. AHF is designed with two types of control scheme. The first performs fast Fourier transforms to calculate the amplitude and phase angle of each harmonic order. The power devices are directed to produce a current of equal amplitude but opposite phase angle for specific harmonic orders. The second method of control is often referred to as full spectrum cancellation in which the full current waveform is used by the controller of the filter, which removes the fundamental frequency component and directs the filter to inject the inverse of the remaining waveform [31–38].

Typically, these filters are sized based on how much harmonic current the filter can produce, normally in amperage increments of 50 Amps. The proper amperage of AHF can be chosen after determining the amount of harmonic cancellation current.

Essentially, the filter consists of a VSD with a special electronic controller which injects the harmonic current onto the system 180 out of phase to the system or drive harmonics. This results in harmonics cancellation. For example, if the VSD created 50 A of 5th harmonic current, and the AHF produced 40 A of 5th harmonic current, the amount of 5th harmonic current exported to the utility grid would be 10 A. The AHF may be classified as a single-phase or three-phase filters.

Also, it could be classified as parallel or series AHF according to the circuit configuration.

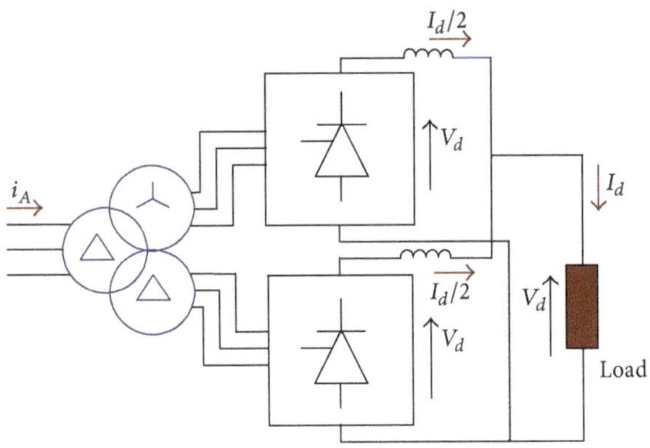

FIGURE 8: Parallel twelve-pulse rectifier connection.

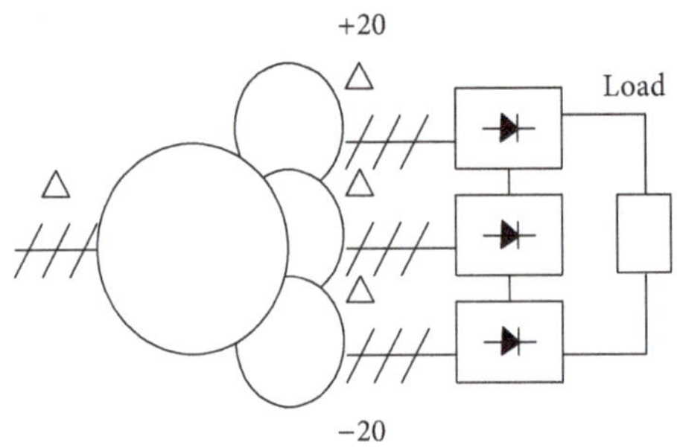

FIGURE 9: 18-pulse rectifier connection.

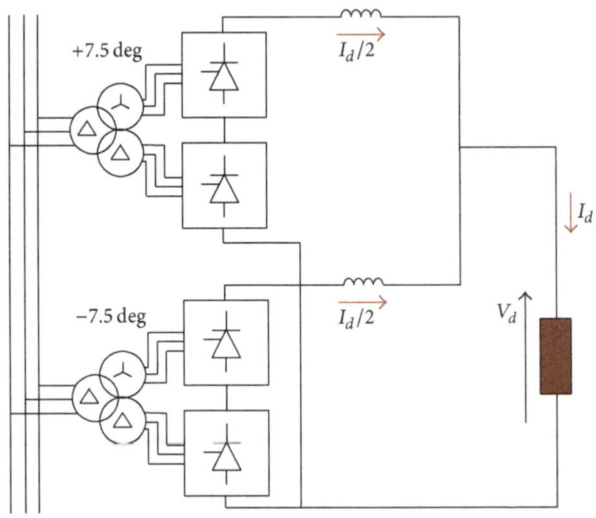

FIGURE 10: 24-pulse rectifier connection.

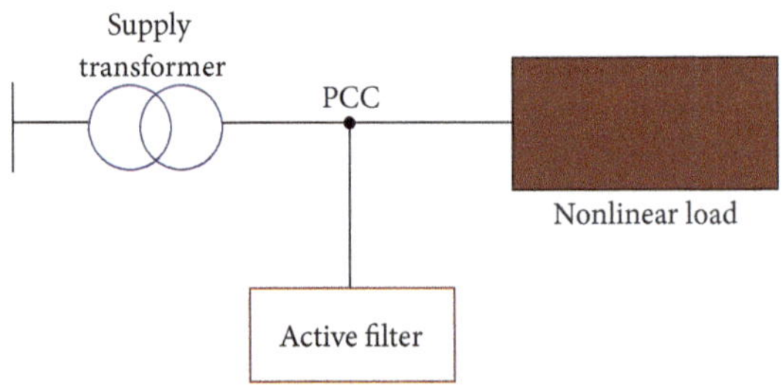

FIGURE 11: Parallel active filter.

3.1. Parallel Active Filters. This is the most widely used type of AHF (more preferable than series AHF in terms of form and function). As the name implies, it is connected in parallel to the main power circuit as shown in Figure 11. The filter is operated to cancel out the load harmonic currents leaving the supply current free from any harmonic distortion. Parallel filters have the advantage of carrying the load harmonic current components only and not the full load current of the circuit [39–44].

AHF can be controlled on the basis of the following methods:

(i) the controller detects the instantaneous load current i_L,

(ii) the AHF extracts the harmonic current i_{Lh} from the detected load current i_L by means of digital signal processing,

(iii) the AHF draws the compensating current i_{AF} (= $-i_{Lh}$) from the utility supply voltage v_S so as to cancel out the harmonic current i_{Lh} [45].

3.2. Series Active Filters. The main circuit configuration for this type of AHF is shown in Figure 12. The idea here is to eliminate voltage harmonic distortions and improve the quality of the voltage applied to the load. This is achieved by producing a sinusoidal pulse width modulated (PWM) voltage waveform across the connection transformer, which is added to the supply voltage to counter the distortion across the supply impedance and present a sinusoidal voltage across the load. Series AHF has to carry the full load current increasing their current ratings and I^2R losses compared with parallel filters, especially across the secondary side of the coupling transformer [43].

Unlike the shunt AHF, the series AHF is controlled on the basis of the following methods:

(i) the controller detects the instantaneous supply current i_S,

(ii) the AHF extracts the harmonic current i_S from the detected supply current by means of digital signal processing,

(iii) the active filter applies the compensating voltage v_{AF} (= $-Ki_{Sh}$) across the primary of the transformer. This will result in a significant reduction in the supply harmonic current (i_{Sh}), when the feedback gain K is set to be high enough [45].

An AHF with both series and parallel (shunt) connected sections, as shown in Figures 11 and 12, respectively, can be used to compensate for both voltage and current harmonics simultaneously [34–36]. In all cases, the critical requirement of any AHF circuit is to calculate the required compensation current accurately and in real time.

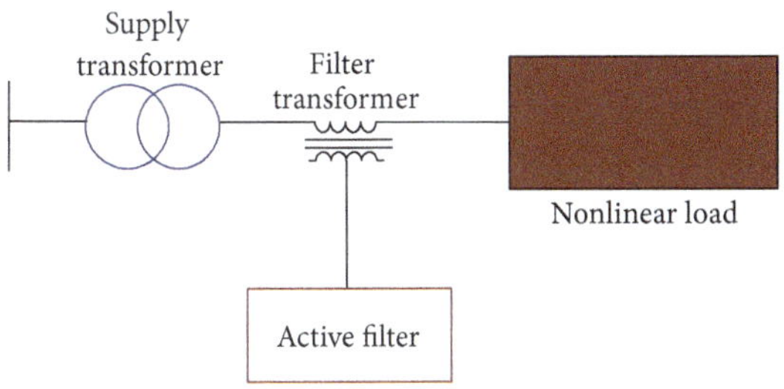

FIGURE 12: Series active filter.

4. Hybrid Harmonic Mitigation Techniques

Hybrid connections of AHF and PHF are also employed to reduce harmonics distortion levels in the network. The PHF with fixed compensation characteristics is ineffective to filter the current harmonics. AHF overcomes the drawbacks of the PHF by using the switching-mode power converter to perform the harmonic current elimination. However, the AHF construction cost in an industry is too high. The AHF power rating of power converter is very large. These bound the applications of AHF used in the power system. Hybrid harmonic filter (HHF) topologies have been developed [46–51] to solve the problems of reactive power and harmonic currents effectively. Using low cost PHF in the HHF, the power rating of active converter is reduced compared with that of AHF. HHF retains the advantages of AHF and does not have the drawbacks of PHF and AHF. Figure 13 shows a number of possible hybrid combinations. Figure 13(a) is a combination of shunt AHF and shunts PHF. Using a combination of PHF will make a significant reduction in the rating of the AHF. As a result, no harmonic resonance occurs, and no harmonic current flows in the supply. In [50], author claimed that in HHF the AHF can improve the filter performance and suppress the harmonic resonance of existing PHF. Figure 13(b) shows a combination of AHF series with the supply and a shunt PHF. The author of reference [46] found that this topology is not suitable for low-frequency interharmonic compensation because the AHF introduces a high compensation voltage which can interfere with downstream phase-controlled nonlinear loads.

Figure 13(c) shows an AHF in series with a shunt PHF. In all cases, it is required that the filters in a hybrid combination share compensation properly in the frequency domain [51]. A lot of improvements and researches have been made on the control strategies of hybrid harmonic filters.

The AHF and PHF are used to generate the equivalent voltage which is related to the mains harmonic current using different methods (i.e., impedance variation method) as shown in Figure 13(c). The mains harmonic current is suppressed by increasing the ratio of effective source impedance to the harmonic components. To achieve a constant DC bus voltage of the AHF, a PI voltage controller is employed. A hysteresis voltage comparator is employed to track the output voltage to perform the equivalent impedance of active converter [48, 49]. HHF is cost effective and becomes more practical in industry applications.

AHF controller mainly is divided into two parts, that is, reference current generation and PWM current controller. The PWM current controller is principally used for providing gating pulse to the AHF. In reference to current generation scheme, reference current is generated by using the distorted waveform. Many control schemes are there for reference current generation, such as p-q theory, deadbeat controller, neuro, adaptive control, wavelet control, fuzzy, delta-sigma modulation, sliding mode control, vector control, repetitive control, and SFX control for improving the steady state and dynamic performance of AHFs [52–59].

4.1. p-q Method. Instantaneous reactive power theory has been published in 1984. Based on this theory, the so-called "p-q method" was applied successfully in the control of AHF. Zero-sequence component is neglected in this method, and because of that the p-q method is not accurate when the three-phase system is distorted or unbalanced.

4.2. d-q Method. Based on the park transformation, the d-q method came. The three-phase load current can be decomposed in positive-sequence, negative sequence and zero-sequence component. The current in the d-q frame i_d and i_q can be transformed from the positive sequence and negative sequence using a PLL (phase locked loop). The division of the AC and DC components can be obtained across a low-pass PHF. The reference current signal can be achieved by the AC component in d-q frame through a countertransformation.

4.3. Direct Testing and Calculating Method (DTC). Separation of the harmonic and reactive components from the load current is the aim of current reference generator. The main characteristic of this method is the direct derivation of the compensating component from the load current, without the use of any reference frame transformation. In fact, this method presents a low-frequency oscillation problem in the AHF DC bus voltage.

4.4. Synchronous Reference Fame Method (SRF). Real currents are transformed into a synchronous reference frame in this method. The reference frame is synchronized with the AC mains voltage and is rotating at the same frequency. In this method, the reference currents are derived directly from the real load currents without considering the source voltages, which represent the most important characteristics of this method. The generation of the reference signals is not affected by distortion or voltage unbalance, therefore increasing the compensation robustness and performance.

4.5. Current Hysteresis Control. The basic principle of this control method is that the switching signals are derived from the comparison of the current error signal with a fixed width hysteresis band. This current control technique exhibits some unsatisfactory features due to simple, extreme robustness, fast dynamic, good stability, and automatic current limited characteristics.

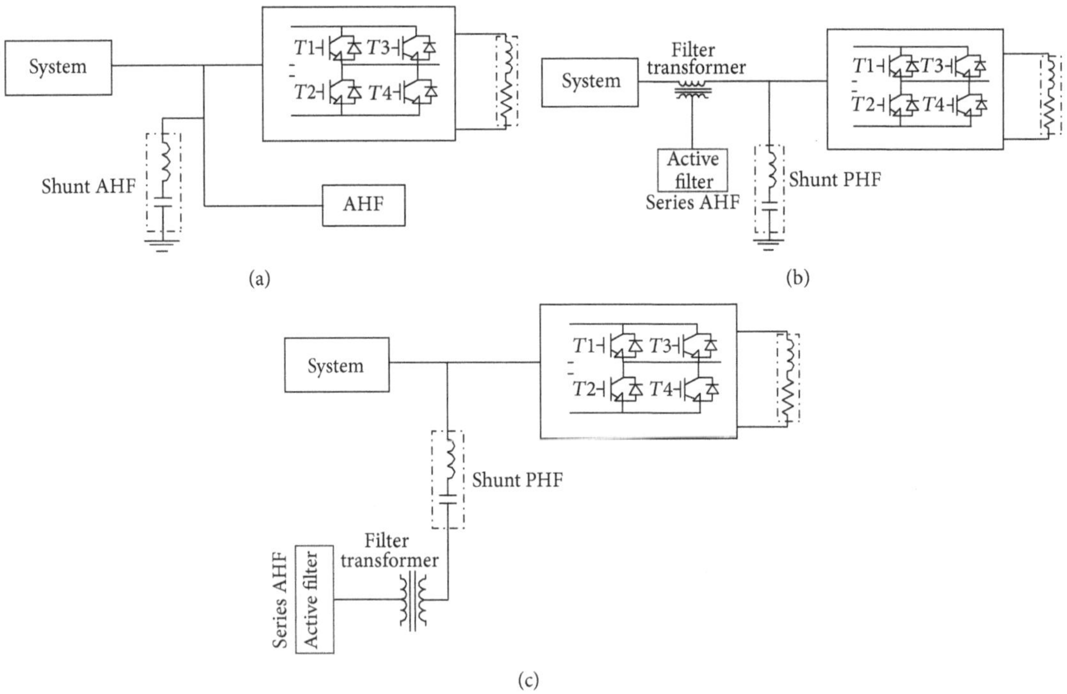

FIGURE 13: Hybrid connections of active and passive filters.

4.6. Triangle-Comparison PWM Control. This control method is also called linear current control. The conventional triangle-comparison PWM control principle is that the modulation signal achieved by a current regulator from the current error signal is intersected with the triangle wave. After that, pulse signals obtained are to control the switches of the converters. With analog PWM circuit, this control method has simple implementation with fast speed of response. Because the modulation frequency equals the triangle frequency, the current loop gain crossover frequency must be kept below the modulation frequency.

4.7. Space Vector Modulation (SVM). The aim of this method is to find the appropriate switching combinations and their duty ratios according to certain modulation scheme. The SVM operates in a complex plane divided in the six sectors separated by a combination of conducting or nonconducting switches in the power circuit. The reference vector is used to locate two adjacent switching-state vectors and compute the time for which each one is active. SVM is of low speed of response caused by the inherent calculation delay, due to the strong antijamming and the good reliability of digital control technique. In order to solve the drawback, the improvement of adopting deadbeat control and a certain oversize of the system reactive components is advised.

Currently, the research trends of the AHF control strategies are mainly towards the optimizing and practical application of the control strategies. At the end, the comparative criteria for PHF, AHF, and HHF could be summarized based on the following:

(i) cost of the equipment and installation,

(ii) harmonic indices (ex. i_h, THD_i, THD_v, TDD, and PWHD),

(iii) life time and failure rate,

(iv) maintenance and engineering.

5. Conclusions

Electrical system reliability and normal operation of electrical equipment rely heavily upon a clean distortion free power supply. Designers and engineers wishing to reduce the level of harmonic pollution on a power distribution network where nonlinear harmonic generating loads are connected have several harmonic mitigation techniques available. Because of the number and variety of available methods, selection of the best-suited technique for a particular application is not always an easy or straightforward process. A broad categorization of different harmonic mitigation techniques (passive, active, and hybrid) has been carried out to give a general viewpoint on this wide-ranging and rapidly developing topic. PHF is traditionally used to absorb harmonic currents because of low cost and simple robust structure. However, they provide fixed compensation and create system resonance. AHF provides multiple functions such as harmonic reduction, isolation, damping and termination, load balancing, PF correction, and voltage regulation. The HHF is more attractive in harmonic filtering than the pure filters from both viability and economical points of view, particularly for high-power applications. It is hoped that the discussion and classification of harmonic mitigation techniques presented in this paper will provide some useful information to help make the selection of an appropriate harmonic reduction method for a given application on an easier task.

References

[1] A. Mansoor, W. M. Grady, A. H. Chowdhury, and M. J. Samotyi, "An investigation of harmonics attenuation and diversity among distributed single-phase power electronic loads," *IEEE Transactions on Power Delivery*, vol. 10, no. 1, pp. 467–473, 1995.

[2] A. Mansoor, W. M. Grady, R. S. Thallam, M. T. Doyle, S. D. Krein, and M. J. Samotyj, "Effect of supply voltage harmonics on the input current of single-phase diode bridge rectifier loads," *IEEE Transactions on Power Delivery*, vol. 10, no. 3, pp. 1416–1422, 1995.

[3] G. Carpinelli, F. Iacovone, P. Varilone, and P. Verde, "Single phase voltage source converters: analytical modelling for harmonic analysis in continuous and discontinuous current conditions," *International Journal of Power and Energy Systems*, vol. 23, no. 1, pp. 37–48, 2003.

[4] E. F. El-Saadany and M. M. A. Salama, "Reduction of the net harmonic current produced by single-phase non-linear loads due to attenuation and diversity effects," *International Journal of Electrical Power and Energy Systems*, vol. 20, no. 4, pp. 259–268, 1998.

[5] T. Key and J. S. Lai, "Analysis of harmonic mitigation methods for building wiring systems," *IEEE Transactions on Power Systems*, vol. 13, no. 3, pp. 890–897, 1998.

[6] M. H. Rashid and A. I. Maswood, "A novel method of harmonic assessment generated by three-phase AC-DC converters under unbalanced supply conditions," *IEEE Transactions on Industry Applications*, vol. 24, no. 4, pp. 590–597, 1988.

[7] R. E. Owen, M. F. McGranghan, and J. R. Vivirito, "Distribution system harmonics: controls for large power converters," *IEEE Transactions on Power Apparatus and Systems*, vol. 101, no. 3, pp. 644–652, 1982.

[8] T. Hoevenaars, K. LeDoux, and M. Colosino, "Interpreting IEEE STD 519 and meeting its harmonic limits in VFD applications," in *Proceedings of the 50th Annual Technical Conference of the Petroleum and Chemical Industry Committee*, pp. 145–150, Houston, Tex, USA, September 2003.

[9] B. Singh, B. N. Singh, A. Chandra, K. Al-Haddad, A. Pandey, and D. P. Kothari, "A review of three-phase improved power quality AC-DC converters," *IEEE Transactions on Industrial Electronics*, vol. 51, no. 3, pp. 641–660, 2004.

[10] "Investigation into execution of harmonic guidelines for household and office electric applications," Japanese IEE of Japan SC77A Domestic Committee Report, 2002.

[11] R. L. Smith and R. P. Stratford, "Power system harmonics effects from adjustable-speed drives," *IEEE Transactions on Industry Applications*, vol. 20, no. 4, pp. 973–977, 1984.

[12] J. C. Das, *Power System Analysis, Short-Circuit Load Flow and Harmonics*, Marcel Dekker, New York, NY, USA, 2002.

[13] D. Alexa, A. Sîrbu, and D. M. Dobrea, "An analysis of three-phase rectifiers with near-sinusoidal input currents," *IEEE Transactions on Industrial Electronics*, vol. 51, no. 4, pp. 884–891, 2004.

[14] E. B. Makram, E. V. Subramaniam, A. A. Girgis, and R. Catoe, "Harmonic filter design using actual recorded data," *IEEE Transactions on Industry Applications*, vol. 29, no. 6, pp. 1176–1183, 1993.

[15] T. S. Key and J. S. Lai, "Comparison of standards and power supply design options for limiting harmonic distortion in power systems," *IEEE Transactions on Industry Applications*, vol. 29, no. 4, pp. 688–695, 1993.

[16] A. R. Prasad, P. D. Ziogas, and S. Manias, "A novel passive waveshaping method for single-phase diode rectifiers," *IEEE Transactions on Industrial Electronics*, vol. 37, no. 6, pp. 521–530, 1990.

[17] J. Yanchao and F. Wang, "Single-phase diode rectifier with novel passive filter," *IEE Proceeding Circuits Devices Systems*, vol. 145, no. 4, pp. 254–259, 1998.

[18] K. Hirachi, T. Iwade, and K. Shibayama, "Improvement of control strategy on a step-down type high power factor converter," *National Conversion Record of Industrial Electronic Engineering Japan*, pp. 4.70–4.71, 1995.

[19] R. Ltoh, K. Lshizaka, H. Oishi, and H. Okada, "Single-phase buck rectifier employing voltage reversal circuit for sinusoidal input current waveshaping," *IEE Proceedings: Electric Power Applications*, vol. 146, no. 6, pp. 707–712, 1999.

[20] C. A. Canesin and I. Barbi, "A novel single-phase ZCS-PWM high-power-factor boost rectifier," *IEEE Transactions on Power Electronics*, vol. 14, no. 4, pp. 629–635, 1999.

[21] P. J. A. Ling and C. J. Eldridge, "Designing modern electrical systems with transformers that inherently reduce harmonic distortion in a PC-rich environment," in *Proceedings of the Power Quality Conference*, pp. 166–178, 1994.

[22] J. C. Read, "The calculation of rectifier and inverter performance characteristics," *Journal of the Institute of Electrical Engineers*, vol. 92, no. 2, pp. 495–509, 1945.

[23] E. J. Cham and T. R. Specht, "The ANSI 49 rectifier with phase shift," *IEEE Transactions on Industry Applications*, vol. 20, no. 3, pp. 615–624, 1984.

[24] R. Hammond, L. Johnson, A. Shimp, and D. Harder, "Magnetic solutions to line current harmonic reduction," in *Proceedings of the Europe by International Power Conversion Conference (PCIM '94)*, pp. 354–364, San Diego, Calif, USA, 1994.

[25] S. Kim, P. N. Enjeti, P. Packebush, and I. J. Pitel, "A new approach to improve power factor and reduce harmonics in a three-phase diode rectifier type utility interface," *IEEE Transactions on Industry Applications*, vol. 30, no. 6, pp. 1557–1564, 1994.

[26] S. Choi, P. N. Enjeti, and I. J. Pitel, "Polyphase transformer arrangements with reduced kVA capacities for harmonic current reduction in rectifier-type utility interface," *IEEE Transactions on Power Electronics*, vol. 11, no. 5, pp. 680–690, 1996.

[27] B. M. Bird, J. F. Marsh, and P. R. McLellan, "Harmonic reduction in multiplex converters by triple-frequency current injection," *Proceedings of the Institution of Electrical Engineers*, vol. 116, no. 10, pp. 1730–1734, 1969.

[28] Y. S. Tzeny, "Harmonic analysis of parallel-connected 12-pulse uncontrolled rectifier without an interphase transformer," *IEE Proceeding Electrical Power Applications*, vol. 145, no. 3, pp. 253–260, 1998.

[29] M. Hink Karl, "18 -Pulse Drives and Voltage Unbalance," http://mtecorp.com/18pulse.html.

[30] T. H. Chen and M. Y. Huang, "Network modelling of 24-pulse rectifier transformers for rigorous simulation of rail transit power systems," *Electric Power Systems Research*, vol. 50, no. 1, pp. 23–33, 1999.

[31] E. F. El-Saadany, R. Elshatshat, M. M. A. Salama, M. Kazerani, and A. Y. Chikhani, "Reactance one-port compensator and

modular active filter for voltage and current harmonic reduction in nonlinear distribution systems: a comparative study," *Electric Power Systems Research*, vol. 52, no. 3, pp. 197–209, 1999.

[32] J. R. Johnson, "Proper use of active harmonic filters to benefit pulp and paper mills," *IEEE Transactions on Industry Applications*, vol. 38, no. 3, pp. 719–725, 2002.

[33] D. Li, Q. Chen, Z. Jia, and J. Ke, "A novel active power filter with fundamental magnetic flux compensation," *IEEE Transactions on Power Delivery*, vol. 19, no. 2, pp. 799–805, 2004.

[34] W. M. Grady, M. J. Samotyi, and A. H. Noyola, "Survey of active line conditioning methodologies," *IEEE Transactions on Power Delivery*, vol. 5, no. 3, pp. 1536–1541, 1990.

[35] M. Takeda, K. Ikeda, and Y. Tominaga, "Harmonic current compensation with active filter," in *Proceedings of the IEEE/IAS Annual Meeting*, pp. 808–815, 1987.

[36] Z. Du, L. M. Tolbert, and J. N. Chiasson, "Active harmonic elimination for multilevel converters," *IEEE Transactions on Power Electronics*, vol. 21, no. 2, pp. 459–469, 2006.

[37] S. Bhattacharya, T. M. Frank, D. M. Divan, and B. Banerjee, "Active filter system implementation," *IEEE Industry Applications Magazine*, vol. 4, no. 5, pp. 47–63, 1998.

[38] I. Takahashi, S. G. Li, and Y. Omura, "Low price and high power active filter," in *Proceedings of the IEEE/IAS Annual Meeting*, pp. 95–98, 1991.

[39] G. W. Chang and T. C. Shee, "A novel reference compensation current strategy for shunt active power filter control," *IEEE Transactions on Power Delivery*, vol. 19, no. 4, pp. 1751–1758, 2004.

[40] H. Akagi, "Control strategy and site selection of a shunt active filter for damping of harmonic propagation in power distribution systems," *IEEE Transactions on Power Delivery*, vol. 12, no. 1, pp. 354–362, 1997.

[41] A. Cavallini and G. C. Montanari, "Compensation strategies for shunt active-filter control," *IEEE Transactions on Power Electronics*, vol. 9, no. 6, pp. 587–593, 1994.

[42] H. Akagi, H. Fujita, and K. Wada, "A shunt active filter based on voltage detection for harmonic termination of a radial power distribution line," *IEEE Transactions on Industry Applications*, vol. 35, no. 3, pp. 638–645, 1999.

[43] H. Fujita and H. Akagi, "A practical approach to harmonic compensation in power systems—series connection of passive and active filters," *IEEE Transactions on Industry Applications*, vol. 27, no. 6, pp. 1020–1025, 1991.

[44] F. Z. Peng, H. Akagi, and A. Nabae, "A new approach to harmonic compensation in power systems—a combined system of shunt passive and series active filters," *IEEE Transactions on Industry Applications*, vol. 26, no. 6, pp. 983–990, 1990.

[45] H. Akagi, "Active harmonic filters," *Proceedings of the IEEE*, vol. 93, no. 12, pp. 2128–2141, 2005.

[46] D. Basic, V. S. Ramsden, and P. K. Muttik, "Hybrid filter control system with adaptive filters for selective elimination of harmonics and interharmonics," *IEE Proceedings: Electric Power Applications*, vol. 147, no. 4, pp. 295–303, 2000.

[47] S. Senini and P. J. Wolfs, "Hybrid active filter for harmonically unbalanced three phase three wire railway traction loads," *IEEE Transactions on Power Electronics*, vol. 15, no. 4, pp. 702–710, 2000.

[48] P. Salmerón, J. C. Montano, J. R. Vázquez, J. Prieto, and A. Pérez, "Compensation in nonsinusoidal, unbalanced three-phase four-wire systems with active power-line conditioner," *IEEE Transactions on Power Delivery*, vol. 19, no. 4, pp. 1968–1974, 2004.

[49] H. Akagi, "New trends in active filters for power conditioning," *IEEE Transactions on Industry Applications*, vol. 32, no. 6, pp. 1312–1322, 1996.

[50] H. L. Jou, J. C. Wu, and K. D. Wu, "Parallel operation of passive power filter and hybrid power filter for harmonic suppression," *IEE Proceedings: Generation, Transmission and Distribution*, vol. 148, no. 1, pp. 8–14, 2001.

[51] X. Zha and Y. Chen, "The iterative learning control strategy for hybrid active filter to dampen harmonic resonance in industrial power system," in *Proceedings of the IEEE International Symposium on Industrial Electronics*, vol. 2, pp. 848–853, 2003.

[52] B. Singh, K. Al-Haddad, and A. Chandra, "A review of active filters for power quality improvement," *IEEE Transactions on Industrial Electronics*, vol. 46, no. 5, pp. 960–971, 1999.

[53] B. R. Lin, B. R. Yang, and H. R. Tsai, "Analysis and operation of hybrid active filter for harmonic elimination," *Electric Power Systems Research*, vol. 62, no. 3, pp. 191–200, 2002.

[54] D. Chen and S. Xie, "Review of the control strategies applied to active power filters," in *Proceedings of the IEEE International Conference on Electric Utility Deregulation, Restructuring and Power Technologies (DRPT '04)*, pp. 666–670, April 2004.

[55] K. Zhou, Z. Lv, A. Luo, and L. Liu, "Control strategy of shunt hybrid active power filter in distribution network containing distributed power," in *Proceedings of the China International Conference on Electricity Distribution (CICED '10)*, pp. 1–10, September 2010.

[56] D. Detjen, J. Jacobs, R. W. De Doncker, and H. G. Mall, "A new hybrid filter to dampen resonances and compensate harmonic currents in industrial power systems with power factor correction equipment," *IEEE Transactions on Power Electronics*, vol. 16, no. 6, pp. 821–827, 2001.

[57] M. Aredes, J. Häfner, and K. Heumann, "Three-phase four-wire shunt active filter control strategies," *IEEE Transactions on Power Electronics*, vol. 12, no. 2, pp. 311–318, 1997.

[58] M. Aredes, L. F. C. Monteiro, and J. M. Miguel, "Control strategies for series and shunt active filters," in *Proceedings of the IEEE Bologna Power Tech Conference*, pp. 1–6, June 2003.

[59] S. Khalid and A. Triapthi, "comparison of sinusoidal current control strategy & synchronous rotating frame strategy for total harmonic reduction for power electronic converters in aircraft system under different load conditions," *International Journal of Advanced Research in Electrical, Electronics and Instrumentation Engineering*, vol. 1, no. 4, pp. 305–313, 2012.

20

Implementation and Control of an AC/DC/AC Converter for Double Wound Flywheel Application

J. G. Oliveira,[1] H. Schettino,[2] V. Gama,[2] R. Carvalho,[2] and H. Bernhoff[1]

[1] *Division for Electricity, Uppsala University, 751 05 Uppsala, Sweden*
[2] *Electrical Engineering Department, Federal University of Juiz de Fora, 36036-900 Juiz de Fora, MG, Brazil*

Correspondence should be addressed to J. G. Oliveira, janaina.goncalves@angstrom.uu.se

Academic Editor: Henry S. H. Chung

An all-electric driveline based on a double wound flywheel, connected in series between main energy storage and a wheel motor, is presented. The flywheel works as a power buffer, allowing the battery to deliver optimized power. It also separates electrically the system in two sides, with the battery connected to the low voltage side and the wheel motor connected to the high voltage side. This paper presents the implementation and control of the AC/DC/AC converter, used to connect the flywheel high voltage windings to the wheel motor. The converter general operation and the adopted control strategy are discussed. The implementation of the AC/DC/AC converter has been described from a practical perspective. Results from experimental tests performed in the full-system prototype are presented. The prototype system is running with satisfactory stability during acceleration mode. Good efficiency and unity power factor could be achieved, based on vector control and space vector modulation.

1. Introduction

Extensive research has been recently done on Electric Vehicles (EVs) [1, 2]. The development of an efficient and robust propulsion system is essential to the feasibility of EVs [3]. But, even though sophisticated engines and advanced electrical power trains exist, the main issue, the long-term energy storage, has not been resolved.

All-electric drivelines based on battery, supercapacitor, flywheel, and combinations of these are being widely discussed and tested, attempting to lower the requirement on power density from the batteries [4, 5].

The propulsion system in development at Uppsala University is based upon a double wound flywheel energy storage device [6]. The flywheel under study is physically divided into two voltage levels through the stator winding. The high voltage side of the flywheel is connected to the wheel motor, whereas the low voltage side is connected to the battery, as shown in Figure 1. The system is bidirectional and the power can either flow from the battery to the wheel motor (acceleration mode) or from the wheel-motor to the battery (regenerative braking) [7].

In acceleration mode, the flywheel function is basically to provide the variant power requested by the wheel motor, so that the battery delivers a smoother output power. In braking mode, the wheel motor acts as a generator, and the flywheel is responsible for storing the regenerated energy.

The system needs a considerable number of power electronics converters and electronic controllers in order to regulate and optimize the power flux between the components [8]. A DC/AC converter is used to connect the battery to the flywheel low voltage side. An AC/DC/AC converter is used to connect the flywheel high voltage side to the wheel motor.

This paper presents the implementation and control of the AC/DC/AC converter to link the flywheel high voltage side to the wheel motor. Although it is a bidirectional converter, only the acceleration mode is studied in the present paper. Section 2 describes the modeling and general operation of the converter. The adopted control strategy

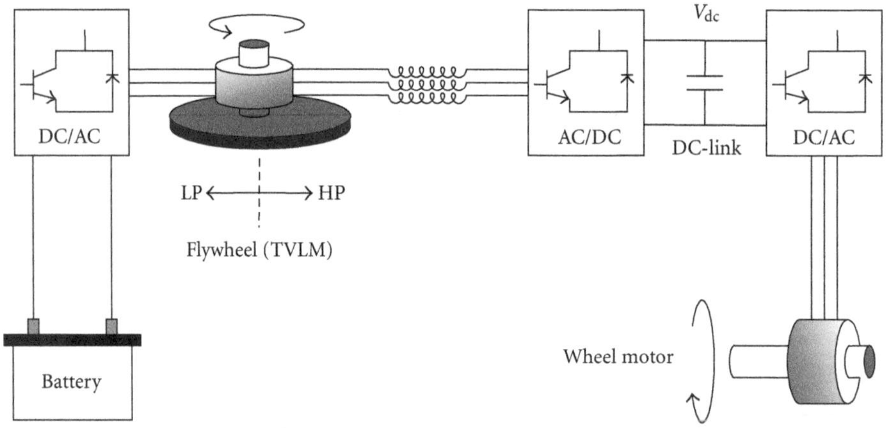

FIGURE 1: All-electric driveline schematics: The power can flow in both directions.

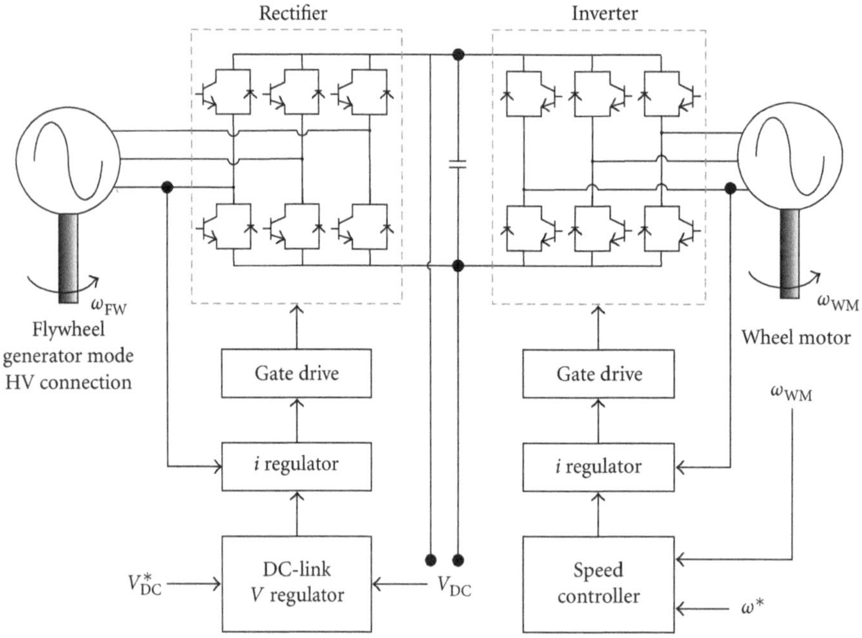

FIGURE 2: The AC/DC/AC converter in acceleration mode.

is discussed in details in Section 3. A prototype system is presented in Section 4, and the experimental results are discussed in Section 5.

2. The AC/DC/AC Converter

Both the wheel motor and the flywheel machine are three-phase synchronous machines in the present system. The angular speed of synchronous machines is proportional to the applied electric frequency in steady state. The energy stored in the flywheel rotor is related to its angular speed. The wheel motor speed must vary according to the vehicle's requirements.

The AC/DC/AC converter is an electrical way to decouple the two frequencies. Two identical three-phase bridges, one capacitor and inductors compose the power circuit. In acceleration mode, the AC/DC/AC converter can be divided into a rectifier and an inverter, as shown Figure 2. This section presents briefly the modeling and functioning of the two modules.

2.1. Rectifier. A forced-commutated three-phase controlled rectifier is required to obtain a desired voltage in the DC-link for different flywheel speeds and different wheel motor loads [9]. A scheme of the forced-commutated rectifier, based on transistors, is shown in Figure 2.

The AC/DC converter must behave like a voltage boost in order to work as a forced-commutated rectifier. In other words, the DC-Link voltage must be larger than the peak DC voltage generated by the rectifying diodes in passive mode. The inductance L performs the boost voltage operation in combination with the capacitor C and acts at the same time as a low-pass filter for the AC line current. Therefore, the choice of L and C values is extremely important for the proper functioning of the rectifier. Since the flywheel has a very small internal inductance, external inductors need to be placed.

Assuming that the source (flywheel machine) is a balanced, sinusoidal three-phase voltage supply with frequency ω, the input voltages can be written as

$$e_a = -E\sin(\omega t),$$
$$e_b = -E\sin\left(\omega t - \frac{2\pi}{3}\right), \quad (1)$$
$$e_c = -E\sin\left(\omega t + \frac{2\pi}{3}\right).$$

In the rotating d-q frame, the input voltage is usually in the quadrature axis [10]. Therefore, the direct component is null

$$e_d = 0,$$
$$e_q = E. \quad (2)$$

The voltage equation in the rotating d-q frame is given by

$$0 = L\frac{di_d}{dt} - \omega L i_q + V_d,$$
$$E = L\frac{di_q}{dt} + \omega L i_d + V_q. \quad (3)$$

The active power supplied from the source is

$$P = \frac{3}{2}\left(e_q i_q + e_d i_d\right) = \frac{3}{2}E i_q. \quad (4)$$

If the i_d current is controlled to zero, the line current is completely represented by i_q component and the active power coming from the source is maximized. Consequently, the input current will be in phase with the input voltage (unity power factor)

$$i_d^* = 0. \quad (5)$$

The idea of the proposed control strategy is to control the DC voltage in combination with the input AC currents so that unity power factor is reached. Instantaneous value of the DC voltage, input currents, and flywheel speed must be known to achieve closed loop control. The control strategy adopted is discussed in detail in Section 3.

2.2. Inverter. The inverter is responsible for driving the wheel motor, controlling its angular speed for different load torques. Since the wheel motor is a synchronous permanent magnetic machine (PMSM), the electromagnetic torque expressed in the d-q rotating frame is given by [11]

$$T_e = \frac{3p}{2}\left[\psi_{PM} i_q + i_q i_d \left(L_d - L_q\right)\right], \quad (6)$$

where p is the number of pole pairs; ψ_{PM} is the flux produced by the permanent magnet; L_d and L_q are, respectively, the direct and quadrature components of the wheel motor inductance.

Considering a nonsalient pole machine (smooth air gap), L_d and L_q are the same, and (5) can be written as

$$T_e = \frac{3p}{2}\psi_{PM} i_q = K i_q. \quad (7)$$

According to (7), only the i_q component can influence the value of the electromagnetic torque produced in the machine. Thus, the maximum torque per ampere occurs when i_q is at its maximum value and, consequently, i_d is equal to zero

$$i_d^* = 0. \quad (8)$$

The implementation of the inverter control requires the feedback signals of the line currents and rotor position instantaneous values, as shown Figure 4. The details of the control strategy are further explained in Section 3.

3. Control Strategy

3.1. DC-Link Control. A simplified diagram of the DC-link control is shown in Figure 3. Due to the characteristic of closed loop, it is necessary to measure the stator currents and rotor angular position. These measurements are carried out by two current sensors, and a hall sensor used to calculate the speed and estimate the rotor position. The instantaneous values of stator currents i_a and i_b are mathematically transformed (Clarke and Park transformations) and then used as the feedback for i_q and i_d control loops.

An outer loop of voltage is connected to a PID regulator. The output of the voltage controller is the reference of the quadrature current i_q. The reference of the direct current i_d is set to zero, in order to obtain unity power factor operation. The output of i_q and i_d PID controllers are transformed into α and β components (Inv Park) and modulated by Space Vector modulation, generating the pulses that are inserted into a three-phase bridge rectifier.

3.2. Vector Control. The Vector control block diagram is similar to the DC-Link diagram, as shown in Figure 4. Current sensors are necessary to capture the instantaneous values of the line currents. In this case, an encoder was used to measure the rotor angular position. The two inner currents loops are maintained in exactly the same way as for the DC-link control. The main difference is in the outer loop, where a speed loop control is implemented. The output of the speed controller is the reference quadrature current i_q, which regulates the torque needed to reach the reference speed. The reference of the direct current i_d is set to zero, in order to obtain maximum torque per ampere in the machine.

4. Prototype System

A scaled experimental test setup has been constructed to investigate the properties of the proposed flywheel system [12]. The experiment allows measurements of complete drive cycles to be performed, improving the understanding of the constituting components and optimization of the complete system.

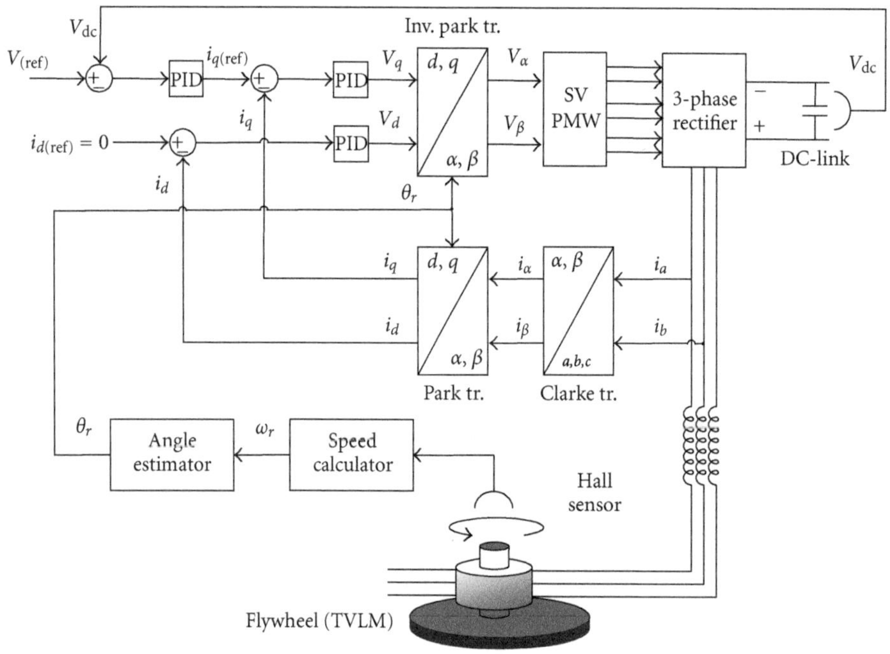

Figure 3: Block diagram of the DC-Link control.

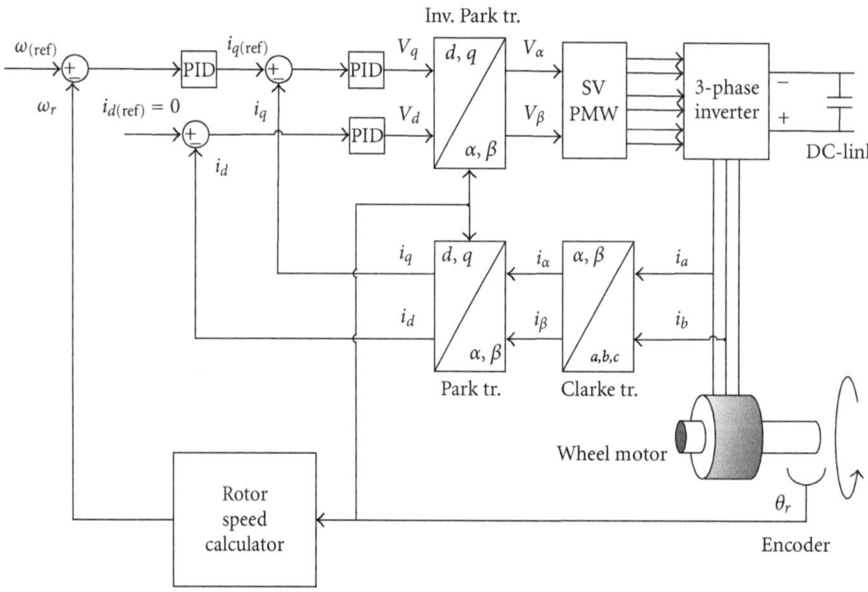

Figure 4: Block diagram of the Vector control.

A prototype of the described AC/DC/AC converter that connects the high voltage of the flywheel machine to the wheel motor has been implemented.

A 3.5 kW/380 V PMSM (manufacturer *Leroy Somer*) is used as wheel motor. The double wound flywheel machine is a prototype developed at Uppsala University [6]. A DC power supply emulates a battery, connected through a DC/AC converter to the low-voltage side of the flywheel (not shown in the diagram).

The high-voltage side of the flywheel is connected to 600 μH inductors per phase, required for the functionality of the forced-commutated rectifier. A 20 mF/350 V capacitor is connected in parallel with the DC-Link.

Two three-phase *MOSFET* bridges were built with discrete *DSEI 2130*. A *snubber* circuit was implemented in order to eliminate the rings presented in the *MOSFET* output.

The control of the AC/DC/AC converter has been performed using two MCUs (Microcontrollers) Piccolo

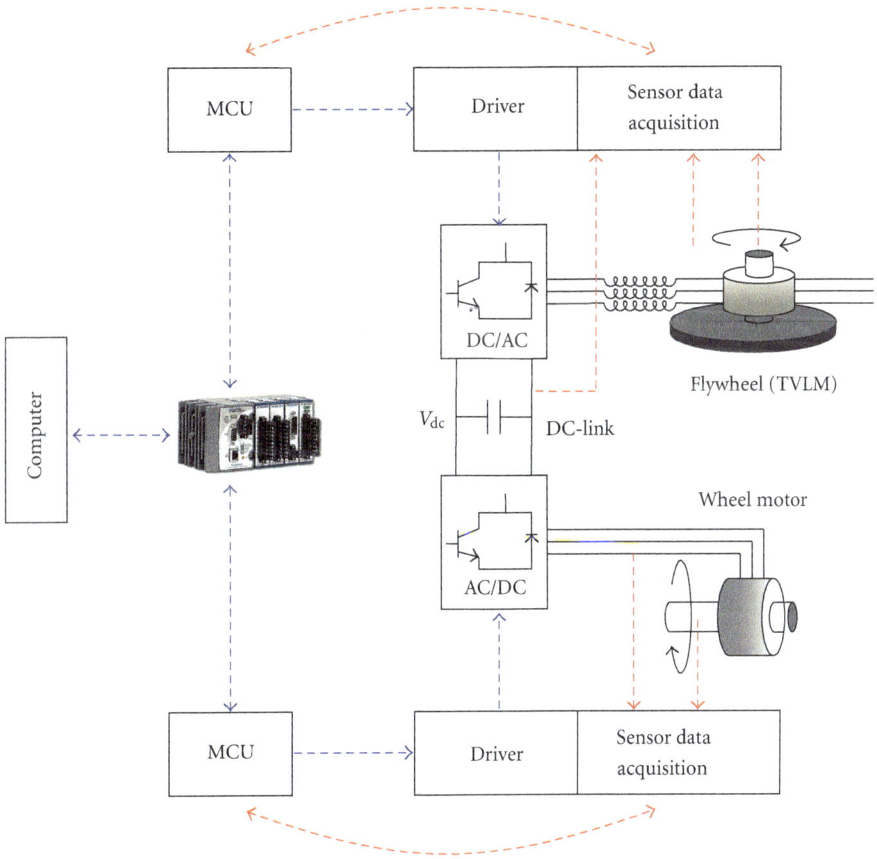

FIGURE 5: Prototype System, the black arrows indicate the sensor signals flux.

TMDSCNCD28035 (manufactured by Texas Instruments) plugged in the TMDSDOCK28035 Experimenter's Kit.

The current setup has two MCU and one computer, which perform real-time debug, through serial communication using a Compact Rio device from National Instruments, as shown in Figure 5.

Two mainboards, one for each module of the converter, contain the driver and sensor circuits. The driver used is an IR2130, from *International Rectifier*. With the purpose of avoiding damage in the low power system, the driver was electrically isolated from the MCU by digital optocouplers (A4800). The system requires a large number of sensors: four current sensors (HAL50-S), one encoder (ERN423), one voltage sensor (resistive divider), and three hall sensors (A1101). The current sensors are used in both boards to capture the instantaneous values of the line AC currents. The hall sensor (flywheel speed capture) and the voltage sensor (DC-Link voltage capture) are used for the rectifier control strategy. The encoder is used to estimate the wheel motor rotor angle for the inverter board. The black arrows in Figure 5 indicate the sensor signals flux.

The rectifier main board with the components indications is shown in Figure 6. Small changes are seen in the inverter mainboard due to the different sensors used.

5. Experimental Results

The experimental waveforms for a 20-second test performed in the full system prototype are shown in Figure 7, Figure 8, and Figure 9. A variant torque is applied to the wheel motor up to $t = 11$ s, as seen in Figure 7.

The approximate input and output power curves can be seen in Figure 8. The mean input power, measured in the terminals of the DC power supply, is 140 W and the mean output power, measured from voltage and current supplied to the wheel motor, is 125 W.

The flywheel speed decreases from 1950 RPM to 1650 RPM in order to supply the power requested by the wheel motor, as seen in Figure 9. When the torque is changed to a small constant value ($t > 11$ s), the flywheel speed increases to the initial value (1950 RPM). While the torque is being applied, a stationary error of about 10% compared to the setpoint values occurs in the motor speed control (Figure 9).

Additional experimental results are shown in Figures 10, 11, and 12.

Different load current and different reference values for the DC-link voltage were applied to the system in order to verify the controllers' response. The rotational speed of

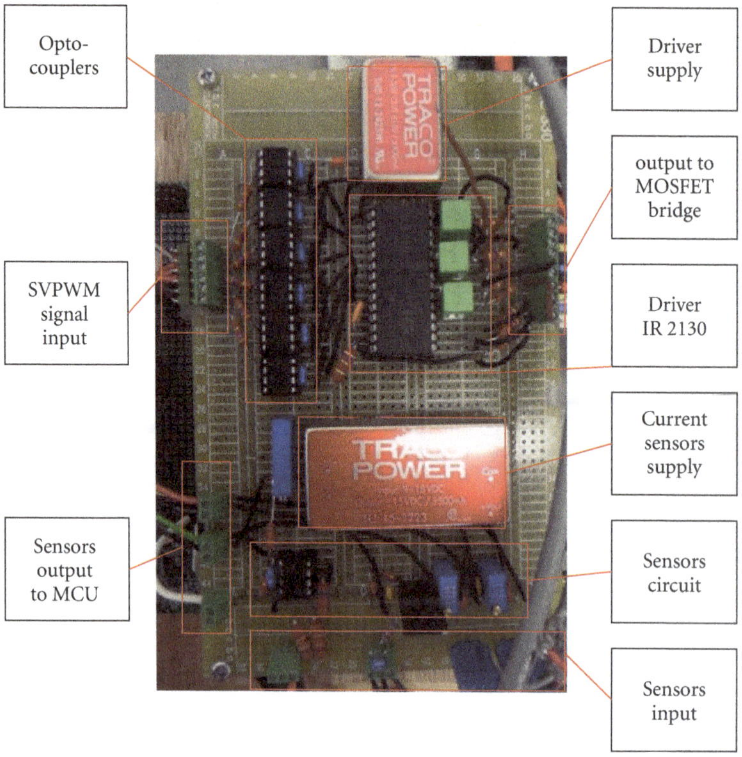

Figure 6: Rectifier main circuit board.

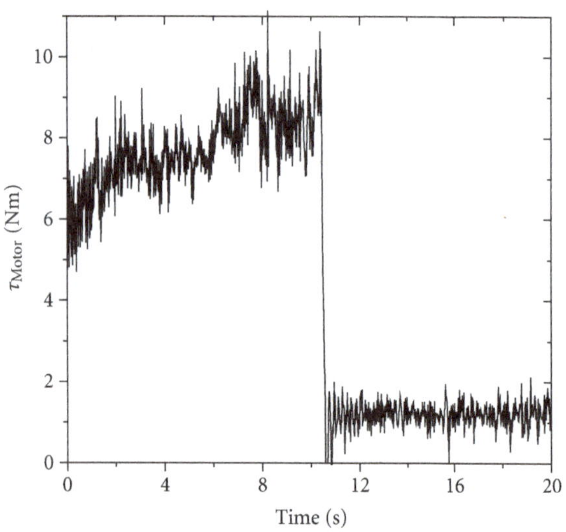

Figure 7: Applied torque.

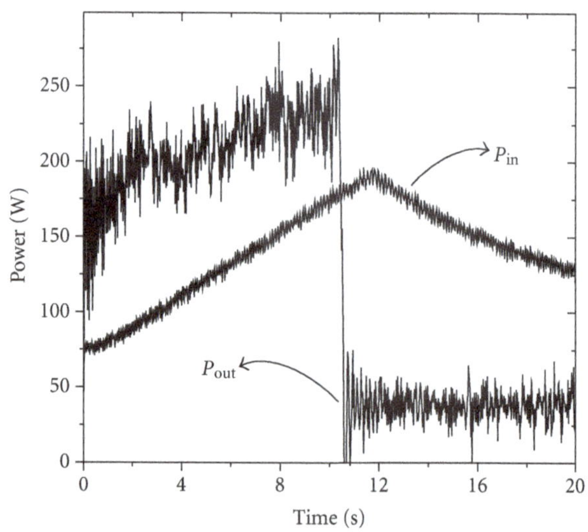

Figure 8: Input and output power.

the flywheel machine varies in order to supply the extra current required on the motor side, as shown in Figure 10(a). Different output currents are shown in Figure 10(b). The voltage on the DC-link is kept constant, with some oscillations when the applied load is changed. Small overshoot can be seen when the reference value was changed. These oscillations might indicate that the PID parameters can be further adjusted.

The flywheel was disconnected from the battery, and the load was kept the same, so the output current of the DC-link would not vary. The speed of the flywheel decreased as shown in Figure 11(a). The voltage in the DC-link was kept constant (80 V) until the speed the flywheel was around 20 rpm, as shown in Figure 11(b). The output current on the flywheel high power side is shown in Figure 11(c). It increased as the speed of the flywheel decreased, to compensate for the falling output voltage.

Output voltage and current on the high voltage of the flywheel machine for different reference values of I_d is shown in Figure 12. As described in Section 2, unity power factor

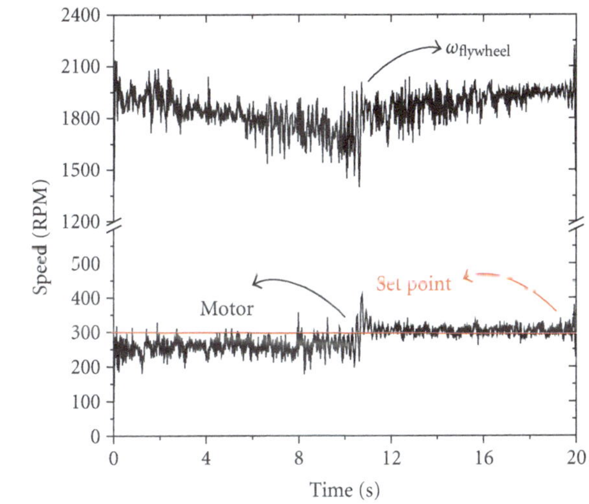

FIGURE 9: Flywheel and wheel motor angular speed.

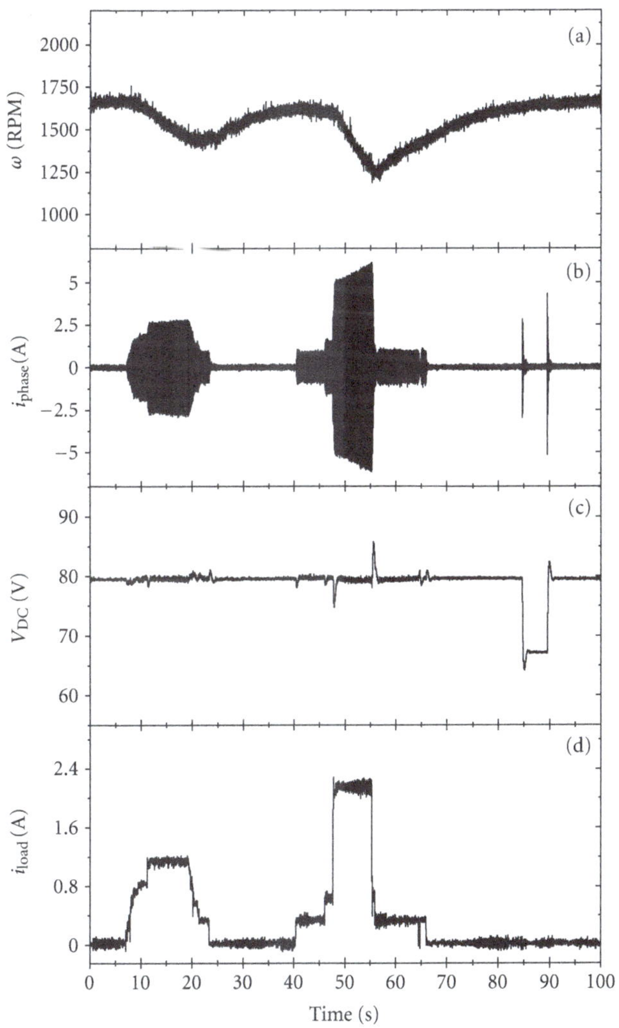

FIGURE 10: (a) Speed of the flywheel, (b) flywheel output current, (c) DC-link voltage, and (d) wheel motor current.

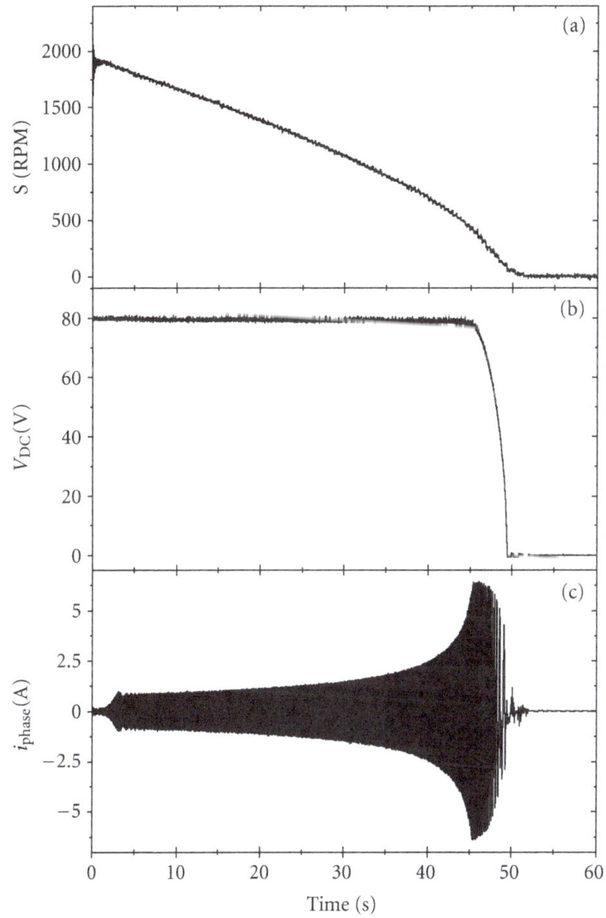

FIGURE 11: (a) Speed of the flywheel, (b) voltage over the DC link, and (c) flywheel output current.

operation is desired on the rectifier side, and this is achieved by making $I_d = 0$ in nonsalient pole machines. The results shown in Figure 12 indicate that the flywheel machine is not purely nonsalient, and due to some small saliency, the required value of I_d in order to reach unity power operation is different from zero (the values are given in p.u. in Figure 12).

6. Conclusion

An AC/DC/AC converter based on MOSFET modules for double wound flywheel application is presented. The prototype system is running with satisfactory stability in acceleration mode. Some changes in the control system must be implemented in order to achieve regenerative braking. Based on vector control theory and space vector modulation, good efficiency and unity power factor could be achieved.

Results have shown some oscillations in the control of the DC-link voltage, which can be suppressed with further adjustment of the PID parameters. However, the control has shown robustness and capability of keeping the DC-link voltage constant for a broad range of input voltages.

Small saliency of the flywheel machine has been shown by the results, but unity power factor operation can be reached by slightly adjusting the value of I_d in the control system.

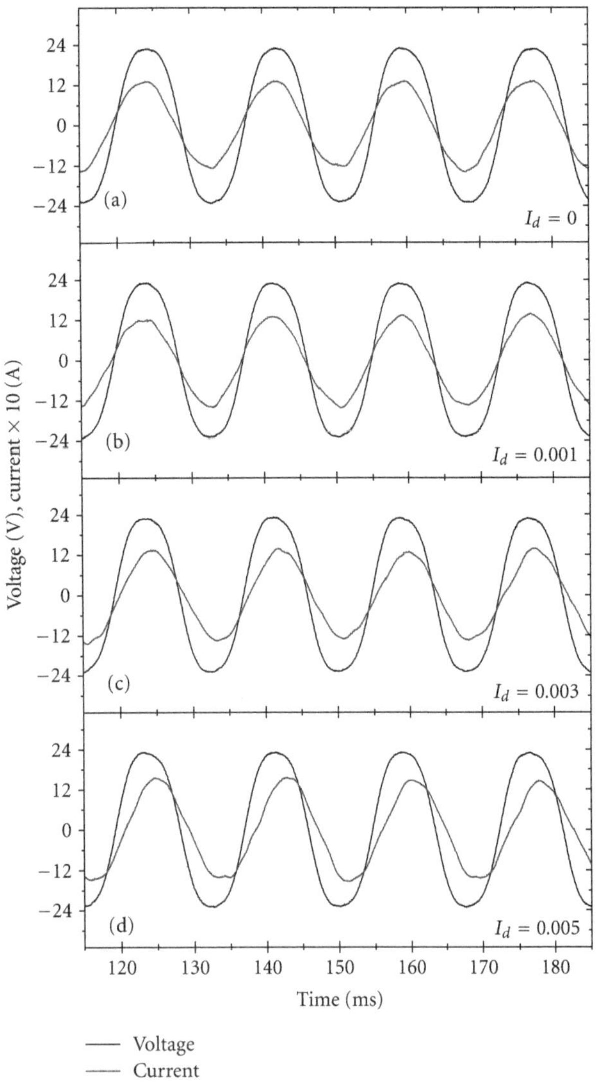

FIGURE 12: Output voltage and current on the high voltage of the flywheel machine for different reference values of I_d.

Future work consists in implementing regenerative braking operation and enlarging the system to a full-scale experimental setup.

Acknowledgments

The authors would like to thank Dr. Juan de Santiago, Licentiate Johan Abrahamsson and Licentiate Johan Lundin for their contribution to this work. Also to Professor Francisco José Gomes and PET/MEC, from the Federal University of Juiz de Fora.

References

[1] J. Dixon, "Energy storage for electric vehicles," in *Proceedings of the IEEE-ICIT International Conference on Industrial Technology (ICIT '10)*, pp. 20–25, March 2010.

[2] P. Fairley, "Speed bumps ahead for electric-vehicle charging," *IEEE Spectrum*, vol. 47, no. 1, pp. 13–14, 2010.

[3] A. Emadi, S. S. Williamson, and A. Khaligh, "Power electronics intensive solutions for advanced electric, hybrid electric, and fuel cell vehicular power systems," *IEEE Transactions on Power Electronics*, vol. 21, no. 3, pp. 567–577, 2006.

[4] J. Van Mierlo, P. Van den Bossche, and G. Maggetto, "Models of energy sources for EV and HEV: fuel cells, batteries, ultracapacitors, flywheels and engine-generators," *Journal of Power Sources*, vol. 128, no. 1, pp. 76–89, 2004.

[5] J. Santiago, J. G. Oliveira, J. Lundin, J. Abrahamsson, A. Larsson, and H. Bernhoff, "Design parameters calculation of a novel driveline for electric vehicles," *World Electric Vehicle Journal*, vol. 3, 2009.

[6] J. De Santiago, A. Larsson, and H. Bernhoff, "Dual voltage driveline for vehicle applications," *International Journal of Emerging Electric Power Systems*, vol. 11, no. 3, article 1, 2010.

[7] J. G. De Oliveira, J. Lundin, J. De Santiago, and H. Bernhoff, "A double wound flywheel system under standard drive cycles: simulations and experiments," *International Journal of Emerging Electric Power Systems*, vol. 11, no. 4, article 6, 2010.

[8] J. G. Oliveira and H. Bernhoff, "Power electronics and control of two-voltage-level flywheel based all-electric driveline," in *IEEE International Symposium on Industrial Electronics*, 2011.

[9] M. H. Rashid, *Power Electronics Handbook*, Academic Press, 2001.

[10] J. W. Choi and S. K. Sul, "Fast current controller in 3-phase AC/DC boost converter using d-q axis cross-coupling," in *Proceedings of the 27th Annual IEEE Power Electronics Specialists Conference (PESC '96)*, vol. 1, pp. 177–182, January 1996.

[11] K. S. Low, M. F. Rahman, and K. W. Lim, "Dq transformation and feedback linearization of a permanent magnet synchronous motor," in *Proceedings of the International Conference on Power Electronics and Drive Systems*, pp. 292–296, February 1995.

[12] J. Abrahamsson, J. De Santiago, J. G. Oliveira, J. Lundin, and H. Bernhoff, "Prototype of electric driveline with magnetically levitated double wound motor," in *Proceedings of the 19th International Conference on Electrical Machines (ICEM '10)*, September 2010.

21

Fault Tolerant Ancillary Function of Power Converters in Distributed Generation Power System within a Microgrid Structure

Antonino O. Di Tommaso, Fabio Genduso, Rosario Miceli, and Giuseppe Ricco Galluzzo

Department of Energy, Informatics and Mathematical Models, University of Palermo, 90128 Palermo, Italy

Correspondence should be addressed to Fabio Genduso; genduso@dieet.unipa.it

Academic Editor: Salem Rahmani

Distributed generation (DG) is deeply changing the existing distribution networks which become very sophisticated and complex incorporating both active and passive equipment. The simplification of their management can be obtained assuming a structure with small networks, namely, microgrids, reproducing, in a smaller scale, the structure of large networks including production, transmission, and distribution of the electrical energy. Power converters in distributed generation systems carry on some different ancillary functions as, for example, grid synchronization, islanding detection, fault ride through, and so on. In view of an optimal utilization of the generated electrical power, fault tolerant operation is to be considered as a suitable ancillary function for the next future. This paper presents a complete modeling of fault tolerant inverters able to simulate the main fault type occurrence and a control algorithm for fault tolerant converters suitable for microgrids. After the model description, formulated in terms of healthy device and leg binary variables, and the illustration of the fault tolerant control strategy, the paper shows how the control preserves power quality when the converter works in the linear range. The effectiveness of the proposed approach and control is shown through computer simulations and experimental results.

1. Introduction

The exponential penetration of distributed generation (DG) is leading to enormous changes in the conception of the electrical system management. Involved networks are no longer to be regarded as passive components but integrate a growing number of functions such as load management, demand side management, and generation curtailment.

Both standalone and grid connected microgrids, locally reproducing the structure of the whole generation and distribution system, may be a viable solution to enhance DG benefits, reducing at the same time the drawbacks of DG itself.

All the future hypothetical scenarios will lead to great transformations in the design of power systems, with interesting implications in different fields of research [1–5].

Some problems which arise immediately are the presence of bidirectional power flow, the need for a different design of the power lines and transformers, the unwanted islanding conditions, a different protection philosophy, and the possibility of a fault tolerant operation to increase the level of continuity.

This paper considers fault tolerant operation as a possible ancillary function. This additional function may be important in autonomous microgrid that cannot benefit from back up interventions of the traditional distribution network [6].

Since about a decade, the fault tolerant converters are the subject of research and development and their concept has now been extended to multilevel converters too because of their large number of devices and their redundancy making them suitable candidates for such an operation.

The present technical literature on the topic is abundant even not still exhaustive because the research fields are still up and many open questions are still waiting for significant answers [7, 8].

Among the various aspects, the most frequently discussed are control problems [8–13], fault diagnosis and protection [6, 7, 9, 14, 15], topologies, reconfiguration of circuits and

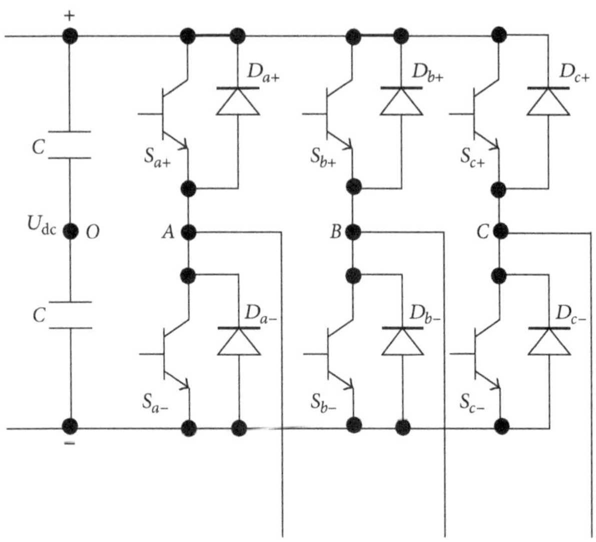

Figure 1: The VSI topology.

controls [4, 8, 13, 16, 17], and effect of fault tolerant operations on the main network [14, 18] and on the loads, with a special reference to evaluation of performance of AC electrical drives [7, 9–11].

This paper presents first a mathematical model able to represent both the behavior of a faulted converter and the fault tolerant operating mode. Subsequently a control strategy that can generate a substantially symmetrical three-phase output and the related implementation algorithm are examined. The fault tolerant operation is therefore considered in the context of a grid converter operation analyzing the system behavior and the operating limits of the same one in order to guarantee an acceptable level of power quality. In this scenario, some immediate indications on levels of transmitted power and performance deviations with respect to a classical grid converter are given. The validation of the control strategy and of the used algorithm is verified through simulations and experimental tests of a prototype built within the SDESLAB at the University of Palermo.

The next sections are organized as follows.

Section 2 deals with the general model of the inverter. Section 3 shows the simulation of the open-circuit fault. Section 4 illustrates the fault tolerant mode and the corresponding control algorithm and displays the simulation results. Section 5 deals with the specific case of a grid converter within a microgrid and outlines some considerations about the transmitted power and the current distortion and the need for a higher level of the DC Link voltage. Section 6 shows the experimental results obtained by tests on the realized prototype. Finally, Section 7 summarizes the conclusions.

2. The General Inverter Model

2.1. The General Model for Healthy Mode. For the general model of the Voltage Source Inverter in faulty and unfaulty, let us consider the circuit of Figure 1 presenting the general topology of a VSI.

The current circulation on free-wheeling diodes is also considered as also illustrated in the references [19–21].

For each controllable device, a switching function is defined in terms of gate pulse. For the jth upper device, it is, for example,

$$S_{j+} = \begin{cases} 1 & \text{if driving pulse is given,} \\ 0 & \text{otherwise } (\text{with } j = \{A, B, C\}). \end{cases} \quad (1)$$

A complementary switching function is defined for the lower device S_{j-}. The effective turning on of the device is conditioned by the current sign. In fact, an upper controllable device can effectively be turned on only if the load current is positive; conversely a lower device can be turned on only if the same current is negative. In general, current not being conducted by controlled devices is conducted by the free-wheeling diodes. Even if diodes are not controllable devices, switching functions may be defined for them also by taking into account the current sign. For example, an upper diode conducting the negative current is turned on only if the lower controllable device is off. In this way, the switching function for the upper diode is

$$D_{j+} = \overline{S_{j-}} \otimes (i_j < 0). \quad (2)$$

In a similar manner for the lower diode, it is

$$D_{j-} = \overline{S_{j+}} \otimes (i_j > 0). \quad (3)$$

By summarizing all the previous considerations, two generalized switching functions may be defined

$$\begin{aligned} \widehat{S}_{j+} &= [S_{j+} \otimes (i_j > 0)] \oplus [\overline{S_{j-}} \otimes (i_j < 0)], \\ \widehat{S}_{j-} &= [S_{j-} \otimes (i_j < 0)] \oplus [\overline{S_{j+}} \otimes (i_j > 0)]. \end{aligned} \quad (4)$$

Considering now the converter topology, output voltage with respect to the O point, as sketched in Figure 1, may be written as

$$v_{jO} = (\widehat{S}_{j+} - \widehat{S}_{j-}) \frac{u_{DC}}{2}; \quad j = \{A, B, C\}. \quad (5)$$

If the load impedances are balanced, the phase to neutral voltages have the well-known expressions

$$v_{jN} = v_{jO} - \frac{\sum_{k=j} v_{kO}}{3}; \quad j = \{A, B, C\}. \quad (6)$$

The load currents can be found from the equations

$$v_{jN} = Ri_j + L\frac{di_j}{dt} + e_{jN}. \quad (7)$$

For a complete inverter model, the equations describing the DC link circuit transient must be considered. In particular the expression of the DC link voltage is

$$C\frac{du_{DC}}{dt} = i_0 - \sum_{k=j} i_k \widehat{S}_{k+} \quad (8)$$

and i_0 is given by

$$u - u_{DC} = R_0 i_0 + L_0 \frac{di_0}{dt}. \quad (9)$$

Equations (4)–(9) constitute the model of a VSI in the healthy mode.

The Section 2.2 considers the modified converter model at faulty mode.

2.2. The General Model for the Faulty Mode. Recent studies have shown as up than the 80% of converter faults are due to single or multiple device failures [22].

Every power structure designer strives to remain well within the operating limits of the power device as specified by the IGBT manufacturer. However, there are component failures either of the power switching device or of supporting components that result in one or more IGBT power device failures. When the IGBT power devices fail, under certain circumstances, the failure can result in the rupture of the power module and extensive damage to the surrounding power components [23].

Typically failures are manifest as counterblow destruction, as in the case of short circuit of silicon packed devices, or by losing driver pulse occurring if driver circuit or its power supply are invalid so that no trigger pulse is sent to the gate [8]. Open-circuit faults generally do not cause shutdown of the system but degrade its performance [10, 21].

Some of the gate drive failure can result from breaking of passive components (resistors and capacity) or from a drift of their values for which the behavior of the driver does not meet the original specifications. With regard to short circuits or defects caused by overcurrent, a crucial parameter is the passing through energy I^2t that, during time, causes a degradation of the materials employed in the construction of the components.

Short circuit faults have a very fast evolution and, in general, no software algorithm is able to detect them in the short time required to interrupt and insulate the faulted zone from the rest of the system. For this reason, they are generally managed directly with hardware additional device or circuits and interrupted with ultrafast fuses, taking care that these last ones should not introduce any additional inductances in the main circuit.

Short circuits and ruptures that do not imply any further damage to the remaining part of the power circuit evolve to an open circuit condition. For this reason only open circuit faults are considered here.

Multiple device faults are less frequent but may be also significant if the inverter leg is an integrated packaged one. However, the presented model will also consider the case of multiple devices damage. Furthermore, thanks to the followed approach, in all the examined cases, an inverter reconfiguration for the fault tolerant mode can be simulated.

In this section, the single device faults for drive failure and also for diode breakdown are considered.

If a controllable upper device goes to a drive failure condition, in this case, the S_{j+} switching function is always zero. A positive current cannot be conducted by any device of the faulted phase. The negative current may instead circulate in the upper diode and in the lower controlled device. Starting from diode conduction if the lower device is turned off, a commutation phenomenon appears, then the diode current turns to zero while the transistor current rises up. In this case, the general switching function becomes

$$\begin{aligned} \widehat{S}_{j+} &= \left[\overline{S_{j-}} \otimes \left(i_j < 0\right)\right], \\ \widehat{S}_{j} &= \left[S_{j-} \otimes \left(i_j < 0\right)\right]. \end{aligned} \quad (10)$$

In particular $\widehat{S}_{j+}$ considers only the upper diode conduction while $\widehat{S}_{j-}$ denotes the lower transistor conduction. Similar consideration applies when the faulted device is the lower. In fact, in this case, the general switching functions become

$$\begin{aligned} \widehat{S}_{j+} &= \left[S_{j+} \otimes \left(i_j > 0\right)\right], \\ \widehat{S}_{j-} &= \left[\overline{S_{j+}} \otimes \left(i_j > 0\right)\right]. \end{aligned} \quad (11)$$

For the general case, the introduction of a *healthy device binary variable (HDBV)* is useful. It is defined as follows:

$$h_{j\pm} = \begin{cases} 1 & \text{if the upper } (+) / \text{lower } (-) \text{ device is healthy,} \\ 0 & \text{otherwise.} \end{cases} \quad (12)$$

The introduction HDBVs makes the definition of the generalized (i.e., in faulty and unfaulty mode) switching function very simple. In fact

$$\begin{aligned} \widehat{S}_{j+} &= \left[h_{j+} S_{j+} \otimes \left(i_j > 0\right)\right] \oplus \left[h_{j-} \overline{S_{j-}} \otimes \left(i_j < 0\right)\right], \\ \widehat{S}_{j-} &= \left[h_{j-} S_{j-} \otimes \left(i_j < 0\right)\right] \oplus \left[h_{j+} \overline{S_{j+}} \otimes \left(i_j > 0\right)\right], \end{aligned} \quad (13)$$

that is, in a general simulation scheme, the fault behavior is modeled simply with the product of $S_{j\pm}$ by $h_{j\pm}$.

The same principle applies when both the transistor and the diode are broken, but the first expression of (10) now contains two incompatible conditions. However, to get a correct simulation result, the signal of (10) may be used to reset the integrators of (7). In other words, when the lower power device is turned off, the load current is forced to zero.

The loss of entire converter leg does not introduce any difference because the approach with healthy device binary variable also allows an easy and affordable modeling in inverter reconfiguration for the management of the converter in fault-tolerant mode. With this aim, the healthy leg binary variables (HLBV) are then introduced as follows:

$$h_j = h_{j+} \oplus h_{j-}. \quad (14)$$

In reconfiguration, after the fault diagnosis, the faulted leg is disabled and the load terminal of the faulted phase is connected to the middle point of the DC link allowing the current circulation by means of bidirectional switches. The reconfigured inverter assumes the topology of a B4 circuit (see Figure 2).

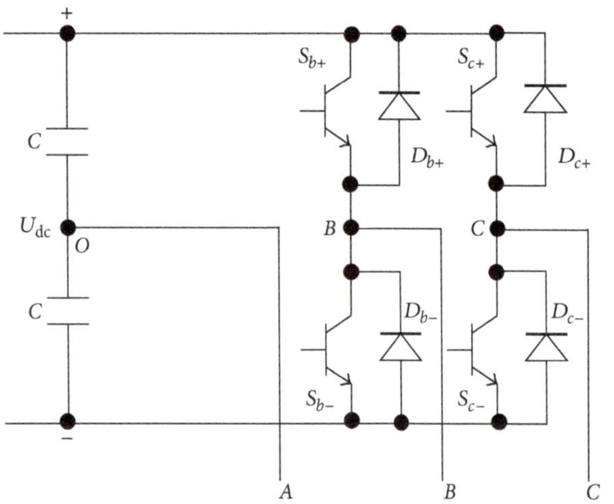

FIGURE 2: The reconfigured inverter after the fault (B4 topology).

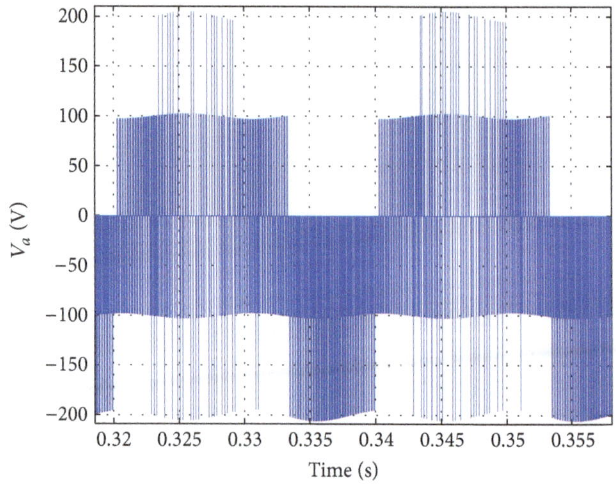

FIGURE 3: Voltage on load faulted phase for upper device drive failure.

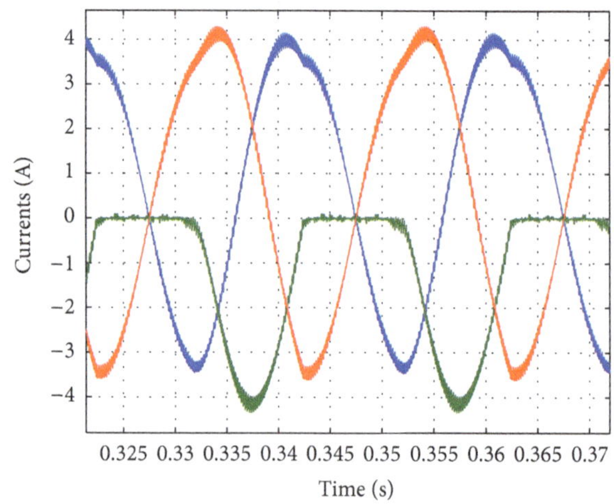

FIGURE 4: Currents on the load after fault.

In previous papers of the authors [14, 18, 21, 24–27], this reconfiguration has been extensively investigated and the general model of the reconfigured inverter has been also presented. This model will consist of the following equations:

$$U_1 = \frac{1}{C}\int_0^t \left(i_0 - \widehat{\mathbf{S}}_{\mathbf{k}}^{\mathbf{t}} \mathbf{H_k} \mathbf{i_k}\right) dt,$$

$$U_2 = \frac{1}{C}\int_0^t \left(i_0 - \left[\widehat{\mathbf{S}}_k^t \mathbf{H}_k - \mathbf{1}^t \mathbf{H}_k\right] \mathbf{i_k}\right) dt, \tag{15}$$

for the partial DC link voltages with

$$\mathbf{H}_k = \begin{pmatrix} h_A & 0 & 0 \\ 0 & h_B & 0 \\ 0 & 0 & h_C \end{pmatrix},$$

$$\mathbf{v}_{kN} = U_{\mathrm{DC}} \mathbf{T} \mathbf{H}_k \widehat{\mathbf{S}}_k - \mathbf{T} \mathbf{H}_k \mathbf{1} U_2, \tag{16}$$

for the inverter output voltages with

$$\mathbf{T} = \frac{1}{3}\begin{pmatrix} 2 & -1 & -1 \\ -1 & 2 & -1 \\ -1 & -1 & 2 \end{pmatrix}. \tag{17}$$

The load (grid) equations remain unchanged.

3. Simulation of the Faults

Simulations of faults are made with the help of the previous model whose equations are implemented thanks to the Matlab-Simulink software package. In the simulation, a VSI with a 300 V DC link voltage is hypothesized. Simulation results are illustrated in the figures below.

Figure 3 shows the output voltage on the faulted phase when a driver failure for an upper device appears, while Figure 4 shows the currents on the load after fault.

Simulations show clearly the positive current interruption and the negative current which flows through the diode.

During the current interruption, the average value of the phase voltage is zero (the instantaneous value is only made up of a series of voltage spikes). Other currents on unfaulty phases become phase opposite while the applied voltages on it are half of the line to line voltage with a different sign.

Figure 5 shows, instead, the voltage on an unfaulted phase.

Figures 6 and 7 show, respectively, the same simulation results when the fault regards both the device and the freewheeling diode.

Voltage is very similar to that of the previous analyzed case except for a higher number of spikes. Currents are instead quite different being composed of a series of negative pulse due to commutation of the lower transistor. The unfaulted phase voltages are very similar to the previous ones and, for this reason, are not reported here.

4. The Control Strategy for the Fault Tolerant Operation

4.1. The Fault Tolerant Control Arrangement. In the case of loss of an entire leg, the only reconnection of the faulted

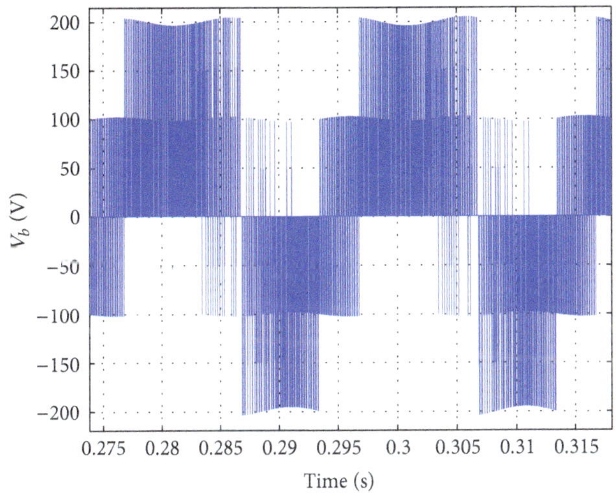

FIGURE 5: Voltage on an unfaulted phase.

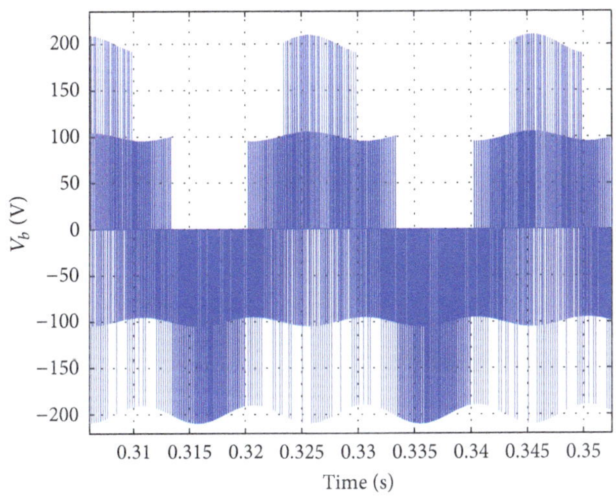

FIGURE 6: Faulty phase voltage after transistor and diode have broken.

TABLE 1: Synthesis of the postfault reference voltages.

Healthy U_{ref}	Phase A fault	Phase B fault	Phase C fault
u_A	0	$-u_C$	u_B
u_B	u_C	0	$-u_A$
u_C	$-u_B$	u_A	0

phase to DC link middle point does not ensure a correct operation. In particular, the current peak amplitude of the faulted phase results lowers by a factor $\sqrt{3}$. In order to obtain a balanced currents system, the voltage references used in the modulation process must be changed.

In the case of a grid converter, its capability to maintain a set of symmetrical voltages is very important for the power quality in view to limit as possible disturbances propagation into the grid avoiding oversized apparatuses to their mitigation.

The new modified set of reference voltages in the fault-tolerant mode should then present a minimal inverse component.

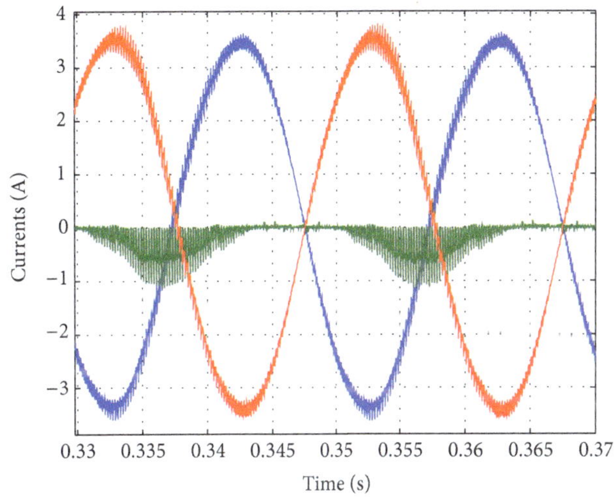

FIGURE 7: Load currents after broken transistor and diode.

The new reference voltages consist of a set of two sinusoidal signals controlling the healthy legs.

Suppose now that the signals corresponding phasors are

$$U_{\text{ref1}} = U_x \exp(j0), \qquad U_{\text{ref2}} = U_x \exp(-j\varphi). \tag{18}$$

For the fault cases, on phases A, B, and C, respectively, the inverse component of this reference system is

$$\begin{aligned} U_{\text{inv}A} &= 0 + \alpha U_{\text{ref1}} + \alpha^2 U_{\text{ref2}}, \\ U_{\text{inv}B} &= U_{\text{ref1}} + \alpha 0 + \alpha^2 U_{\text{ref2}}, \\ U_{\text{inv}C} &= U_{\text{ref1}} + \alpha U_{\text{ref2}} + \alpha^2 0. \end{aligned} \tag{19}$$

By calculating the module of the inverse components, it follows that

$$|U_{\text{inv}}| = \sqrt{2} U_x \sqrt{1 + \cos\left(\varphi \pm \frac{2\pi}{3}\right)}, \tag{20}$$

in which the sign "+" is for a fault on phase A and C while the sign "−" is for a fault on phase B. In both cases, the minimum inverse factor is reached for $\varphi \pm 2\pi/3 = \pi$, that is, for $\varphi = \pm\pi/3$ and it results just $|U_{\text{inv}}| = 0$.

Hence, in a fault tolerant converter, the new references have equal amplitude and a 60-degree phase displacement.

Furthermore, the new references can be built following the suggestion of Table 1.

Therefore, the general expression of the reference voltages (u_k^* with $k = A, B, C$) for healthy and faulted conditions is

$$\begin{aligned} u_A^* &= u_A h_A h_B h_C + \sqrt{3}\left(\underline{h}_C u_B - \underline{h}_B u_C\right), \\ u_B^* &= u_B h_A h_B h_C + \sqrt{3}\left(\underline{h}_A u_C - \underline{h}_C u_A\right), \\ u_C^* &= u_C h_A h_B h_C + \sqrt{3}\left(\underline{h}_B u_A - \underline{h}_A u_B\right), \end{aligned} \tag{21}$$

where u_j is the set of normal symmetrical three-phase reference voltages.

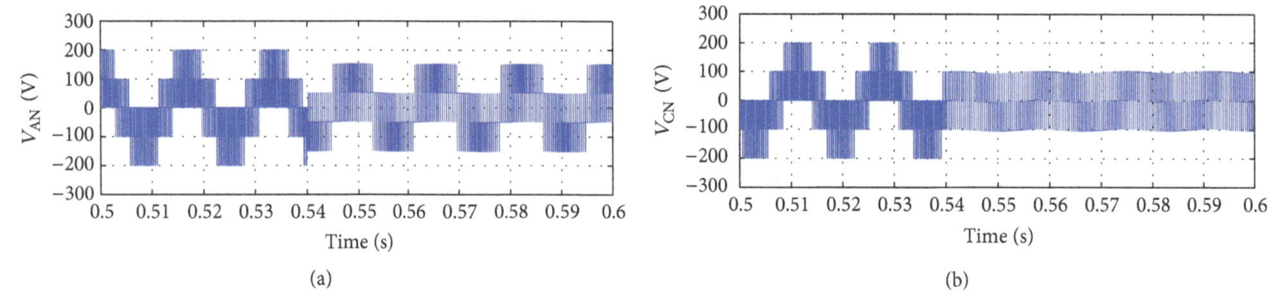

FIGURE 8: Phase voltages at unfaulted and faulted converter leg.

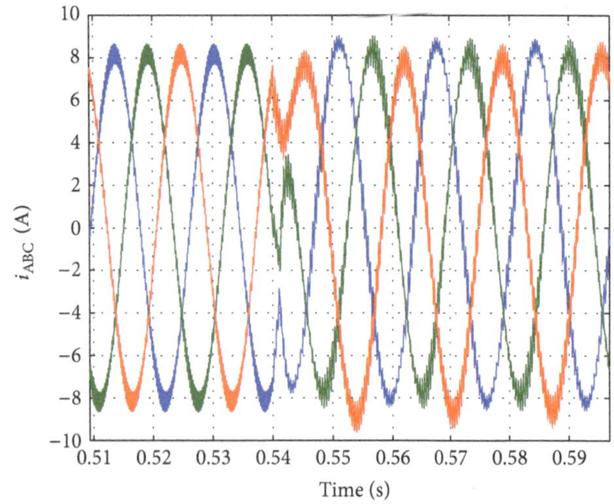

FIGURE 9: Load current before and after faults.

Equation (21) may be also written in vector notation

$$\mathbf{u}^* = \lambda \mathbf{u} + \sqrt{3}\mathbf{u} \times \underline{\mathbf{h}}, \tag{22}$$

where $\lambda = h_A h_B h_C = \det(\mathbf{H}_k)$.

The modified reference voltage are very simple to be implemented on microprocessors and diagnostic signals on the converters can be used to synthesize the new postfault reference voltages.

In (21), the general term $\lambda\mathbf{u}$ is the reference for the healthy case while the vectors cross product $\sqrt{3}(\mathbf{u} \times \underline{\mathbf{h}})$ is the new reference in the faulted case. The factor $\sqrt{3}$ guarantees the identity of the voltage and current amplitude, even after the fault has occurred.

Note that the cross product term $\sqrt{3}(\mathbf{u}\times\underline{\mathbf{h}})$ should be $\pi/2$ clockwise rotated if the converter is a grid connected, thus maintaining the synchronization between converter and grid voltages. In this case, the reference voltages system, expressed in vector form, results

$$\mathbf{u}^* = \lambda \mathbf{u} + \sqrt{3}\left(\mathbf{u} \times \underline{\mathbf{h}}\right) e^{-j(\pi/2)}. \tag{23}$$

The first advantage of (23) is a close integration of the diagnostic signals directly in the control for the inverter reconfiguration, another advantage is to make the fault and reconfiguration transparent to the main control system. In fact, any regulating action of external variables can provide a three-phase voltage reference that will be then manipulated according to (23) in order to generate the correct reference. In other words, the only reason for which the control system can be affected by the fault is the unavoidable DC link voltage ripple (as discussed later) and its effect on the current.

4.2. The Control Simulation. Control simulation results are illustrated in the following figures showing how the fault condition is managed by the model with the modified control strategy in order to guarantee the minimal derating in power quality.

Figure 8 shows the converter phase voltage measured at unfaulted (a) and faulted leg (b) after the faults occurred, resulting their first harmonics a symmetrical system even after the fault.

The voltage waveforms are quite different from those with unfaulted converter condition and the faulted leg voltage exhibits the greatest difference.

Figure 9 shows converter currents before and after fault occurrence with the instantaneous modification of the converter references. The residual current imbalance after fault is due to the alternating components in the DC-link voltage depending on the current amplitude and on the output frequency. In general, the voltage oscillation in single DC capacitors does not affect significantly the entire value of the DC link voltage and the general converter performance.

Postfault currents exhibit a phase jump due to reconfiguration unless this is compensated. Figure 10 shows instead the effect on load currents with the phase jump correction introduced in (23).

The linear region of the converter during the fault is, in each case, restricted. Linear range is quite important in order to achieve minimal currents asymmetry.

Figure 11 shows a higher load current distortion and unbalance during overmodulation (i.e., with an amplitude modulation index higher than 0.5, in the faulted case).

Figure 12 shows the voltage imbalance on the two DC link capacitors.

5. Simulation of a Grid Converter with Fault Tolerant Operation

The illustrated simulation was made for a standalone converter. In the case of a grid converter fed by a DG source,

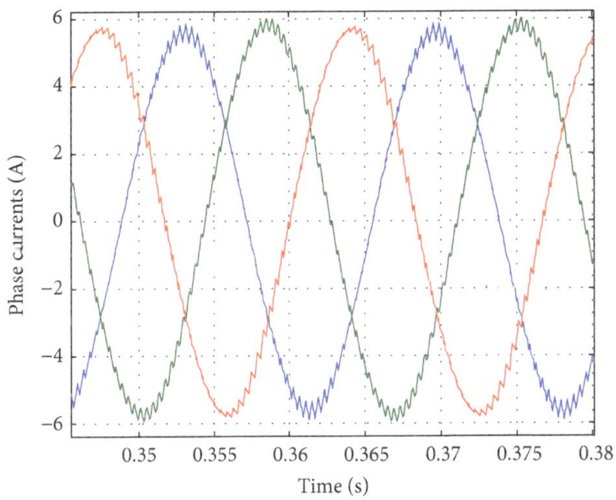

FIGURE 10: Load currents before and after fault with phase jump correction.

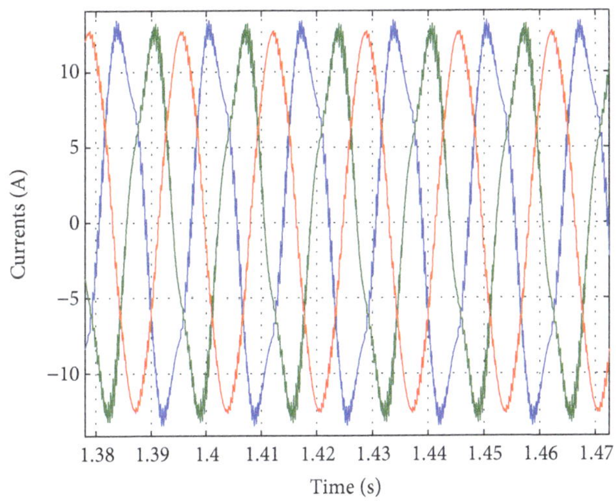

FIGURE 11: Distorted and unbalanced load currents with higher distortion during faulted condition in overmodulation mode.

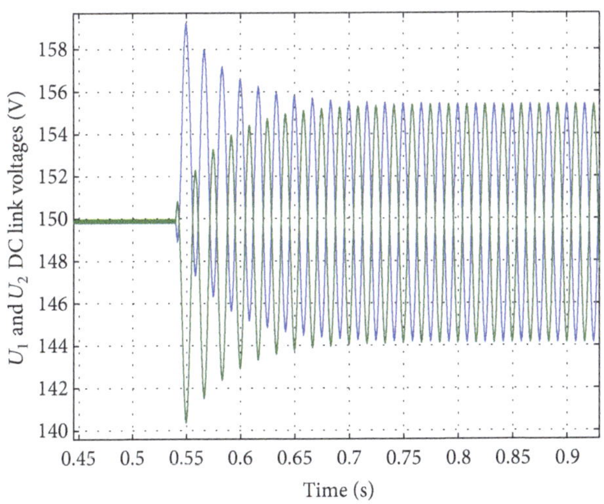

FIGURE 12: Voltage unbalance through DC link ca pacitors.

the system and the control are more complicated. However, no modification of the classical control structure of a grid converter is needed. Figure 13 shows the simplified scheme of the control structure for a grid converter control in which active and reactive power are managed via current control loop and DC link voltage loop. A complete description of the grid converter control system and of some other different solutions is given in [28, 29].

The control scheme allows active P and reactive power Q to be managed separately by decoupling DQ channels for the current control, so that P depends on i_D and Q depends on i_Q via previous proper grid synchronization via PLL. The DC link voltage control is also performed.

In the example of this section, the grid filter is not optimally designed with the aim to put better in evidence some aspect of the fault tolerant grid converter behavior.

The simulation assumes the occurrence of a fault, the fast reconfiguration of the converter, and the decrease of the active power injected into the grid. Subsequently the reactive power transferred to the grid is vanished. It is evident that despite the active power decrease, there is a greater distortion of the output current, as it is seen in Figures 14 and 15 (also with reference to the i_D and i_Q components) due to the increased voltage ripple on the DC link. This oscillation is natural and the control system can only fix its mean value forced by the set point. The simulation shows well as vanishing the delivered reactive power, the DC link voltage oscillation decreases with benefit for both the input and the grid current that exhibit a lower distortion as evident from Figures 16, 17, 18, 19, and 20.

It is worthwhile to remark that for avoiding overmodulation, the fault tolerant converter has a DC link voltage 25–30% greater than that of a traditional grid converter.

6. Experimental Results

Experimental results were made available in the case of the fault tolerant mode where the reconfiguration makes the converter operation more safe for a normal laboratory test. It was not possible to test the converter as a grid-converter one, so the tests on the fault tolerant operation have been made by feeding an induction motor and registering output voltages and currents.

For the experimental test, a customary benchmark based on an INFRANOR power converter and a dSPACE control board has been built and set up.

The converter used in the test bench is a commercial frequency converter whose control board has been replaced by a customary one designed to be driven by the dSPACE control board and converting the TTL gating pulses to the "line drive" type of the INFRANOR converter (see Figure 21).

In the practical realization of a fault tolerant converter, some issues are mandatory to be taken into account: Triacs for the current path closure may suffer for a dv/dt not sufficient immunity, having, therefore, unwanted switchings. A solution to this problem can be the choice of a different bidirectional controllable switch. In particular, the combination of power diodes in a bridge configuration with an auxiliary

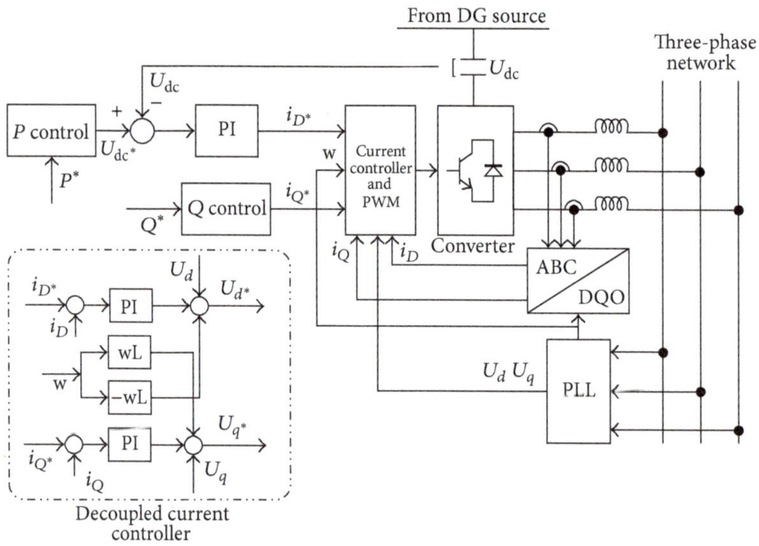

FIGURE 13: The control structure of a grid converter.

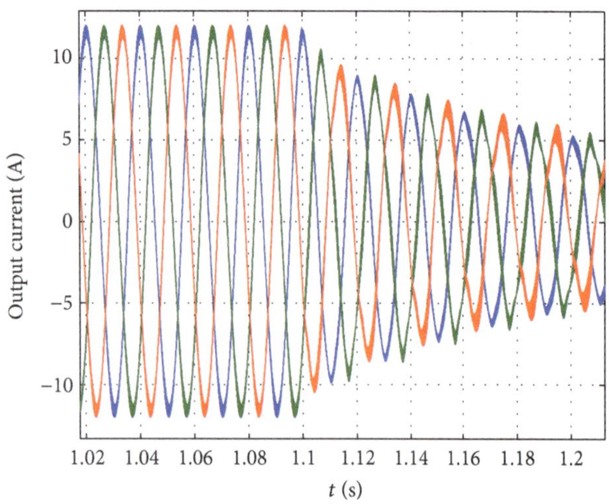

FIGURE 14: Current decrease after fault and reconfiguration.

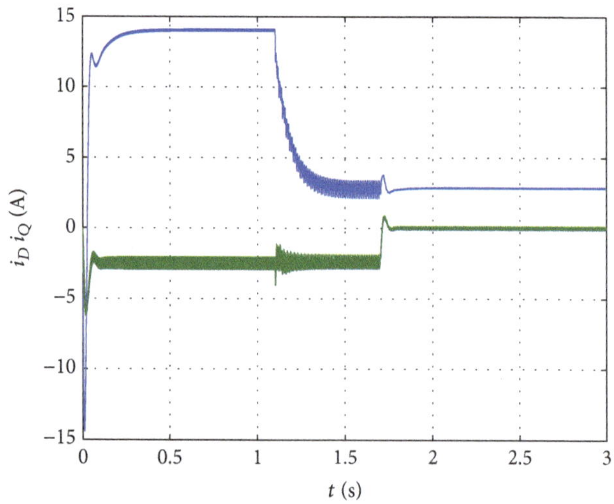

FIGURE 15: i_D (blue) and i_Q (green) current components.

IGBT is able to reach bidirectional current circulation (see Figure 22 for topology details). A fault detection algorithm based on the elaboration of load current signal has also been implemented. An extensive technical literature (see e.g., [4, 5]) exists on fault detection, but its discussion here is out of the scopes of this paper.

The fed induction motor is a two-pole three-phase squirrel cage whose nameplate is reported in Table 2.

A digital oscilloscope with two differential probes (up to a 25-MHz bandwidth) and 3% accuracy has been used to register voltage and current waveforms. Measurements data were stored in the computer hosting the dSPACE board with the help of "Octave" software.

Fault emulation have been realized with the help of solid state relays and connectors being driven from the dSPACE board and software interface. In this way, it was possible to test the behavior of the fault-tolerant converter in all possible condition without a real disruption of the power devices.

Figures 23, 24, and 25 shows, respectively, the measurements of voltage on faulty and unfaulty phases and the measurement of the currents, before and after fault and reconfiguration.

This final comparison validates the proposed general model of VSI in healthy and faulty mode implemented algorithm.

7. Conclusions

The use of fault-tolerant systems is essential for the optimal development of electrical power production by renewable sources as well as for the achievement of high reliability

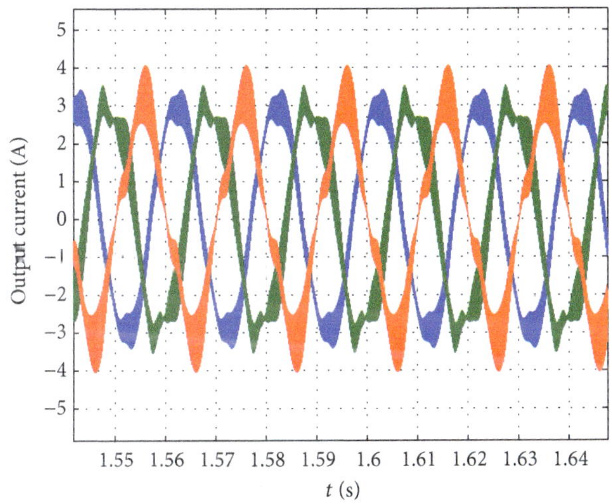

FIGURE 16: After fault current zoom.

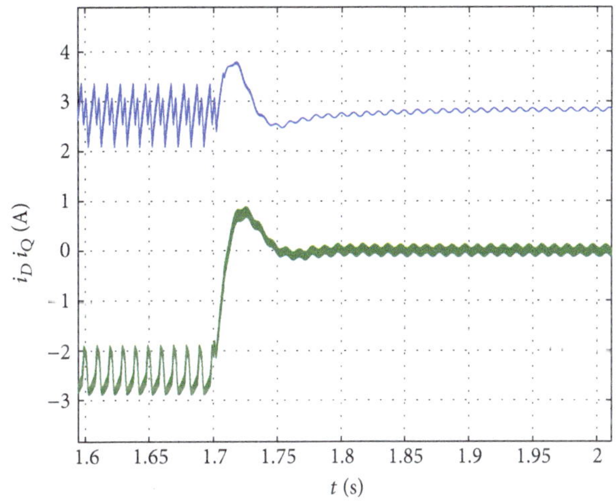

FIGURE 19: i_D and i_Q transient after vanishing Q.

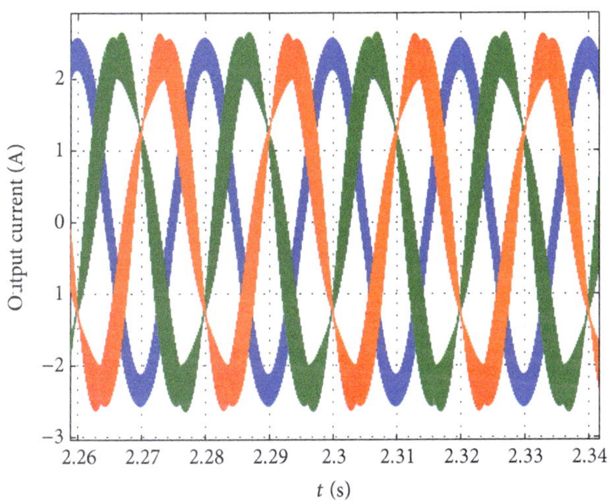

FIGURE 17: After fault current with no reactive power.

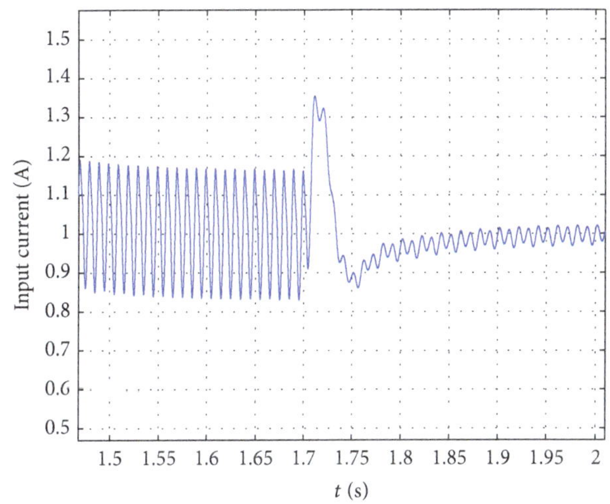

FIGURE 20: Transient on i_0 after vanishing Q.

TABLE 2: Test bench induction motor nameplates.

Ratings	Value
Power	5.5 kW
Speed	2870 r.p.m
Frequency	50–60 Hz
Torque	18.3 Nm
Voltage	400 V
Current	13 A
Poles	2

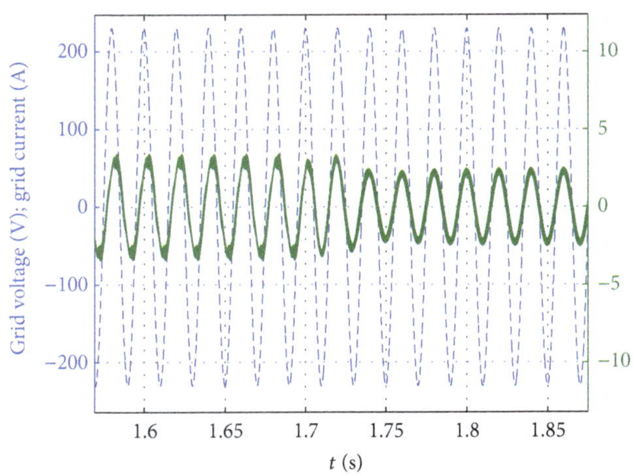

FIGURE 18: Phase voltage (blue dashed) and current (green continuous) with and without Q transmission.

and power quality fundamental for both end users and grid manager.

Fault tolerant converters require intelligent real-time detection algorithms of incoming failures easy to implement even on low cost hardware, such as microcontrollers or Pegasus.

In this paper, the authors have considered the fault tolerant operation of an inverter for distributed generation

FIGURE 21: Picture of the realized test bench.

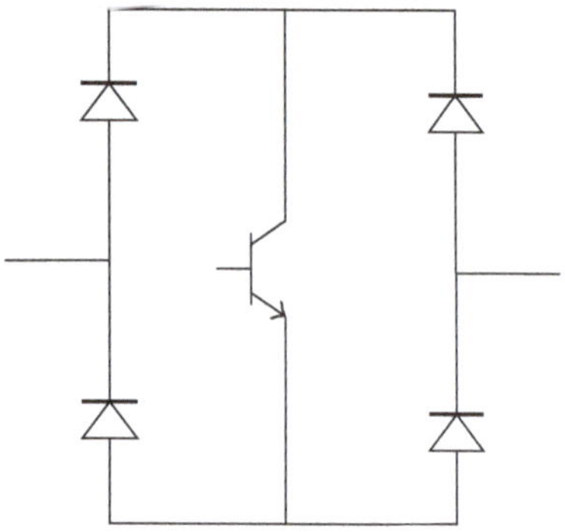

FIGURE 22: A bidirectional switch realized with diode and IGBT.

system seen as an ancillary complementary function in addiction to the classical ones already developed. The general inverter model here presented in the paper is suitable and useful for the simulation of faults and fault-tolerant operation of a VSI. The presented model is very helpful to predict transient phenomena due to faults occurrence. Switching function definition has been extended in order to cover the condition of diode turning on and open fault occurrence. The introduction of HDBVs and HLBVs allows for the construction of a unified model suitable for simulations with equation solver tools. This makes the simulations affordable and at the same time fast in their operation.

The implementation algorithm is very suitable in large context as grid connected converter for DG because it is transparent to the action of the main power control and synchronization.

Numerical simulations presented in the paper have considered both a standalone and a grid converter. In order to guarantee minimal power quality degradation, the following considerations apply in the converter control:

(i) over modulation must be avoided;

(ii) for this reason both a reduction of the transmitted active power and vanishing the reactive power are mandatory;

(iii) a 25–30% greater DC link voltage with respect to traditional value must be chosen.

The control has been also verified with experimental results in the case of reconfigurable operation. Comparison of

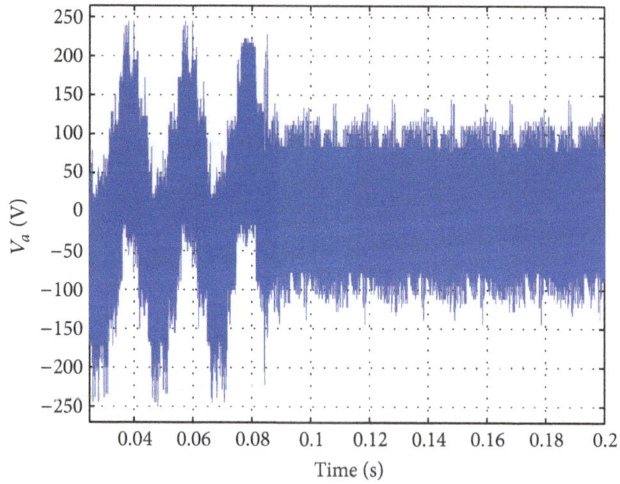

FIGURE 23: Measurement of faulty phase voltage before and after the fault.

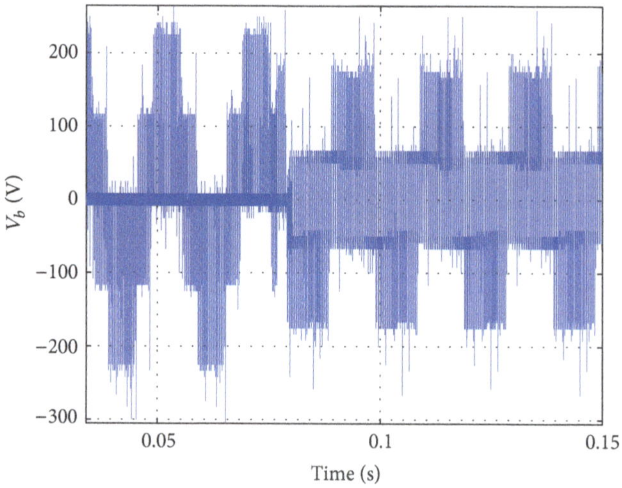

FIGURE 24: Measurement of unfaulty phase voltage before and after the fault.

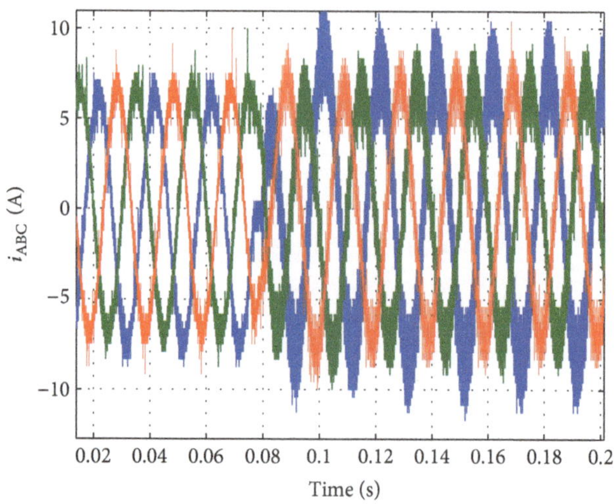

FIGURE 25: Measurement of phase currents before and after the fault.

simulations and experimental results, with the clear good accordance between them, confirms the validity of the proposed model and encourages the proposed implementation algorithm to include fault tolerance as an additional ancillary function in the next generation of grid converters.

Abbreviations

$j = \{A, B, C\}$: Phase index
$v_{jO(N)}$; i_j: Phase voltages and currents on the AC side
e_{jN}: Back e.m.f. of the load (or grid voltages)
$h_{j\pm}$: Upper(+)/lower(−) healthy device binary variables
h_j: Healthy leg binary variables
$S_{j\pm}$: Upper(+)/lower(−) side transistor switching function for phase j
$\widehat{S}_{j\pm}$: Upper(+)/lower(−) side transistor generalized switching function for phase j
$D_{j\pm}$: Upper(+)/lower(−) side diode switching function for phase j
$\otimes$: Logical AND operator
$\oplus$: Logical OR operator
U_{DC}: Whole inverter capacitor bank voltage
U_1, U_2: Partial capacitor bank voltage
i_0: DC side inverter input current
R_0, L_0: DC side wires parasitic circuital parameters
C: Value of the single DC link capacitor
R, L: Load circuital parameter (or grid filter resistance and inductance)
$[\cdots]^t$: Transposition of a matrix (a vector)
$\mathbf{1} = [111]^t$: Unitary three-dimensional vector
$\mathbf{u}_{ref}$: Vector of the reference voltage in PWM control
$\mathbf{u}^*_{ref}$: Vector of the reference voltage in PWM control after a fault
$\mathbf{v_k}$: Vector of the converter output voltages
$\mathbf{h}$: Vector of the healthy leg binary variables
$\underline{\mathbf{h}}$: NOT($\mathbf{h}$) vector
$\mathbf{i_k}$: Vector of the load currents
$\widehat{\mathbf{S}}_\mathbf{k}$: Vector of the switching functions $[\widehat{S}_A, \widehat{S}_B, \widehat{S}_C]^t$
φ: Angle displacement
U_x: Phasor voltage peak
$\alpha = \exp(j(2\pi/3))$: Fortescue operator for symmetrical Components
$\mathbf{v}_1 \times \mathbf{v}_2$: Cross product between generic vectors $\mathbf{v}_1$ and $\mathbf{v}_2$

References

[1] F. Abrahamsen, F. Blaabjerg, and C. Klumpner, "Low inductive fuses in DC Link inverter application," in *Proceedings of the Power Conversion Intelligent Motion Conference (PCIM '00)*, Boston, Mass, USA, March 2000.

[2] H. Chen, X. Zhang, S. Liu, and S. Yang, "Research on control strategies for distributed inverters in low voltage micro-grids," in *Proceedings of the 2nd International Symposium on Power Electronics for Distributed Generation Systems (PEDG '10)*, pp. 748–752, Hefei, China, June 2010.

[3] F. Blaabjerg, S. Freysson, H. Hansen, and S. Hansen, "A new optimized space-vector modulation strategy for a component-minimized voltage source inverter," *IEEE Transactions on Power Electronics*, vol. 12, no. 4, pp. 704–714, 1997.

[4] B. Wang, J. Wang, X. Sun, J. Wu, and W. Wu, "Phase multilevel inverter fault diagnosis and tolerant control technique," in *Proceedings of the CES/IEEE 5th International Power Electronics and Motion Control Conference (IPEMC '06)*, pp. 1–5, Shanghai, China, August 2006.

[5] S. Duong, J. Schanen, C. Schaeffer, F. Sarrus, and J. de Palma, "IGBT protection by fuse calculation, behaviour and connection," in *Proceedings of the Conference Record of the IEEE Industry Applications Conference: 32nd IAS Annual Meeting (IAS '97)*, pp. 1228–1235, New Orleans, La, USA, 1997.

[6] F. Genduso, R. Miceli, and S. Favuzza, "A Perspective on the future of distribution: smart grids, state of the art, benefits and research plans," *Energy and Power Engineering*, vol. 5, no. 1, pp. 36–42, 2013.

[7] J. Salmon, "Current overload protection features of hybrid inverter drives," in *Proceedings of the International Conference on Power Electronics and Motion Control*, vol. 1, pp. 470–476, San Diego, CA, USA, November 1992.

[8] M. Ma, L. Hu, A. Chen, and X. He, "Reconfiguration of carrier-based modulation strategy for fault tolerant multilevel inverters," *IEEE Transactions on Power Electronics*, vol. 22, no. 5, pp. 2050–2060, 2010.

[9] L. Zhou and K. Smedley, "A fault tolerant control system for hexagram inverter motor drive," in *Proceedings of the 25th Annual IEEE Applied Power Electronics Conference and Exposition (APEC '10)*, pp. 264–270, Palm Springs, CA, USA, February 2010.

[10] D. Sun and Y. He, "A modified direct torque control for PMSM under inverter fault," in *Proceedings of the 8th International Conference on Electrical Machines and Systems (ICEMS '05)*, vol. 3, pp. 2473–2473, Nanjing, China, 2005.

[11] M. A. Parker, C. Ng, and L. Ran, "Fault-tolerant control for a modular generatorconverter scheme for direct-drive wind turbines," *IEEE Transactions on Industrial Electronics*, vol. 58, no. 1, pp. 305–315, 2011.

[12] O. Wallmark, L. Harnefors, and O. Carlson, "Post-fault operation of fault-tolerant inverters for PMSM drives," in *Proceedings of the European Conference on Power Electronics and Applications*, pp. 1–10, Dresden, Germany, September 2005.

[13] P. Correa, M. Pacas, and J. Rodriguez, "Modulation strategies for fault-tolerant operation of H-bridge multilevel inverters," in *Proceedings of the IEEE International Symposium on Industrial Electronics (ISIE '06)*, pp. 1589–1594, Montreal, Quebec, Canada, July 2006.

[14] F. Genduso, R. Miceli, and G. R. Galluzzo, "Flexible power converters for the fault tolerant operation of micro-grids," in *Proceedings of the 19th International Conference on Electrical Machines (ICEM '10)*, Rome, Italy, September 2010.

[15] B. Xu, D. Yang, and X. Wang, "Neural network based fault diagnosis and reconfiguration method for multilevel inverter," in *Proceedings of the Chinese Control and Decision Conference (CCDC '08)*, pp. 564–568, Yantai, Shandong, China, July 2008.

[16] S. Bolognani, M. Zordan, and M. Zigliotto, "Experimental fault-tolerant control of a PMSM drive," *IEEE Transactions on Industrial Electronics*, vol. 47, no. 5, pp. 1134–1141, 2000.

[17] C. Cecati and N. Rotondale, "A double-PWM strategy for improved electric drive reliability," in *Proceedings of the Symposium on Power Electronics, Electrical Drives, Automation & Motion (SPEEDAM '02)*, vol. 1, pp. 25–30, Ravello, Italy, 2002.

[18] A. O. di Tommaso, S. Favuzza, F. Genduso, R. Miceli, and G. R. Galluzzo, "Development of diagnostic systems for the fault tolerant operation of Micro-Grids," in *Proceedings of the International Symposium on Power Electronics, Electrical Drives, Automation and Motion (SPEEDAM '10)*, pp. 1645–1650, Pisa, Italy, June 2010.

[19] L. Li-Chun, P. Ching-Tsai, and T. Lang-Jong, "Switching flow graph modeling technique for three phase inverter," *IEEE Transactions on Industrial Electronics*, vol. 55, no. 4, pp. 1603–1613, 2008.

[20] S. Cuk and R. Middlebrook, "A general unified approach to modeling switching converters power stages," in *Proceedings of the IEEE Power Electronics Specialist Conference*, pp. 18–34, 1976.

[21] F. Genduso, R. Miceli, and G. R. Galluzzo, "Experimental validation of a general model for three phase inverters operating in healthy and faulty modes," in *Proceedings of the International Symposium on Power Electronics, Electrical Drives, Automation and Motion (SPEEDAM '12)*, pp. 1–6, May 2012.

[22] P. Chen, Z. Ying, and D. Enzeng, "Fault detection algorithm of inverter based on singular value decomposition nonlinear filtering," in *Proceedings of the IEEE International Conference on Computer Science and Automation Engineering (CSAE '11)*, vol. 2, pp. 85–89, 2011.

[23] D. Braun, D. Pixler, and P. LeMay, "IGBT module rupture categorization and testing," in *Proceedings of the IEEE Industry Applications Conference 32nd IAS Annual Meeting*, pp. 1259–1266, October 1997.

[24] C. Cecati, F. Genduso, R. Miceli, and G. R. Galluzzo, "A suitable control technique for fault-tolerant converters in distributed generation," in *Proceedings of the International Symposium on Industrial Electronics (ISIE' 11)*, pp. 107–112, Gdansk, Poland, June 2011.

[25] F. Genduso and R. Miceli, "A general mathematical model for nonredundant fault-tolerant inverters," in *Proceedings of the IEEE International Electric Machines & Drives Conference (IEMDC '11)*, pp. 705–710, 2011.

[26] F. Genduso and R. Miceli, "Power quality savings with controlled fault-tolerant power converters," in *Proceedings of the International Conference on Theory and Application of Modeling and Simulation in Electrical Power Engineering (ELECTRIMACS '11)*, pp. 1–6, Cergy Pontoise, France, June 2011.

[27] A. D. Tommaso, F. Genduso, R. Miceli, and G. R. Galluzzo, "A general mathematical model for the simulation of common faults in three phase voltage source inverters," *International Journal of Automation and Power Engineering*, vol. 2, pp. 1–11, 2013.

[28] F. Blaabjerg, R. Teodorescu, M. Liserre, and A. V. Timbus, "Overview of control and grid synchronization for distributed power generation systems," *IEEE Transactions on Industrial Electronics*, vol. 53, no. 5, pp. 1398–1409, 2006.

[29] M. Liserre, P. Rodriguez, and R. Teodorescu, *Grid Converter for Photovoltaic and Wind Power Systems*, Wiley, IEEE press, 2011.

22

Improved Expression for Estimation of Leakage Inductance in *E* Core Transformer Using Energy Method

Sivananda Reddy Thondapu,[1] Mangesh B. Borage,[2]
Yashwant D. Wanmode,[1] and Purushottam Shrivastava[1]

[1] *Pulse High Power Microwave Section, Raja Ramanna Centre for Advanced Technology, Indore 452013, India*
[2] *Power Supplies and Industrial Accelerator Division, Raja Ramanna Centre for Advanced Technology, Indore 452013, India*

Correspondence should be addressed to Sivananda Reddy Thondapu, sivananda@rrcat.gov.in

Academic Editor: Pavol Bauer

This paper proposes a simpler and more accurate expression for estimation of leakage inductance in *E* core transformer, which is the most widely used transformer structure. The derived expression for leakage inductance accounts for the flux extending into air. The finite element method (FEM) analysis is made on the secondary shorted transformer to observe the *H*-field pattern. The results obtained from FEM analysis are used for approximating the field that is extending into air to derive an expression for leakage inductance. This expression is experimentally validated on prototype transformers of different core dimensions.

1. Introduction

Transformer is one of the basic building blocks of many power converters. The following are some of the cases where accurate estimation of leakage inductance is required.

(i) Different resonant converter topologies, discussed in [1–5], use parasitics of transformer as a part of resonant tank network. For designing power converter with such topologies, one requires accurate estimation of leakage inductance.

(ii) In hard switched converters, in every cycle the energy stored in the parasitics appears as loss in converter. In estimation of efficiency of such converters, one needs to estimate leakage inductance before hand.

(iii) For designing snubber circuits to limit device voltage during turn-off transients [6–8], one needs to estimate leakage inductance. These turn-off transients mainly occur due to energy stored in the leakage inductance of the transformer.

Methods that are usually employed for estimation of leakage inductance are (i) energy method [8–13] and (ii) method of mutual fluxes.

In energy method, the energy stored in magnetic field of the secondary shorted transformer is calculated and equated to $(1/2)L_{\text{leak}}I_p^2$ where L_{leak} is the leakage inductance of the transformer when referred to primary, and I_p is current flowing through primary.

The H-profile inside the coil is calculated using Ampere's law. The energy stored in magnetic field is calculated by evaluating the volume integral in (1):

$$E_{\text{stored}} = \frac{\mu}{2}\iiint H^2 dv = \frac{1}{2}L_{\text{leak}}I_p^2. \quad (1)$$

The expression derived for leakage inductance using energy method is independent of frequency. Hence, it does not consider any frequency-dependent effects on leakage inductance. The energy method is used for comparing leakage inductance, in different winding configurations.

On the other hand, method of mutual fluxes uses Maxwell's equations to predict the leakage inductance more accurately at high frequencies. As this method accounts for frequency-dependent effects like eddy current losses and altered flux pattern due to eddy currents, it gives more accurate results, particularly at high frequencies. In [14], a frequency-dependent formula is presented to find leakage inductance in a toroidal core transformers.

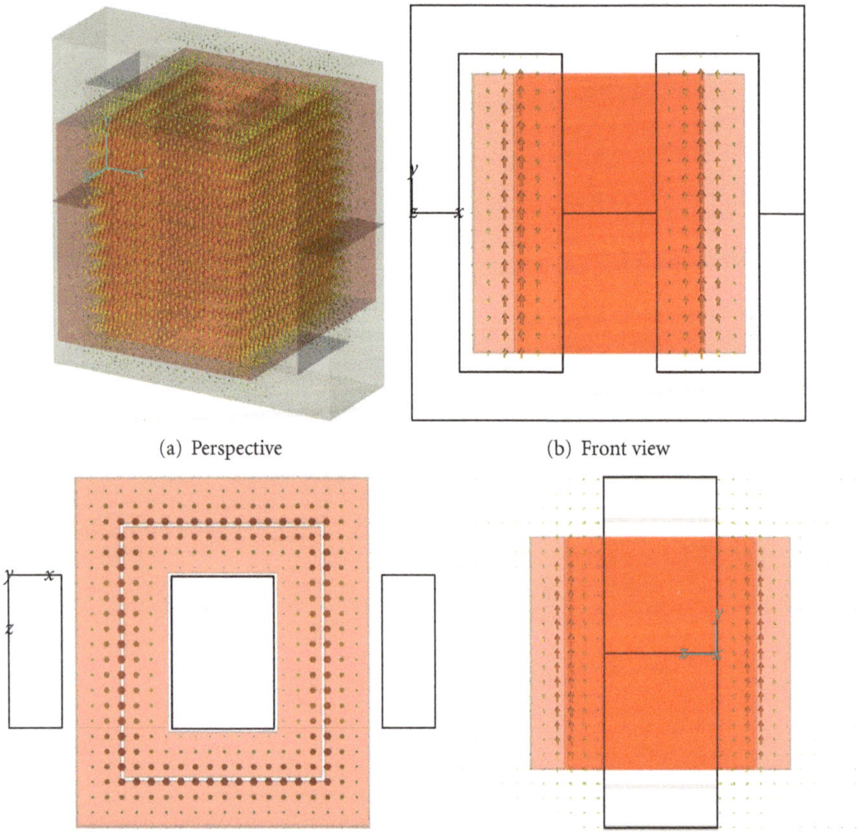

(a) Perspective

(b) Front view

(c) Top view

(d) Side view

Figure 1: Simulation result of FEM analysis.

In this paper, an expression for leakage inductance is derived using energy method. Expression derived accounts for the flux that extends into air in a secondary shorted transformer, which is not considered earlier [8, 10, 11]. Therefore, the estimated leakage inductance using this expression has better accuracy.

2. Estimation of Leakage Inductance on *E* Core Transformer

FEM analysis on secondary shorted transformer is made using CST Ver 2008.5 (with magneto-static solver). The simulation result of FEM analysis is shown in Figure 1. The simulation result is the H-field pattern of secondary shorted E core transformer. This gives an idea of profile of H-field from the surface of the core to the outer surface of the whole winding. The profile of H-field of a E core transformer and its winding configuration is shown in Figures 2 and 3, respectively.

Energy stored in the magnetic field in the secondary shorted transformer is equated to energy stored in leakage inductor, which is given in (1).

The volume represented with purple color in Figure 4 is considered for volume integration. This constitutes (i) the volume occupied by the coils and (ii) the volume formed by extrusion of the coils along the centre limb of the core, excluding the volume of the core. The flux extends into air shown in Figure 5. The volume shown in Figure 4 is divided into four parts, whose top view is shown in Figure 3. The profile of H-field is calculated using Ampere's law, the same profile is observed in FEM analysis. The H-profile is expressed mathematically in (2):

$$H(z) = \begin{cases} H_m \dfrac{z}{h_1} & 0 < z < h_1, \\ H_m & h_1 < z < h_1 + t, \\ H_m \left(\dfrac{h_1 + h_2 + t - z}{h_2} \right) & h_1 + t < z < h_1 + h_2 + t. \end{cases} \tag{2}$$

Using H-profile given in (2) to evaluate volume integral given in (1), we get expression of the leakage inductance:

$$L_{\text{leak}} = \frac{1}{3} \frac{\mu_o N_1^2}{F^2} (h + 2t)[FC + B(E + 2h)], \tag{3}$$

where $h = h_1 + h_2 + t$ is total thickness of the winding measured from the surface of the core to the outer surface of the outer winding, t is total thickness of the insulation used between the layers. B, C, E, F is are the dimensions of the E core shown in Figure 6.

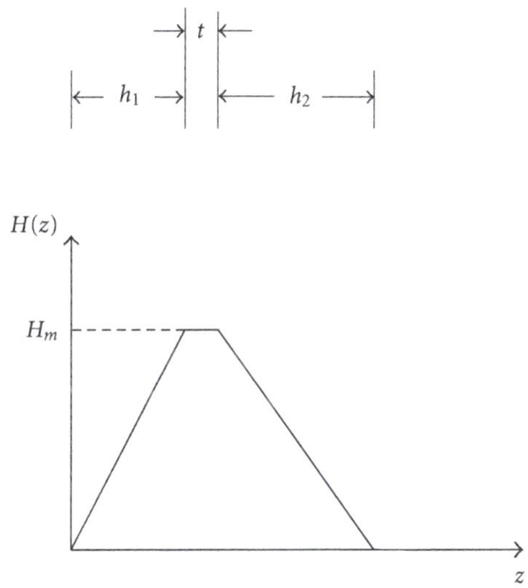

FIGURE 2: Profile of $H(z)$ from the surface of the core.

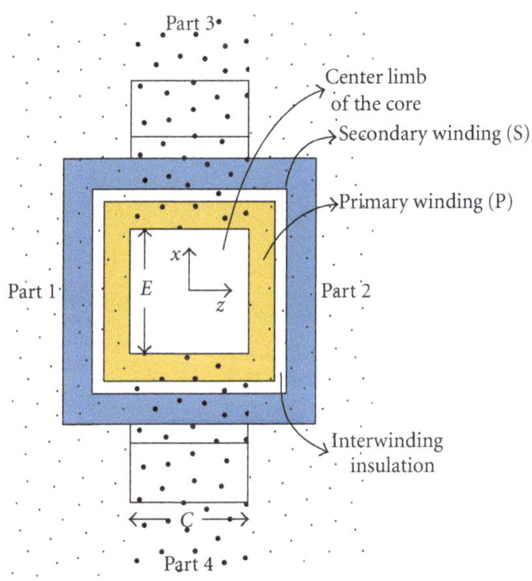

FIGURE 3: Top view of E core transformer with windings.

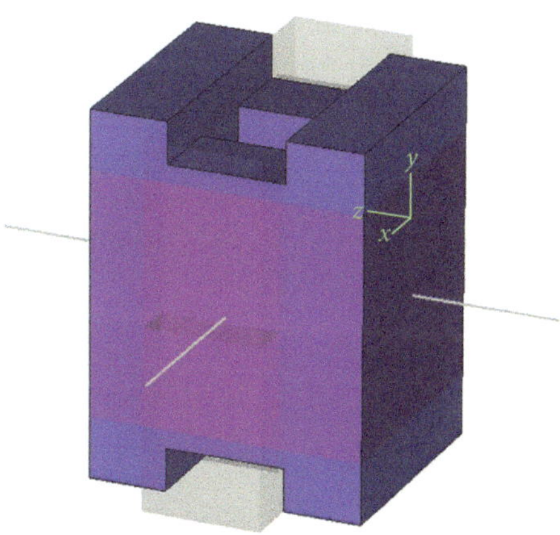

FIGURE 4: Volume around the core considered for integration.

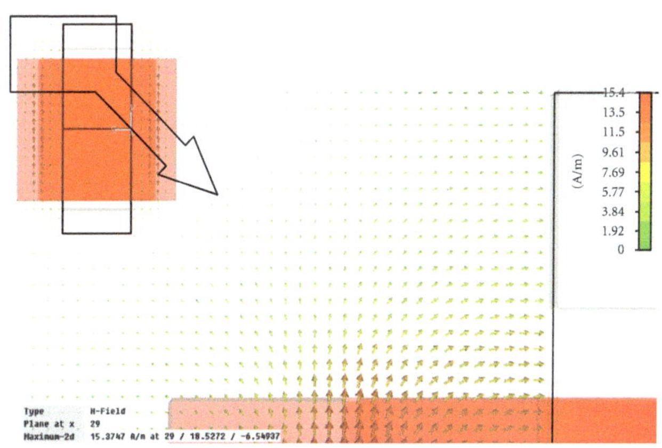

FIGURE 5: Scaled side view to show the flux that extends into air.

Table 1: EE core transformer with interwinding insulation.

Sample no.	1	2	
Core type	EE42/21/15	EE65/38/13	
F	14.45 mm	22.65 mm	
C	15.20 mm	13.45 mm	
B	21.10 mm	32.59 mm	
E	12.05 mm	19.77 mm	
h_1 thickness of primary	3.20 mm	3.81 mm	
h_2 thickness of secondary	1.90 mm	1.55 mm	
t thickness of insulation	1.27 mm	2.00 mm	
N_1	34	48	
N_2	17	24	Winding configuration
Photograph			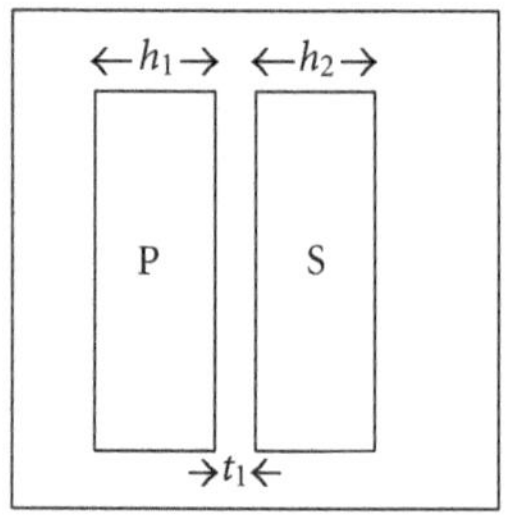
Leakage inductance $L_{\text{leak,earlier}}$ from (5)	11.91 μH	22.52 μH	
Leakage inductance L_{leak} from (3)	15.32 μH	28.21 μH	
Leakage inductance L_{exp} at 10 kHz measured	14.13 μH	26.76 μH	
% deviation of L_{exp} from calculated $L_{\text{leak,earlier}}$	15.71%	15.84%	
% deviation of L_{exp} from calculated L_{leak}	8.38%	5.12%	

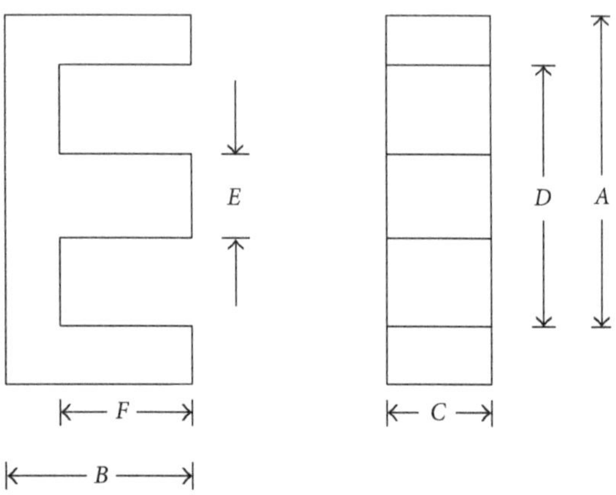

Figure 6: E core dimensions given in data sheets.

It is seen in [8, 11–13], by sandwiching the windings, leakage inductance reduces by p^2 times, where p is the number of interfaces between primary and secondary [8]:

$$L_{\text{leak}} = \frac{1}{3p^2}\frac{\mu_o N_1^2}{F^2}(h+2t)[FC+B(E+2h)]. \quad (4)$$

3. Experimental Results

Three different samples are made to validate the derived expression. The results are compared with the expressions available in literature. The expression given in [8, 10, 11] is replaced with the dimensions of the core (shown in Figure 6) and the dimensions of windings will give (5):

$$L_{\text{leak,earlier}} = \frac{1}{3p^2}\frac{\mu_o N_1^2}{F^2}(h+2t)F(C+E+2h). \quad (5)$$

Sample no. 1 is made with E core (EE42/21/15) wound with stranded conductor made with three conductors of 24 SWG. With interwinding insulation of thickness 1.27 mm is used. Standard winding configuration as shown in Table 1 is used to wind primary and secondary.

Sample no. 2 is made with E core (EE65/38/13) wound with stranded conductor made with three conductors of 24 SWG. With interwinding insulation of thickness 2.00 mm. Standard winding configuration as shown in Table 1 is used to wind primary and secondary.

Sample no. 3 is made with E core (EE42/21/15) wound with stranded conductor made with two conductors of 24 SWG. With total interwinding insulation of thickness 0.46 mm + 0.26 mm = 0.72 mm is used. Sandwiched winding

TABLE 2: EE core transformer with sandwiched winding (P/2, S, P/2).

Sample no.	3	
Core type	EE42/21/15	
F	14.45 mm	
C	15.20 mm	
B	21.10 mm	
E	12.05 mm	
h_1 thickness of primary half	1.41 mm	
h_2 thickness of secondary	2.93 mm	
h_3 thickness of primary half	1.52 mm	
t_1 thickness of insulation	0.46 mm	
t_2 thickness of insulation	0.26 mm	
N_1	46	
N_2	44	Winding configuration
Photograph		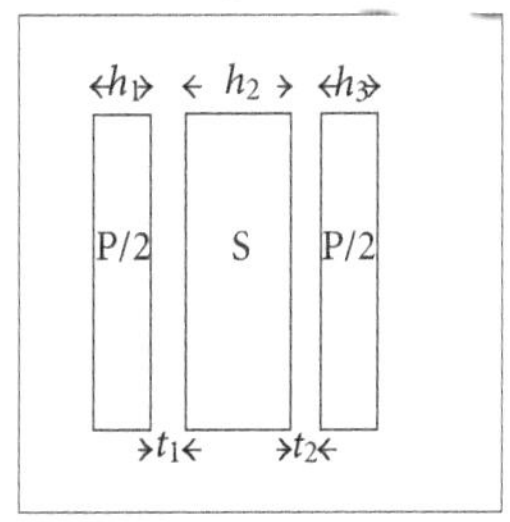
Leakage inductance $L_{leak,earlier}$ from (5)	4.94 μH	
Leakage inductance L_{leak} from (4)	6.37 μH	
Leakage inductance L_{exp} at 10 kHz measured	5.91 μH	
% deviation of L_{exp} from calculated $L_{leak,earlier}$	16.41%	
% deviation of L_{exp} from calculated L_{leak}	7.82%	

configuration as shown in Table 2 is used to wind primary and secondary.

All the measurements for leakage inductance are made by shorting the secondary. The measurements are made with LCR meter (make: Gw instek Model: LCR-8101) at 10 kHz.

Tables 1 and 2 tabulate experimentally measured values, estimated values using the expression derived in this paper and expression derived earlier for leakage inductance and their corresponding errors when compared with the experimentally measured values.

4. Conclusion

In this paper an improved expression for leakage inductance has been derived, using energy method. To observe the H-field pattern, FEM analysis has been carried out on a secondary shorted transformer, with primary excited. Experimental results show that the formula derived in this paper has better accuracy. It is observed that the leakage inductance estimated using improved expression L_{leak} has less deviation from L_{exp} when compared with leakage inductance estimated from earlier expression $L_{leak,earlier}$. The accuracy of the estimated value is improved due to consideration of flux that is extending into air.

References

[1] S. D. Johnson, A. F. Witulski, and R. W. Erickson, "Comparison of resonant topologies in high-voltage dc applications," *IEEE Transactions on Aerospace and Electronic Systems*, vol. 24, no. 3, pp. 263–274, 1988.

[2] A. K. S. Bhat, A. Biswas, and B. S. R. Iyengar, "Analysis and design of (LC)(LC)-type series-parallel resonant converter," *IEEE Transactions on Aerospace and Electronic Systems*, vol. 31, no. 3, pp. 1186–1193, 1995.

[3] B. Yang, F. C. Lee, A. J. Zhang, and G. Huang, "LLC resonant converter for front end DC/DC conversion," in *Proceedings of the 17th Annual IEEE on Applied Power Electronics Conference and Exposition Conference (APEC '02)*, vol. 2, pp. 1108–1112, Institute of Electrical and Electronics Engineers, 2002.

[4] J. A. Sabate, V. Vlatkovic, R. B. Ridley, F. C. Lee, and B. H. Cho, "Design considerations for high-voltage high-power full-bridge zero-voltage-switched pwm converter," in *Proceedings of the 5th Annual Applied Power Electronics Conference and Exposition (APEC '90)*, pp. 275–284, Institute of Electrical and Electronics Engineers, 1990.

[5] J. F. Lazar and R. Martinelli, "Steady-state analysis of the LLC series resonant converter," in *Proceedings of the 16th Annual Applied Power Electronics Conference and Exposition (APEC '01)*, vol. 2, pp. 728–735, Institute of Electrical and Electronics Engineers, 2001.

[6] P. C. Todd, "Snubber circuits: theory, design and application," in *Unitrode-Power Supply Design Seminar*, 1993.
[7] William McMurray, "Selection of snubbers and clamps to optimize the design of transistor switching converters," *IEEE Transactions on Industry Applications*, vol. IA-16, no. 4, pp. 513–523, 1980.
[8] N. Mohan, T. M. Undeland, and W. P. Robbins, *Power Electronics: Converters, Applications, and Design*, vol. 1, John Wiley & Sons, New York, NY, USA, 2003.
[9] R. Doebbelin, M. Benecke, and A. Lindemann, "Calculation of leakage inductance of core-type transformers for power electronic circuits," in *Proceedings of the 13th Power Electronics and Motion Control Conference (EPE-PEMC '08)*, pp. 1280–1286, September 2008.
[10] A. A. Dauhajre, *Modelling and estimation of leakage phenomena in magnetic circuits*, Ph.D. thesis, California Institute of Technology, Pasadena, Calif, USA, 1986.
[11] W. T. McLyman and C. W. T. McLyman, *Transformer and Inductor Design Handbook*, Marcel Dekker, New York, NY, USA, 2004.
[12] Z. Ouyang, O. C. Thomsen, and M. Andersen, "The analysis and comparison of leakage inductance in different winding arrangements for planar transformer," in *Proceedings of the International Conference on Power Electronics and Drive Systems (PEDS '09)*, pp. 1143–1148, November 2009.
[13] R. Doebbelin, R. Herms, C. Teichert, W. Schaetzing, and A. Lindemann, "Analysis methods and design of transformers with low leakage inductance for pulsed power applications," in *Proceedings of the European Conference on Power Electronics and Applications*, pp. 1–7, September 2007.
[14] W. G. Hurley and D. J. Wilcox, "Calculation of leakage inductance in transformer windings," *IEEE Transactions on Power Electronics*, vol. 9, no. 1, pp. 121–126, 1994.

Analysis and Minimization of the Oscillatory Currents in Multibranch Thyristor-Switched Capacitors

Mohamed M. Saied

Electrical Engineering Department, College of Engineering and Petroleum, Kuwait University, P.O. Box 5969, Kuwait 13060, Kuwait

Correspondence should be addressed to Mohamed M. Saied, mmsaied2@yahoo.com

Academic Editor: Neville Watson

This paper addresses the switching transients in multibranch thyristor-switched capacitors (TSCs). The current transients following the addition of a branch to a group of already connected ones are analyzed. Expressions for both its fundamental and its oscillatory components are given in terms of the power network voltage, frequency, short-circuit level, and the switching angle. The relations include also the compensator parameters such as its total reactive power rating, total number of branches, the number of already connected branches, and the initial voltage on the capacitor involved in the switching transient. An expression for the distortion of the supply current is also given. A minimization procedure is presented for identifying the optimal switching angle leading to the least magnitude of the oscillatory current. Switching when the instantaneous supply voltage is equal to the initial voltage will result in the least oscillatory current only in the two special cases of a single-branch compensator, or in the switching of the first branch of a multi-branch TSC. The effect of both the total number of branches and the branch switching steps on the oscillatory current and on the optimal switching angle is also discussed. The advantage of the suggested procedure is demonstrated by investigating several case studies.

1. Introduction

There are four main practical applications of static Var compensators (SVCs) in electric power systems which can lead to considerable economical technical and/or economical advantages: to maintain the voltage at or near a constant level, to improve the power system stability, to improve the power factor, and, finally, to correct phase unbalances [1–11]. Most SVC's have branches comprising thyristor-switched capacitors (TSCs). A TSC includes several usually identical branches connected in parallel. By controlling the thyristors, the total capacitance and hence the capacitive reactive power delivered to the network can be changed. There are several factors affecting the choice of the number of branches in a TSC compensator [2, 4]. In most practical applications, the number of branches, and hence of the control steps, lies between one and four [1]. Due to the switching-on and -off of these branches, the TSC terminal current includes unrequired high-frequency oscillatory components in addition to the useful capacitive current of the fundamental frequency. These components can lead to serious system disturbances. Similar problems of capacitor switching transients related to the operation of distribution and transmission networks have been recognized in earlier investigations such as [6–10]. Reference [6] presented a comparative study of several suggested 600-V or less 3-phase switching configurations for minimal transients, while the study [7] was focusing on the transient overvoltages at the switched capacitors and other network locations in addition to the expected implications on the duties of the surge arresters and current limiting reactors.

The amplitudes and frequencies of the currents in a the multi-branch compensator depend mainly on the parameters of the power network supplying the TSC (primarily its operating voltage and short-circuit level), the total reactive power rating of the compensator, the total number of its branches, the switching time point of the TSC branch under consideration, and the number of already connected branches. A preliminary approach to the calculation of these currents and the identification of the optimal switching conditions for least oscillatory components was presented in the study [4]. The reference came to the main conclusion

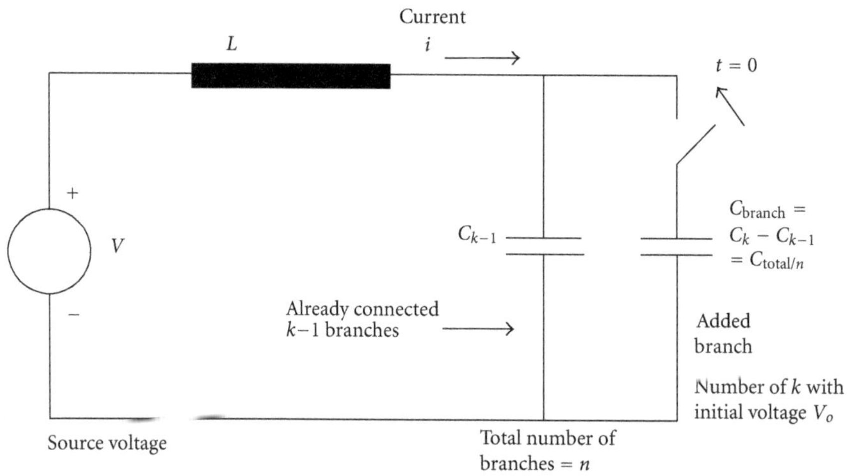

FIGURE 1: The equivalent circuit.

that the conditions for minimum oscillatory currents are close to the simple and nonideal condition of gating the thyristors when the instantaneous value of the supply voltage reaches the capacitor initial voltages. This statement implies the need for a more rigorous and generalized approach to the problem. Moreover, in order for the results to be more useful for practicing engineers, the current amplitudes, the time waveforms of the TSC total current and terminal voltage following each switching step, and the current distortion factor should be presented in a normalized way. They will be given in terms of the network's short-circuit level S_{sc} rather than its equivalent internal inductance L and the TSC Var rating (preferably in per unit of S_{sc}) instead of its capacitance C. This will help recognize some general trends and implementation guidelines from the presented results.

2. Method of Analysis

Consider the equivalent circuit shown in Figure 1 representing one phase of a 3-phase TSC. It includes a voltage source given by $V_m \sin(\omega_o t + \alpha)$, where V_m, ω_o, and α are its maximum phase value, angular frequency and its phase angle, respectively. It follows that at the switching time point $t = 0$, the supply voltage assumes the instantaneous value $V_m \sin\alpha$. It can be shown that

$$L = \frac{3V_m^2}{(2S_{sc}\omega_o)}. \tag{1}$$

It should be noted that the mentioned inductance is not a physically existing element. The equivalent inductance L (having the power frequency reactance $\omega_o L$) in the circuit shown in Figure 1 represents the series connection of both the power network's short-circuit reactance (e.g., of a feeder or a cable section) as well as the leakage reactance of the eventually existing step-down power transformer.

S_{sc} is the network's 3-phase short-circuit power at the compensator terminals. The capacitor C_{k-1} is the total capacitance of all already connected branches, while $(C_k - C_{k-1}) = C_{branch}$ is the capacitance of the branch to be switched at $t = 0$. If the total number of the identical branches is n, then each branch has a capacitance $C_{branch} = C_{total}/n$. All these capacitance and inductance values are referred to the source side of the transformer, if any. It can be shown that

$$C_{total} = \frac{2Q_{total}}{(3V_m^2\omega_o)}, \tag{2}$$

where Q_{total} represents the total 3-phase reactive power delivered by the compensator if operated at V_m and ω_o. For convenience, the parameter $\gamma = Q_{total}/S_{sc}$ will be referred to as the per unit compensator reactive power rating. Each of the n switching steps is given a number k denoting the addition of the branch number k in parallel with the $(k-1)$ already connected branches. From the sinusoidal steady state analysis, the following expressions for the initial values of the supply current, $I_{initial}$ and the voltage, $V_{initial}$ across the $(k-1)$ branches can be derived as follows:

$$\begin{aligned} I_{initial} &= \frac{V_m}{[1/\omega_o C_{k-1} - \omega_o L]} \sin\left[\alpha + \frac{\pi}{2}\right], \\ V_{initial} &= \frac{V_m(1/\omega_o C_{k-1})}{[1/\omega_o C_{k-1} - \omega_o L]} \sin\alpha, \end{aligned} \tag{3}$$

where

$$C_k = \frac{2\gamma S_{sc} k}{3nV_m^2\omega_o}. \tag{4}$$

Neglecting losses, the switching transient is described by the following differential equations subject to the above current initial condition:

$$\begin{aligned} V_m \sin(\omega_o t + \alpha) &= L\frac{di(t)}{dt} + v_{C_k}(t), \\ i(t) &= C_k \frac{dv_{C_k}(t)}{dt}. \end{aligned} \tag{5}$$

The initial voltage across the capacitor C_k can be determined from the initial voltage $V_{initial}$ across C_{k-1} and the initial voltage V_o across the switched new branch as $(V_{\text{initial}}C_{k-1} + V_o C_{\text{total}}/n)/C_k$. It will be referred to as V_{eq}. The following solution for $i(t)$ and $v_{C_k}(t)$ could be obtained

$$i(t) = \frac{1}{\sqrt{L}(-1+\omega_o^2 LC_k)} \times \Big[\sqrt{L}[I_{\text{initial}}(-1+\omega_o^2 LC_k) + C_k\omega_o V_m \cos\alpha] \times \cos\omega_{n,k}t - \sqrt{C_k} \times \Big\{\sqrt{C_k L}\omega_o V_m \cos\alpha \cos\omega_o t + \left(-V_{\text{eq}} + C_k L\omega_o^2 V_{\text{eq}} + V_m \sin\alpha\right) \times \sin\omega_{n,k}t - \sqrt{C_k L}\omega_o V_m \sin\alpha \sin\omega_o t\Big\}\Big], \quad (6)$$

$$v(t) = \frac{1}{\sqrt{C_k}(-1+LC_k\omega_o^2)} \times \Big[-\sqrt{C_k}V_m \sin\alpha \cos\omega_o t + \sqrt{C_k}\big[(-1+LC_k\omega_o^2)V_{\text{eq}} + V_m\big]\cos\omega_{n,k}t - I_{\text{initial}}\sqrt{L}\sin\omega_{n,k}t + C_k I_o L^{3/2}\omega_o^2 \sin\omega_{n,k}t + C_k\sqrt{L}\omega_o V_m \cos\alpha \sin\omega_{n,k}t - \sqrt{C_k}V_m \cos\alpha \sin\omega_o t\Big]. \quad (7)$$

Both the current and the capacitor voltage will contain higher frequency parts of the angular frequency $\omega_{n,k} = 1/\sqrt{LC_k}$ in addition to the power frequency components.

Focusing on the current, its power frequency component has the waveform $I_{\text{fund,max}}\,\text{Cos}(\omega_o t + \alpha)$. Its amplitude is $I_{\text{fund,max}} = \omega_o V_m C_k/(1 - \omega_o^2 LC_k)$. It can be seen that the oscillatory component follows the equation:

$$i_{\text{osc}}(t) = A\sin\omega_{n,k}t + B\cos\omega_{n,k}t, \quad (8)$$

where A and B can be easily determined from the expression of the current waveform in (6) as:

$$A = -\sqrt{\frac{C_k}{L}}\Big(-V_o + L\omega_o^2 V_o C_k + V_m \sin\alpha - \big[\big(C_{k-1}(-1+L\omega_o^2 C_k)(-V_o + V_o L\omega_o^2 C_{k-1}) + V_m C_k \sin\alpha\big)/(-1+L\omega_o^2 C_{k-1})C_k\big]\Big)(-1+L\omega_o^2 C_k)^{-1},$$

$$B = \frac{\omega_o V_m (C_{k-1} - C_k)}{(-1+L\omega_o^2 C_{k-1})(-1+L\omega_o^2 C_k)}\cos\alpha, \quad (9)$$

the amplitude of the oscillatory current is therefore $I_{\text{osc}} = \sqrt{A^2 + B^2}$.

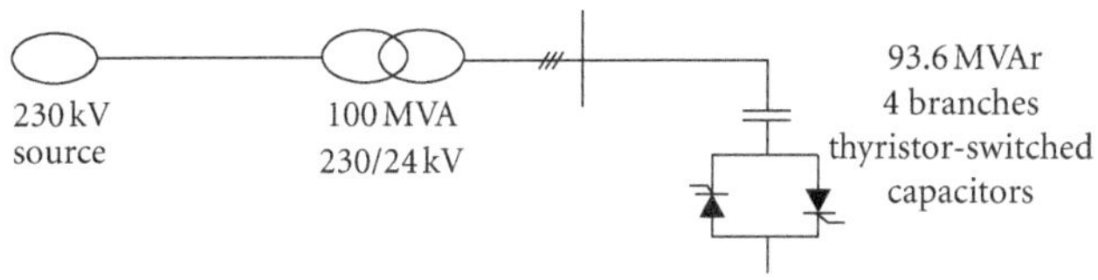

Figure 2: Sample network.

From (9), the main factors affecting the amplitude of the oscillatory current are the series inductance, the compensator reactive power rating, the total number of its branches n, the initial voltage V_o of the switched branch of the number k, and the switching angle α.

The results of applying the above expressions to the special case of a single branch ($n = 1$) agree with those available in the literature.

The optimal switching angle α leading to the least amplitude of the oscillating current I_{osc} can be determined from

$$\frac{d(A^2 + B^2)}{d\alpha} = 0. \quad (10)$$

3. Sample Results

The procedure is applied to the network shown in Figure 2, with the following data [1, 4].

Rated line voltage = 230 kV.

Rated frequency = 60-Hz.

Total compensator capacitive reactive power = 93.6 Mvar.

Number of identical branches = 4.

Assumed short-circuit level at the TSC terminals, taking into account the transformer reactance S_{sc} = 900 MVA.

The results depicted in Figure 3 are computed for an initial branch capacitor voltage equal to the amplitude of the rated network phase voltage V_m and a switching angle of $\alpha = 90°$. Only integer values on the x-axes (switching step k) are meaningful. The plots show the conditions after switching the kth branch. Plot (a) indicates that the amplitude of the fundamental current is almost proportional to the number k denoting the switching step. The slight deviation from the linear relationship is due to the voltage drop across the series inductance, which depends on k. Plots (b) and (c) give the amplitudes and frequencies (as multiples of the supply frequency 60-Hz) of the oscillatory current component following each switching step. Beyond $k = 2$, the oscillatory current increases less than proportional with k. Its frequency drops from 6.2 times the supply frequency upon connecting the first branch ($k = 1$) to half this value (3.1 per unit) for $k = 4$, while adding the last (fourth) branch to the already connected three. This is because the capacitance included in the equivalent circuit in the first case is one fourth of that in the case. Plot (d) illustrates the current distortion defined as the oscillatory current amplitude in

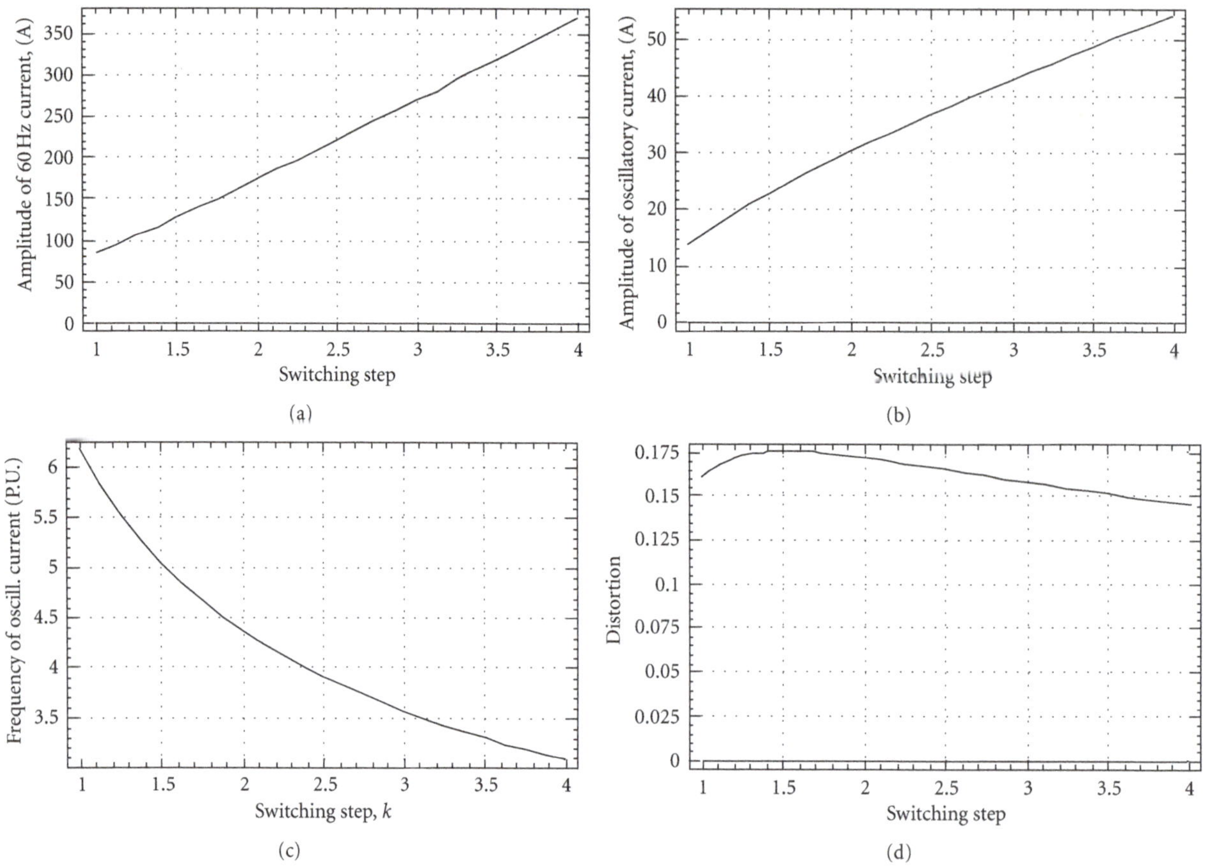

FIGURE 3: The fundamental and oscillatory currents for the different switching steps k. The initial branch capacitor voltages $V_o = V_m$ and the switching angle $\alpha = 90°$.

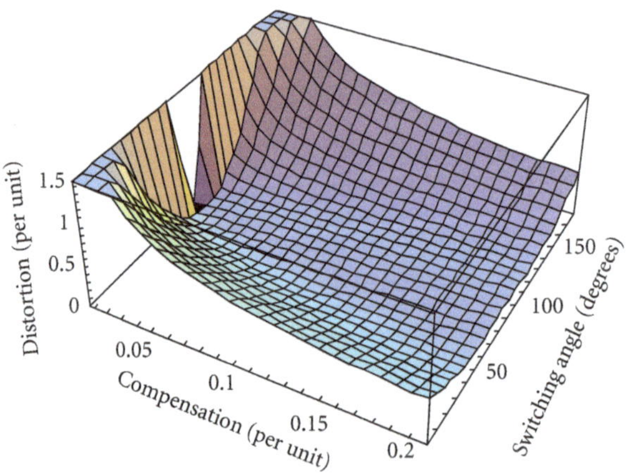

FIGURE 4: The current distortion as a function of both the per unit compensator rating and the switching angle α.

per unit of the fundamental current, that is, $I_{osc}/I_{fund,max}$. Its largest value of about 0.170 occurs while switching the second branch, and the least value is 0.146 and corresponds to $k = 4$.

The 3-dimensional plot in Figure 4 shows the dependence of the distortion factor of the last switching step ($k = 4$) on both the total per unit reactive power rating of the compensator γ and the switching angle α, assuming an initial voltage $V_o = V_m$ for the branch capacitors. It indicates the general trend that the current distortion decreases with the compensator rating. It further shows that for small compensator ratings the least distortion occurs if the switching takes place at a switching angle close to $\alpha = 90°$, while for larger ratings the minimum distortion can occur over a wider range of α.

The impact of the initial voltage V_o together with the switching angle α on the current distortion is depicted in Figure 5. It shows that, for small initial capacitor voltages,

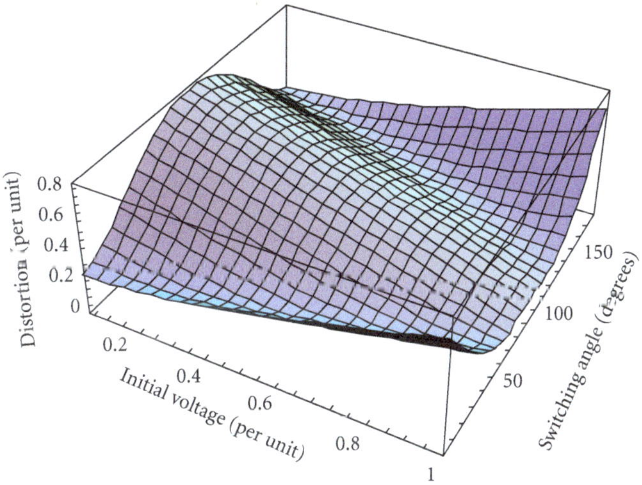

FIGURE 5: The current distortion versus the per unit value of the initial capacitor voltage $z = V_o/V_m$ and the switching angle α.

the least distortion occurs at the switching angle $\alpha = 0$, and the largest distortion results if $\alpha = 90°$. For initial capacitor voltages close to $V_o = V_m$, however, the largest distortion occurs at $\alpha = 0$. The least distortion will result if $\alpha = 67.2°$.

The optimization procedure based on (10) will be demonstrated by considering the example of a 4-branch compensator having a per unit compensation level $\gamma = 0.104$. By direct differentiation with respect to α, the following expression can be obtained for the switching of the branch number k to the already connected $(k-1)$ branches with an initial capacitor voltage of $V_o = z \cdot V_m$ as follows:

$$\begin{aligned}\left(\frac{1}{I^2_{\text{fund,max}}}\right)\frac{d(A^2+B^2)}{d\alpha} &= \left(4.257\times10^{-69}(1.8\times10^9-4.68\times10^7k)^2\right.\\ &\quad\times\cos\alpha\cdot(-1.0597\times10^{39}(-39.4615+k)\\ &\quad\times(-38.4603+k)(7.786\times10^{15}+k)\\ &\quad\times(1479.34-76.9243k+k^2)z\\ &\quad+2.0134\times10^{26}(-1.576\times10^{30}+k)\\ &\quad\times(-38.4614+k)(1479.3-76.9232k+k^2)\\ &\quad\times\sin\alpha))/\left((-39.4615+k)^2(-38.4615+k)^4\right.\\ &\quad\left.\times(1\times10^{-25}+k)(1\times10^{-6}+k)^2\right).\end{aligned} \tag{11}$$

Equating this first derivative to zero will give the following three roots:

$$\begin{aligned}\alpha &= \pm\frac{\pi}{2},\\ \alpha &= \sin^{-1}10^{13}\\ &\quad\cdot\left[z\cdot(3.37\times10^{22}-3.48\times10^{21}k+1.35\times10^{20}k^2\right.\\ &\quad-2.32\times10^{18}k^3+1.50\times10^{16}k^4+1.93k^5)\\ &\quad\times(3.28\times10^{35}-2.56\times10^{34}k+6.66\times10^{32}k^2\\ &\quad\left.-5.77\times10^{30}k^3+3.66k^4)^{-1}\right].\end{aligned} \tag{12}$$

As will be seen later, the third root will result in the global minimum of the current distortion. It can be seen that, for all switching steps k, the value of $\sin\alpha$ is proportional to the per unit initial voltage of the switched capacitor z.

The three plots (a), (b), and (c) in Figure 6 give the optimal switching angle α for the total compensator ratings: $\gamma = 0.03$, 0.104 and 0.208 per unit, respectively. The four curves in each plot give the optimal angle α for the switching of the first branch $k = 1$ (the upper curve), $k = 2$, $k = 3$, and $k = 4$ (the lowest curve). The upper curves for switching the first branch $(k = 1)$ in the three plots (a), (b), and (c) are identical, and they follow the known relation of a single-branch TSC: $\sin\alpha = V_o/V_m$. They all indicate that the optimal α is 90° for $V_o = V_m$. For the switching of the three other branches, the optimal angle yielding the least distortion will be smaller than $\sin^{-1}(V_o/V_m)$. The deviation increases with k and with the capacitor initial voltage $V_o = z \cdot V_m$.

Table 1 gives the optimal switching angle α in degrees corresponding to the four switching steps k, with the capacitor initial voltages $V_o = V_m$, that is, $z = 1$. They are calculated for three different values of the per unit compensator rating γ: 0.030, 0.104, and 0.208. The results show that the optimal switching angle decreases with the total per unit rating of the compensator γ and with the number k denoting the switching step. As known from single-branch TSC, the optimal value of α for the first step $(k = 1)$ is 90° for all per unit compensator ratings γ.

Table 2 shows the fundamental and oscillatory current components as well as the current distortion factor if the optimal values of the switching angle are used for the four switching steps for a compensator per unit rating $\gamma = 0.104$. Both current components increase approximately proportional to the step number k. The distortion factor ranges between 0.163 per unit for $k = 2$ and 0.132 for the last step $k = 4$. These values are less than the corresponding ones (0.173 per unit for $k = 2$ and 0.146 for the last step $k = 4$) in Table 3 presenting the results if the switching angle $\alpha = 90°$ is applied. By comparing Tables 2 and 3, it can be realized that using the optimal values of α will result in

Table 1: The optimal switching angles α ($z = 1$).

Compensator step rating →	$\gamma = 0.030$ per unit	$\gamma = 0.104$ per unit	$\gamma = 0.208$ per unit
$k = 1$	90	90	90
$k = 2$	83	77	71.44
$k = 3$	80	71.5	63.64
$k = 4$	78	67.2	57.57

Table 2: The fundamental and oscillatory currents and the current distortion factor while applying optimal switching angles α ($\gamma = 0.104$, $z = 1$).

Switching step	Fundamental current, A	Oscillatory current, A	Current distortion, per unit
$k = 1$	85.287	13.752	0.161
$k = 2$	175.253	28.543	0.163
$k = 3$	270.292	39.346	0.146
$k = 4$	370.847	49.089	0.132

Table 3: The fundamental and oscillatory currents and the current distortion factor while applying switching angle $\alpha = 90°$ ($\gamma = 0.104, z = 1$).

Switching step	Fundamental current, A	Oscillatory current, A	Current distortion, per unit
$k = 1$	85.287	13.752	0.161
$k = 2$	175.253	30.240	0.173
$k = 3$	270.292	42.858	0.159
$k = 4$	370.847	54.220	0.146

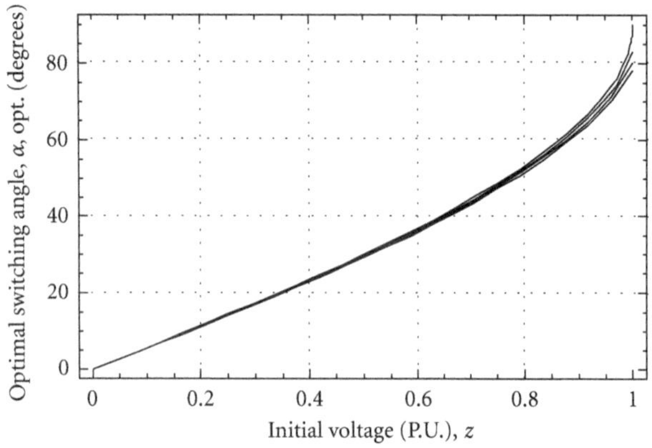

(a) Compensator rating $\gamma = 0.03$ per unit

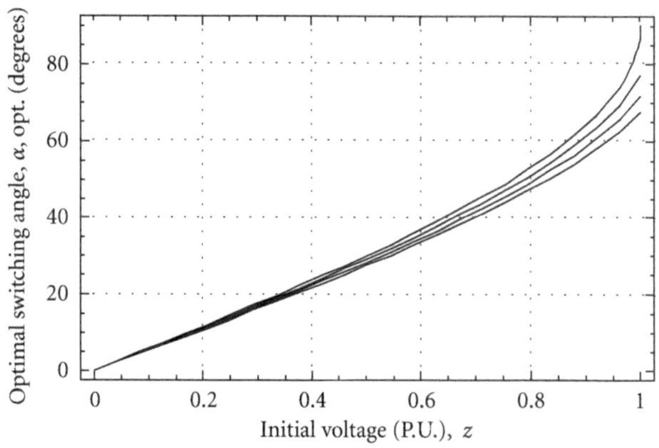

(b) Compensator rating $\gamma = 0.104$ per unit

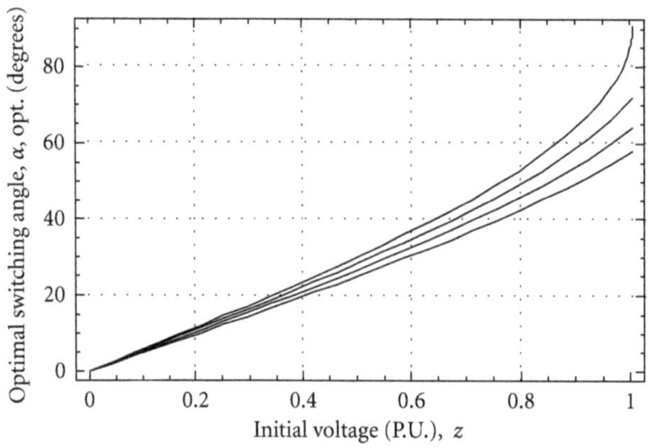

(c) Compensator rating $\gamma = 0.208$ per unit

Figure 6: The optimal switching angle α as a function of the per unit capacitor initial voltage $z = V_o/V_m$, for switching each of the four branches $k = 1, 2, 3$ and 4. Plots for three different per unit total compensator ratings are shown. In each plot, the top curve corresponds to $k = 1$, and the bottom curve applies for $k = 4$.

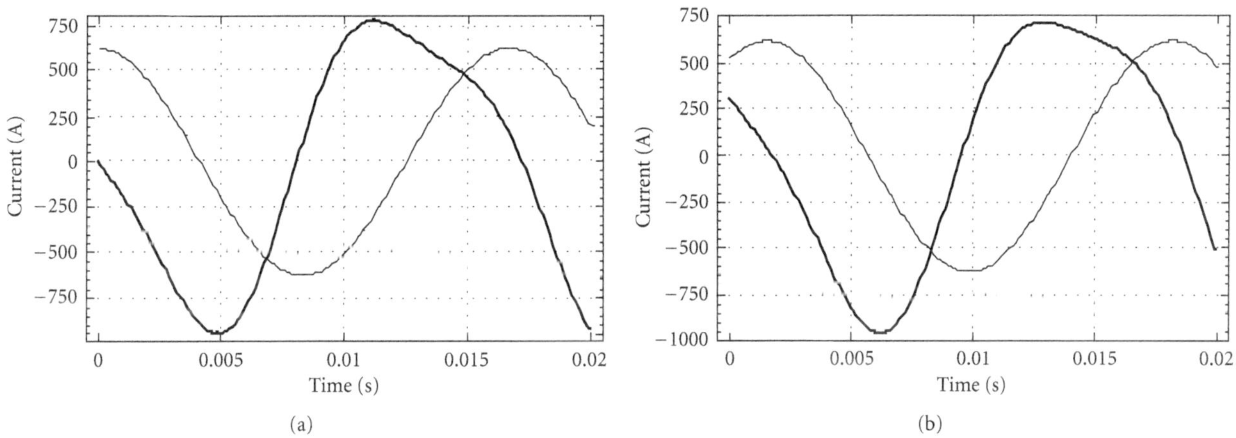

FIGURE 7: Waveforms of the total current (thick curves) for the switching of the fourth branch ($k = 4$). The initial voltage is $V_o = V_m$, $y = 0.208$ per unit. The thinner curves represent (to a reduced scale of 1/300) the waveform of the source phase voltage. (a) $\alpha = 90°$ and (b) for the optimal switching angle $\alpha = 57.57°$.

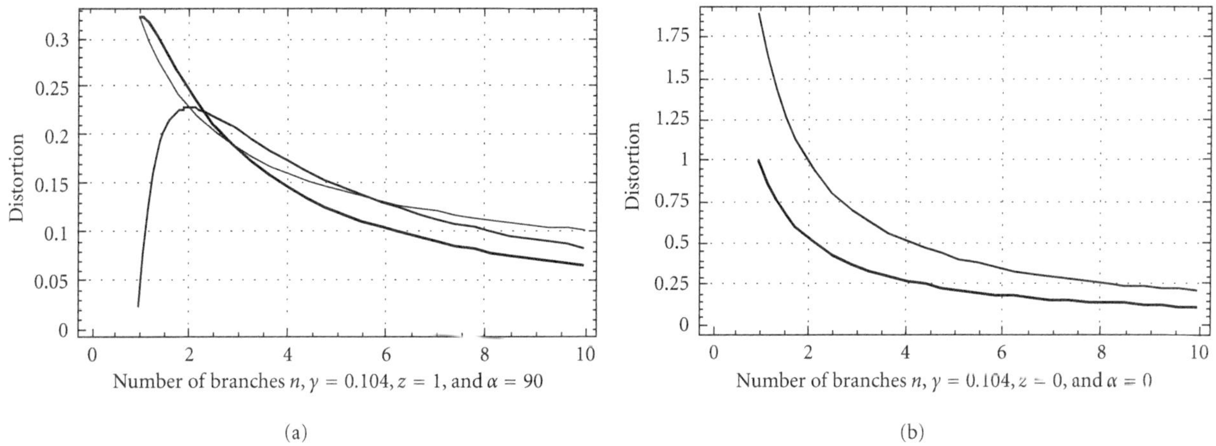

FIGURE 8: The current distortion factor in per unit versus the total number of branches, n for two cases: (a) initial voltage $V_o = V_m$, that is, $z = 1$ and the switching angle $\alpha = 90°$. (b) initial voltage $V_o = 0$, that is, $z = 0$ and the switching angle $\alpha = 0°$. In both cases, the compensator size is $y = 0.104$ per unit.

a reduction in the oscillatory current and the corresponding distortion factor in all switching steps except the first. This improvement is 5.78%, 8.176%, and 9.589% for the steps $k = 2, 3$, and 4, respectively. These observations are also valid for other per unit compensator ratings y. As an example, if $y = 0.208$, and considering the last switching step ($k = 4$), the amplitude of the fundamental current will be 839.1 A. If $\alpha = 90°$ is used, the oscillatory current will be 180.686 A leading to a distortion factor of 0.215336. If the optimal angle $\alpha = 57.57°$ is applied, however, the corresponding values will be 164.083 A for the oscillatory current and 0.19555 for the distortion factor. Again, there a reduction of 9.19%. This improvement is recognizable in the two plots (a) and (b) in Figure 7 pertinent to the fourth switching step ($k = 4$) for an initial voltage $V_o = V_m$, that is, $z = 1$. They illustrate the waveform of the total current $i(t)$, over 20 millisecond, for the two cases: (a) $\alpha = 90°$ and (b) for the corresponding optimal switching angle $\alpha = 57.57°$.

The two plots (a) and (b) of Figure 8 illustrate the effect of the total number of branches n on the oscillatory current component. Each plot includes 3 curves. The thinnest one corresponds to the switching of the first branch (i.e., $k = 1$). The intermediate one applies for switching the branch of number $k = n/2$ (this is valid only for even values of n). The thickest curve is for switching the last branch (i.e., $k = n$). In both plots, the per unit compensator size is $y = 0.104$. In plot (a) the initial voltage on the switched capacitor is $V_o = V_m$, that is, $z = 1$ and the switching angle $\alpha = 90°$. In plot (b) the initial voltage is assumed $V_o = 0$, that is, $z = 0$ and the switching angle $\alpha = 0°$. It should be noted that the thinnest curve in plot (b) is the flat horizontal line of a distortion = 1 per unit, as expected from the analysis of a single-branch TSC. It can be recognized that, with the exception of the case of switching the first capacitor $k = 1$ of the case in plot (b), the current distortion decreases steadily with the total number of branches n. For example, increasing the number of branches from $n = 2$ to $n = 6$ will lead to a reduction in the current distortion by a factor of about 2 in case (a) and by a factor of about 3 in case (b).

4. Conclusions

(1) Expressions are derived for the amplitudes of the fundamental current, the oscillatory current and the terminal voltage in a TSC, in terms of the AC network parameters, the TSC total reactive power rating, its total number of branches n, the switching step number k, and the initial voltage V_o of the switched branch as well as on the switching angle α.

(2) The procedure is applied to a 3-phase, 60-Hz, 24-kV, 4-branch, and 93.6 MVAr TSC connected to a 230-kV network of a 900 MVA short-circuit level via a 230/24-kV transformer.

(3) The results for the four switching steps assuming initial capacitor voltages equal to the peak supply phase voltage ($V_o = V_m$) and a switching angle $\alpha = 90°$ indicate that the amplitude of the fundamental current is almost proportional to the switching step number k.

(4) Beyond the second switching step, the oscillatory current increases less than proportional with k. Its frequency drops from 6.2 times the supply frequency when the first branch is connected ($k = 1$) to half of this value for $k = 4$. The largest value of the current distortion is about 0.170 per unit. It occurs when switching the second branch, and the least value is 0.146 and corresponds to $k = 4$.

(5) Results of the last switching step ($k = 4$) show the general trend that the current distortion decreases with the compensator rating. For small compensator ratings, the least distortion occurs if the switching takes place at a switching angle close to $\alpha = 90°$, upon for larger ratings, the minimum distortion can occur over a wider range of α.

(6) For small initial capacitor voltages, the least current distortion occurs at the switching angle $\alpha = 0$, while the largest distortion results if $\alpha = 90°$. For initial capacitor voltages close to V_m, however, the largest distortion occurs at $\alpha = 0$. The least distortion will result if $\alpha = 67.2°$.

(7) The dependence of the optimal switching angle on both the compensator rating and the per unit capacitor initial voltage $z = V_o/V_m$, for switching each of the four branches, is illustrated. The results for switching the first branch ($k = 1$) are the identical for all compensator ratings, and they follow the relation: $\sin\alpha = V_o/V_m$. They indicate that the optimal α is 90° for $V_o = V_m$. For the switching of the three other branches, the optimal angle will be smaller than $\sin^{-1}(V_o/V_m)$. The deviation increases with both the number k switching step and the capacitor initial voltage V_o.

(8) Results for the optimal α corresponding to the four switching steps k, with $V_o = V_m$ for three different values of the per unit compensator rating y: 0.030, 0.104, and 0.208 are presented. The optimal α decreases with the TSC rating and with the number k of the switching step.

(9) The values of the fundamental and oscillatory currents as well as the current distortion factor are discussed, if the optimal values of α are used for the four switching steps. Both current components increase approximately proportional to the step number k. The distortion factor ranges between 0.163 per unit for $k = 2$ and 0.132 for the last step $k = 4$. These values are less than the corresponding ones (0.173 per unit for $k = 2$ and 0.146 for the last step $k = 4$) resulting from applying $\alpha = 90°$.

(10) Using the optimal values of α will result in a reduction in the oscillatory current and the corresponding distortion factor in all switching steps except the first. This improvement is 5.78%, 8.18%, and 9.59% for the steps $k = 2$, 3, and 4, respectively. This advantage increases with the compensator TSC rating.

(11) The effect of the total number of branches n on the oscillatory current is presented for two cases: (a) $V_o = V_m$ and $\alpha = 90°$, and (b) $V_o = 0$ and $\alpha = 0°$. With the exception of the case of switching the first capacitor $k = 1$ of the case (b), the current distortion decreases steadily with the total number of branches n. For example, increasing the number of branches from $n = 2$ to $n = 6$ will lead to a reduction in the current distortion by a factor of about 2 in case (a) and about 3 in case (b).

References

[1] T. J. E. Miller, *Reactive Power Control in Electric Systems*, chapter 4, John Wiley & Sons, New York, NY, USA, 1982.

[2] The IEEE: Tutorial Course, "Applications of staticVAR systems for system dynamic performance," IEEE Publication 87TH0187-5-PWR, 1987.

[3] L. Gyugyi, R. Otto, and T. H. Putman, "Principles and applications of static thyristor-controlled shunt compensators," *IEEE Transactions on Power Apparatus and Systems*, vol. 97, no. 5, pp. 1935–1945, 1978.

[4] M. M. Saied, "Generalised relations for the current waveforms and the condition for minimum oscillatory currents in multi-branch Thyristor-Switched Capacitors," *Journal of Electrical and Electronics Engineering, Australia*, vol. 12, no. 1, pp. 42–48, 1992.

[5] J. Dixon, L. Morán, J. Rodríguez, and R. Domke, "Reactive power compensation technologies: state-of-the-art review," *Proceedings of the IEEE*, vol. 93, no. 12, pp. 2144–2164, 2005.

[6] G. Olivier, I. Mougharbel, and G. Dobson-Mack, "Minimal transient switching of capacitors," *IEEE Transactions on Power Delivery*, vol. 8, no. 4, pp. 1988–1994, 1993.

[7] S. Mikhail and M. F. McGranaghan, "Evaluation of switching concerns associated with 345 kV shunt capacitor applications," *IEEE Transactions on Power Apparatus and Systems*, vol. 1, no. 2, pp. 221–230, 1986.

[8] A. Gelen and T. Yalcinoz, "The behavior of thyristor switched capacitor (TSC) installed in an infinite bus system," in *Proceedings of the IEEE International Conference on Computer*

as a Tool (EUROCON '09), pp. 614–617, Petersburg, Russia, May 2009.

[9] J. Peltola, P. Halonen, and A. Kähkönen, "Static var compensator for power oscillation damping," in *Proceedings of the IEEE PES Transmission and Distribution Conference and Exposition*, New Orleans, La, USA, April 2010.

[10] W. Guo, D. Xu, J. Wu, and L. Wang, "United system of TSC and SVG for reactive power compensation," in *Proceedings of the 2nd International Symposium on Power Electronics for Distributed Generation Systems (PEDG '10)*, pp. 507–511, Hefei, China, June 2010.

[11] P. Chopade, M. Bikdash, I. Kateeb, and A. D. Kelkar, "Reactive power management and voltage control of large transmission system using SVC (static VAR compensator)," in *Proceedings of the IEEE Southeastcon*, pp. 85–90, Nashville, Tenn, USA, March 2011.

Permissions

The contributors of this book come from diverse backgrounds, making this book a truly international effort. This book will bring forth new frontiers with its revolutionizing research information and detailed analysis of the nascent developments around the world.

We would like to thank all the contributing authors for lending their expertise to make the book truly unique. They have played a crucial role in the development of this book. Without their invaluable contributions this book wouldn't have been possible. They have made vital efforts to compile up to date information on the varied aspects of this subject to make this book a valuable addition to the collection of many professionals and students.

This book was conceptualized with the vision of imparting up-to-date information and advanced data in this field. To ensure the same, a matchless editorial board was set up. Every individual on the board went through rigorous rounds of assessment to prove their worth. After which they invested a large part of their time researching and compiling the most relevant data for our readers. Conferences and sessions were held from time to time between the editorial board and the contributing authors to present the data in the most comprehensible form. The editorial team has worked tirelessly to provide valuable and valid information to help people across the globe.

Every chapter published in this book has been scrutinized by our experts. Their significance has been extensively debated. The topics covered herein carry significant findings which will fuel the growth of the discipline. They may even be implemented as practical applications or may be referred to as a beginning point for another development. Chapters in this book were first published by Hindawi Publishing Corporation; hereby published with permission under the Creative Commons Attribution License or equivalent.

The editorial board has been involved in producing this book since its inception. They have spent rigorous hours researching and exploring the diverse topics which have resulted in the successful publishing of this book. They have passed on their knowledge of decades through this book. To expedite this challenging task, the publisher supported the team at every step. A small team of assistant editors was also appointed to further simplify the editing procedure and attain best results for the readers.

Our editorial team has been hand-picked from every corner of the world. Their multi-ethnicity adds dynamic inputs to the discussions which result in innovative outcomes. These outcomes are then further discussed with the researchers and contributors who give their valuable feedback and opinion regarding the same. The feedback is then collaborated with the researches and they are edited in a comprehensive manner to aid the understanding of the subject.

Apart from the editorial board, the designing team has also invested a significant amount of their time in understanding the subject and creating the most relevant covers. They scrutinized every image to scout for the most suitable representation of the subject and create an appropriate cover for the book.

The publishing team has been involved in this book since its early stages. They were actively engaged in every process, be it collecting the data, connecting with the contributors or procuring relevant information. The team has been an ardent support to the editorial, designing and production team. Their endless efforts to recruit the best for this project, has resulted in the accomplishment of this book. They are a veteran in the field of academics and their pool of knowledge is as vast as their experience in printing. Their expertise and guidance has proved useful at every step. Their uncompromising quality standards have made this book an exceptional effort. Their encouragement from time to time has been an inspiration for everyone.

The publisher and the editorial board hope that this book will prove to be a valuable piece of knowledge for researchers, students, practitioners and scholars across the globe.

List of Contributors

Bassem Omri, Kaicar Ammous and Anis Ammous
Power Electronics Group, Universite de Sfax, ENIS-Sfax, BP. 1173, 3038 Sfax, Tunisia

Monika Jain, Sushma Gupta and Gayatri Agnihotri
Electrical Engineering Department, Maulana Azad National Institute (M.A.N.I.T), Deemed University, Bhopal (M.P.) 462052, India

Deepika Masand
Electrical and Electronic Department of Engineering, Oriental college of science and technology (O.I.S.T), R.G.P.V University, Bhopal (M.P.) 462021, India

Dionysios V. Spyropoulos and Epaminondas D. Mitronikas
Electromechanical Energy Conversion Laboratory, Department of Electrical and Computer Engineering, University of Patras, 26504 Rio, Patras, Greece

Moulay Tahar Lamchich
Department of Physics, Faculty of Sciences Semlalia, Cadi Ayyad University, 40000 Marrakech, Morocco

Carlos Gonzalez-Garcia and Jorge Pleite
Electronic Technology Department, Carlos III University of Madrid, Avenida de la Universidad 30, Leganés, 28911 Madrid, Spain

Rickard Ekström, Senad Apelfröjd and Mats Leijon
Division for Electricity, Swedish Centre for Renewable Electric Energy Conversion, Uppsala University, P.O. Box 534, 751 21 Uppsala, Sweden

M. F. Bulacio, T. A. Gonzalez, G. Marinelli, R. Alonso and H. E. Tacca
LABCATyP, Department of Electronics, Faculty of Engineering, University of Buenos Aires, Av. Paseo Colon 850, C1063ACV, Buenos Aires, Argentina

Alexander Abramovitz
Department of Electrical Engineering and Computer Science, University of California, Irvine, CA 92697, USA

Pradyumn Chaturvedi
Electrical Engineering Department, SATI, Vidisha 464001, India

Shailendra Jain
Electrical Engineering Department, MANIT, Bhopal 462051, India

Pramod Agarwal
Electrical Engineering Department, IIT, Uttaranchal, Roorkee 247667, India

Amin Shabanpour and Ali Reza Seifi
School of Electrical & Computer Engineering, University of Shiraz, Shiraz 7134851151, Iran

Tourkia Lajnef, Slim Abid and Anis Ammous
Power Electronic Group (PEG), ENIS, BPW, 3038 Sfax, Tunisia

Branko Blanuša
Faculty of Electrical Engineering, University of Banja Luka, Patre 5, 78000 Banja Luka, Bosnia and Herzegovina

Bojan Knezevic
Faculty of Mechanical Engineering, University of Banja Luka, Bulevar Stepe Stepanovica 75, 78000 Banja Luka, Bosnia and Herzegovina

L. Charaabi
L.S.E., Ecole Nationale dIngenieur de Tunis, BP 37, 1002 Le Belvedere, Tunisia

Takahiko Yamamoto and Kohji Koshiji
Department of Electrical Engineering, Faculty of Science and Technology, Tokyo University of Science, 2641, Yamazaki, Noda, Chiba 278-8510, Japan

Bahram Ashrafi and Mehdi Niroomand
Department of Electrical Engineering, University of Isfahan (UI), Isfahan 81746-73441, Iran

Hamid Radmanesh
Department of Electrical Engineering, Islamic Azad University, Takestan Branch, Takestan 1995755681, Iran
Electrical Engineering Department, Amirkabir University of Technology, Tehran, Iran

G. B. Gharehpetian and Hamid Fathi
Electrical Engineering Department, Amirkabir University of Technology, Tehran, Iran

Alexander Abramovitz
Department of Electrical Engineering and Computer Science, University of California, Irvine, CA 92697-2625, USA

Mangesh Borage and Sunil Tiwari
Power Supplies and Industrial Accelerator Division, Raja Ramanna Centre for Advanced Technology, Indore 452013, India

Hussein A. Kazem
Faculty of Engineering, Sohar University, P.O. Box 44, 311 Sohar, Oman

J. G. Oliveira and H. Bernhoff
Division for Electricity, Uppsala University, 751 05 Uppsala, Sweden

H. Schettino, V. Gama and R. Carvalho
Electrical Engineering Department, Federal University of Juiz de Fora, 36036-900 Juiz de Fora, MG, Brazil

Antonino O. Di Tommaso, Fabio Genduso, Rosario Miceli and Giuseppe Ricco Galluzzo
Department of Energy, Informatics and Mathematical Models, University of Palermo, 90128 Palermo, Italy

Sivananda Reddy Thondapu, Yashwant D. Wanmode and Purushottam Shrivastava
Pulse High Power Microwave Section, Raja Ramanna Centre for Advanced Technology, Indore 452013, India

Mangesh B. Borage
Power Supplies and Industrial Accelerator Division, Raja Ramanna Centre for Advanced Technology, Indore 452013, India

Mohamed M. Saied
Electrical Engineering Department, College of Engineering and Petroleum, Kuwait University, P.O. Box 5969, Kuwait 13060, Kuwait